Terra and Silva in the Pannonian Plain

Opovo agro-gathering in the Late Neolithic

Ksenija Borojević

BAR International Series 1563

2006

Published in 2016 by
BAR Publishing, Oxford

BAR International Series 1563

Terra and Silva in the Pannonian Plain

ISBN 978 1 84171 993 1

© K Borojević and the Publisher 2006

The author's moral rights under the 1988 UK Copyright, Designs and Patents Act are hereby expressly asserted.

All rights reserved. No part of this work may be copied, reproduced, stored, sold, distributed, scanned, saved in any form of digital format or transmitted in any form digitally, without the written permission of the Publisher.

BAR Publishing is the trading name of British Archaeological Reports (Oxford) Ltd. British Archaeological Reports was first incorporated in 1974 to publish the BAR Series, International and British. In 1992 Hadrian Books Ltd became part of the BAR group. This volume was originally published by Archaeopress in conjunction with British Archaeological Reports (Oxford) Ltd / Hadrian Books Ltd, the Series principal publisher, in 2006. This present volume is published by BAR Publishing, 2016.

Printed in England

BAR
PUBLISHING

BAR titles are available from:

BAR Publishing
122 Banbury Rd, Oxford, OX2 7BP, UK
EMAIL info@barpublishing.com
PHONE +44 (0)1865 310431
FAX +44 (0)1865 316916
www.barpublishing.com

Table of Contents

Preface

Plant macro remains from the late Neolithic site of Opovo are the topic of this book. Opovo is dated from 4700 to 4500 B.C., and culturally to the late phase of the Vinča culture, which is considered one of the most prominent Neolithic cultures of the Balkans.

The Opovo site is located on the southern edge of the Pannonian Plain, in the Banat region that is part of the modern province of Voyvodina in Serbia. The Pannonian Plain, often referred to as the Carpathian Basin, is a large plain in central/south-eastern Europe, left after the prehistoric Pannonian Sea dried out.

The site of Opovo was excavated (1983–1989) by an international archaeological team from the University of Novi Sad, Serbia, and from the University of California Berkeley, U.S.A. The preliminary results from the Opovo excavations were published by Tringham et al. (1985, 1992) in the *Journal of Field Archaeology*. During the project, I was in charge of the retrieval and analysis of plant remains.

Through analysis of plant macro remains from the Opovo site, I provide information about such important issues as vegetation reconstruction, plant use, subsistence, husbandry, wild plant procurement, and intra-site plant distribution at this late Vinča culture site. Relevant data from the artifactual analyses are incorporated in the description of the context from which the plant remains were recovered, with special emphasis on the integration of floral and faunal data. Systematic recovery of floral and faunal remains during the excavations of the Opovo site and subsequent analyses of animal bones (Russell 1993) and plant remains (Borojević 1998) allowed such comparisons to be made. Studies of house structures provided additional information for contextual interpretation of the Opovo plant remains (Stevanović 1996, 1997a, Tringham 1994).

The Opovo site served as a case study of diverse subsistence strategies practiced within the general cultural context of the Vinča culture during the late Neolithic in the Balkans. In reconstructing past subsistence strategies, I explicitly used ethnohistoric data throughout the book to offer possible analogies for prehistoric activities, drawing selectively on a diverse range of limited analogies, and using multiple sources "to suggest what can and what cannot be assumed about the quite distant and unfamiliar subject" (Wylie 2002:153).

In the interpretation, I tried always to keep in mind that plants were gathered, grown, and consumed by humans and animals, and were primarily deposited by humans who once lived in the houses that we revealed in the Opovo site. I used the term "Opovo inhabitants" when referring to the people that once occupied the prehistoric site that we excavated during the 1980s, and I placed emphasis on reconstructing the taphonomic process and contexts of the plants found.

A term used in the title, *agro-gathering,* refers to a combination of crop cultivation and wild plant procurement on the one hand, and hunting and raising domestic animals on the other, and points to a mixed strategy labeled *terra et silva* (meaning land and woods). The Latin expression *terra et silva*, frequently used in the Latin documents from the sixth to tenth century, is adopted from Montanari's description of a similar mixed strategy of subsistence during the early Middle Ages (Montanari 1994). I believe the term can appropriately be used to describe the Opovo subsistence during the late Neolithic.

In the concluding chapters, I discuss relevant models about social context and land use during the late Neolithic period in the southern Pannonian Plan and the Balkans. In any kind of historic and particularly prehistoric reconstructions, we make many assumptions, and I often discuss more than one plausible interpretation. Some of those interpretations might be testable in the future, while others can provide relevant models for future anthropological endeavors.

Acknowledgments

Completion of this book would have been impossible without the cooperation and support from many individuals and institutions. This book was accomplished in three stages. First, I collected archaeobotanical material during the archaeological excavations of the late Neolithic site Opovo Ugar-Bajbuk, Voyvodina, Yugoslavia, 1983–1989. Dr. Bogdan Brukner, from the University in Novi Sad, Serbia (the former Yugoslavia) and Dr. Ruth Tringham from the University of California in Berkeley were the principal investigators of the Opovo Archaeological Project. The late Dr. Ljubomir Bukvić, the curator at the National Museum of Pančevo, was in charge of the organizational logistics. I am grateful to them for inviting me to participate in the project and to all the members of archaeological team for helping me with collection of samples at the site and with flotation. Second, I analyzed the majority of the archaeobotanical material during my graduate studies at Washington University, where I have received assistance from the faculty and graduate students, in particular from Dr. Gayle Fritz, Dr. Patty Jo Watson, Dr. Fiona Marshall, and Dr. Milica Banjanin. Third, I accomplished additional analysis and writing of the book while an assistant professor at the University of Birmingham at Alabama. Dr. Brian Hesse, the former Chair of the Department of Anthropology, suggested the title of the book and encouraged me to publish it in this series. Final editing was completed during my first year at Boston University.

Dr. Neal Lopinot, director of the Center for Archaeological Research, Southwestern Missouri State University in Springfield, kindly identified wood charcoal from six samples. Dr. Helmut Kroll, from the University of Kiel in Germany, helped me with the botanical identification of the ambiguous specimens using scanning electron microscopy (SEM) photographs. Dr. Manfred Rösch, Elske Fischer, Germany; Dr. Elena Marinova, Bulgaria; Dr. Sultana Valamoti, Greece; Dr. Ferenc Gyulai; and Dr. Brigitta Berzsényi, Hungary; provided invaluable information for the section "Opovo Plant Remains in Regional Perspective" (Chapter 3).

With Dr. Alyson Wylie, Columbia University, New York, I discussed the use of ethnohistoric data as offering possible analogies for prehistoric activities.

Dr. Stanija Parabućski from the Department of Botany at the University of Novi Sad helped with the reconstruction of potential vegetation. The late Drs. Miloje Sarić and Milorad Janković from the University of Belgrade, Dr. Ljubo Pavičevic, and Stevo Lopičić from Montenegro furnished additional information about water chestnut exploitation.

SEM photographs were done by Michael Vieth from the Department of Biology, Washington University, St. Louis and Wendy Carden, Vista Engineering in Birmingham. Dr. Ruth Tringham, Dr. Barbara Voytek, and Dr. Mirijana Stevanović, University of California at Berkeley, provided majority of the photographs of the Opovo archaeological features and some artifacts. Eugene Hoefel of St. Louis, drew the cover vignette; Kenyon Ross and David Fisher, of Birmingham, assisted with drawing of figures; and Mirela Butrić of Belgrade, helped with drawing the maps. Dr. Slobadan Marković of Novi Sad, supplied the topographic profile of Opovo. Dean Tenant McWilliams, University of Alabama at Birmingham, provided partial financial support for the illustrations.

Language editors LynnMarie Cook and James Lee Griner corrected the text. Dr. Justin Lev-Tov read the manuscript and together with Dr. Gabriele Faßback provided useful comments. Mihály Buzás, Novi Sad, did the computer layout and Kevin Mullen of Boston University made final corrections to the typeset manuscript. Trina Arpin of Boston University proofread several chapters.

My mother, Katarina Borojević, and my late father, Slavko Borojević—agronomists and plant geneticists, professors emeritii at the University of Novi Sad—introduced me into the scientific world and offered endless professional and material support throughout my career. My mother financed the final technical production of this book.

I am indebted and grateful to all of them. The greatest thanks go to my husband, Bogdan Milošević, and our son, Marko, for their help, emotional support, and patience throughout all these years.

CHAPTER 1. INTRODUCTION

A hundred years have passed since Vasić (1906) introduced the Vinča culture to the archaeological public. The type-site Vinča Belo-Brdo (Figure 1.1) became a renowned Neolithic site among numerous others of the Vinča culture discovered since then. The Vinča culture (Figures 1.2–1.6), one of the most outstanding Neolithic cultures of Southeast Europe, continues to attract the interest of many prehistorians and has been the subject of many studies, including this one.

Figure 1.1. The profile of the type-site Vinča-Belo Brdo and the Danube River, summer 2001.

Investigations of the Vinča culture traditionally have focused on the retrieval and analyses of "hard" artifacts such as pottery and figurines, and less on the retrieval of ecofacts, especially plant remains. Although plant macro remains (e.g., seeds, spikelet forks, fruit stones, nut shells, and wood fragments) were identified from several Vinča culture sites, most archaeobotanical endeavors concentrated on the identification of Old World crops, especially various kinds of cereals (see reviews by Kroll 1991a; Renfrew 1979). Less attention has been paid to the study of other plant species that were domesticated, such as oil and fiber crops, fruit trees and nuts, vegetables and tubers, condiments, and dyes, and even less to the study of plants collected in the wild. For decades, it was often assumed that Neolithic cultivators, including the Vinča culture settlers, did not depend much on wild plant resources once main staples were introduced. These premature assumptions, in addition to the focus of general interest on the origins of crop cultivation and domestication in the Balkans, influenced archaeologists to collect only visible grains of cereal and pulses. Archaeologists were limited by the sparse information about plant remains when discussing broad issues of the beginnings of agriculture and the role of the Balkan Neolithic cultures in it (e.g., Barker 1985; Benac 1978; Dennell 1992; Garašanin 1980; Glišić 1968; Halstead 1996; Kotsakis 2001; Srejović 1978; Todorova and Vaysov 1993; Tringham 2000). Another aspect of the prehistoric plant exploitation that has been taken into consideration by archaeologists, despite limited archaeobotanical data, are some features of the agriculture, such as location of the fields in terms of type of soils and distance from the settlements and the scale of cultivation of the Vinča farmers (see Barker 1985; Chapman 1981, 1990; Garašanin 1979; Tringham and Krstić 1990).

Unfortunately, systematic recovery of plant remains is not yet a standard procedure of many archaeological sites in the region. Even after implementation of water flotation for retrieval of plant macro remains at a few Vinča culture

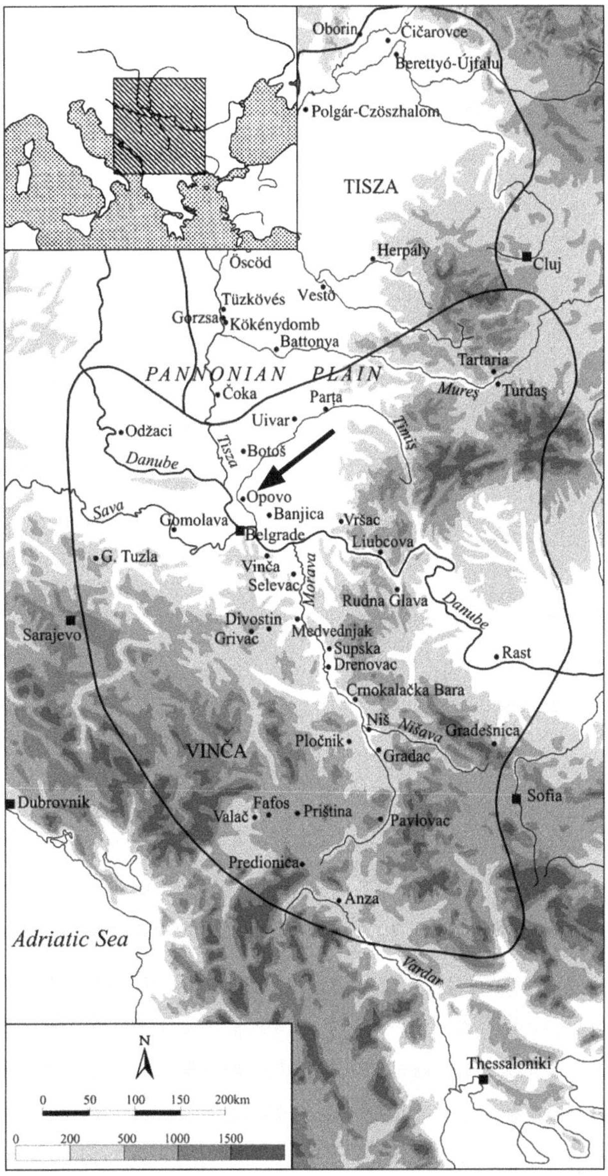

Figure 1.2. Map of the Vinča and Tisza culture showing important sites (modified after Gimbutas 1991 and Garašanin 1984). The Opovo site is indicated with an arrow.

sites, the archaeobotanical analyses were usually limited to the precise taxonomic identification of the plants, their relative abundance, and possible change through time (McLaren and Hubbard 1990; Renfrew 1974, 1976; van Zeist 2001/2002).

The goal of this work was to fill the gap in the studies of various aspects of human exploitation of plants on the Pannonian Plain during the Neolithic, and to provide an example of the potential of comprehensive studies of plants within the wider archaeological context of the Central Balkans and Southeast Europe.

The excavations (1983–1989) of the late Neolithic site of Opovo (Figure 1.3) and systematic recovery of plant and animal remains, fortunately, provided direct information on plants and animals used and deposited at the Vinča culture site. Analysis of plant macro remains and the context in which they were found, served as a basis for this study of taphonomic processes, crop processing areas, crop husbandry, wild procurement strategies, and social implications of the diverse subsistence practices at the Opovo site in the cultural setting of the Vinča culture.

The Opovo Site and the Vinča Culture Context

The Opovo site is located on the southern edge of the Pannonian Plain, 30 km northwest of Belgrade in Voyvodina, Serbia. Culturally, the Opovo site is located on the northern border of the Vinča culture domain, close to the southern limits of the Tisza culture sphere (Figures 1.2–1.4). The site of Opovo is a late Neolithic settlement of the Vinča culture and is dated on the basis of pottery typology to the Vinča-Pločnik I and IIa (Vinča-Pločnik C) period, which would place it ca. 4700–4500 B.C. (Table 1.1). According

Figure 1.3. Excavations at the Opovo site, 1983.

to the available radiocarbon dates from other sites (Gimbutas 1991:451), the occupation of the site would be partially contemporaneous with other Vinča sites such Vinča-Belo Brdo (layers 4.1 m–6 m), Gomolava (Ib), Banjica, and Gornja Tuzla.

The Vinča Culture

The Vinča culture is one of the most outstanding Neolithic cultures and also one of the southeast European prehistoric cultures extensively studied by national and international archaeological teams and individuals, starting with the seminal work of Vasić (1906, 1930, 1932, and 1936). In this brief review of relevant cultural history of the Vinča culture, I have relied primarily on information compiled from several syntheses of the Neolithic in general, and the Vinča culture in particular, including publications of Brukner et al. (1974), Chapman (1981), Garašanin (1979), Gimbutas (1991), Renfrew (1970), Srejović and Tasić (1990), Tasić and Petrović (1988), Tringham and Krstić (1990), Vinča exhibition catalogue (Srpska Akademija

Table 1.1. Periodization and chronology of the Vinča culture.

Date b.c. radiocarbon years (Tringham and Krstić 1990)	Date B.C. calender years (Tringham and Krstić 1990)	Date B.C. calender years (Gimbutas 1991)	Elevations–depth at Vinča–Belo Brdo (Garašanin 1984)	Garašanin (1984)	Milojčić (1949)	Shier (1996)	Date B.C. calender years (Shier 1996, 2004)
4500	5000	5400	pit houses at 8 m	Vinča–Tordoš I	Vinča A	Vinča A1–A3	5200 (A3/B1)
4300			8 m–ca.6.5 m	Vinča –Tordoš II (possibly 2 phases)	Vinča B1, B2/C	Vinča B1a, B1b, B2	
4100			6.5–6 m	Gradac Phase	Vinča C	Vinča C1	4950/4900 (C1/C2)
4000			6–4.1 m	Vinča–Pločnik I	Vinča C	Vinča C2, C3	4700/4600? (C3/D1)
3950			4.1 m	Vinča–Pločnik IIa	Vinča D	Vinča D1	
3850–3300	4500–4000	4000	3.48 m	Vinča–Pločnik IIb			

Nauka i Umetnosti, *Vinča u praistoriji i srednjem veku* 1984) and others.[1] Although much general information about the Vinča culture is well known, the genesis of the Vinča culture is still disputed (e.g., Benac 1990; Garašanin 1984; Özdoğan 1993, 1997).

Setting and Chronology

The Vinča culture complex occupied the central Balkans, including central Serbia. Northward it expanded to the Banat region on the Serbian and Romanian side, eastwards to the western area of Bulgaria, southward to the northern area of Macedonia, and westward to the area of eastern Bosnia (Figures 1.2 and 1.4). There are many regional facies (variants) of the Vinča culture complex. These regional facies did not last equally long. The classical core of the group is considered to be in Serbia and Voyvodina, where there is evidence for the most intensive development of the Vinča culture.

The Vinča culture covers approximately the same territory of the preceding Starčevo culture from which it may have developed (Figure 1.4). Based primarily on the different pottery types and ornamentation techniques, as well as the type of dwellings, some authors favor a migration model, arguing that the Vinča complex is so different from the Starčevo culture that it represents a combination of slow migratory influx and acculturation to the so-called Balkan-Anatolian complex (Garašanin 1984). More recently, Özdoğan (1993, 1997) revived the idea of the central and western Anatolian origins of the Vinča culture, primarily based on new results from excavations in northwest Turkey.

Based on the change of pottery styles and settlement types, the disintegration and collapse of the Vinča culture are interpreted as due to the arrival of culturally different groups from the east for whom a pastoral way of life and metallurgy were the main activities, replacing Vinča traditional farming communities (Garašanin 1984). As with any society, general internal economic and cultural decline usually precedes and facilitates collapse and foreign intrusion. Chapman (1981) believes that intensive cultivation of fertile soil for two millennia could have exhausted the soil, especially on marginal lands, and would have affected the degree of settlement nucleation and the availability of ritual surplus. According to the author, the focus of innovations shifts from the Vinča culture to the east, i.e. to the smaller settlements of the Gumelnitsa and Cucuteni-Tripolye cultures (Chapman 1981).

The relative chronology of the Vinča culture is established on the basis of stratigraphy and typological ceramic change at the type site of Vinča-Belo Brdo (Figures 1.1). The cultural layers at this tell are in some parts over 9 m thick. The Vinča culture layers extend variably from 8 m to 3.5 m or 1 m in depth. Commonly, two corresponding chronological periodizations are used interchangeably (Garašanin 1951; Milojčić 1949), although new ones are becoming available (Schier 1996). The early phases of the Vinča culture are Vinča-Tordoš I and II, corresponding to Vinča A and B. The late phases are called Vinča-Plocnik I and II, corresponding to Vinča C and D. Between the early and late phases is an intermediate period called the Gradac phase or Vinča B2, which is important for the beginnings of metallurgy (Jovanović 1982). Because of the evidence of the indigenous copper metallurgy, the later phases of the Vinča culture are also included in the Eneolithic period (Table 1.1).

The Vinča culture lasted from 5300 to 4000 B.C., based on radiocarbon dates (Gimbutas 1991:451). However, the duration of various phases within the Vinča culture is uncertain because some radiocarbon dates indicate a long occupation of sites, which is not corroborated by stratigraphic data and vice versa (Garašanin 1984; Stevanović and Jovanović 1996). Nevertheless, the Vinča culture lasted more than one millennium and is considered to be a typical Neolithic culture.[2]

Vinča culture material record exhibits similarities with several contemporaneous cultures that are differently classified by southeastern archaeologists to the Middle Neolithic, Late Neolithic/Chalcolithic, or Eneolithic cultures of Southeast Europe (Figure 1.4). To overcome the terminological complexity of the local culture designation, Chapman (2000) has suggested a broad categorization into five stages of the Balkan Mesolithic, Neolithic, and Copper Age: late Mesolithic (8000–6000 B.C.), early farmers (6000–5000 B.C.), mature farmers (5000–4000 B.C.), climax Copper Age (4500–3500 B.C.), and post-climax Copper Age (3500–2500 B.C.). The earlier part of the Vinča culture is included in the "mature farmers" and the late Vinča culture in the "climax Copper Age" stages (Chapman 2000:3).

Settlements and Cemeteries

A total of ca. 650 Vinča sites are registered, in contrast to only two known cemeteries, Botoš and Gomolava (Figure 1.2). Most of the sites are flat, although the tell settlements of Gomolava and Vinča-Belo Brdo are also known. The site territories range in size from less than a hectare to more than 100 ha (e.g., Potporanj). Many of the sites were occupied for several generations, with occupation especially continuous on tell sites (Chapman 1981). The type site of Vinča-Belo Brdo has yielded over 7 m of stratified Vinča layers (Figure 1.1). Ditches were uncovered in several sites, including the tell site Vinča-Belo Brdo, but also in flat sites such as Jakovo-Kormadin. In most cases, the ditches were dated to the later phase of the Vinča culture period; one exception was the first ditch at the type site of Vinča-Belo Brdo which was built at the end of the early Vinča phase. The remains of the defensive ditches around some of the settlements point to periods of unrest during the late Vinča-Pločnik period (Garašanin 1979). Recently,

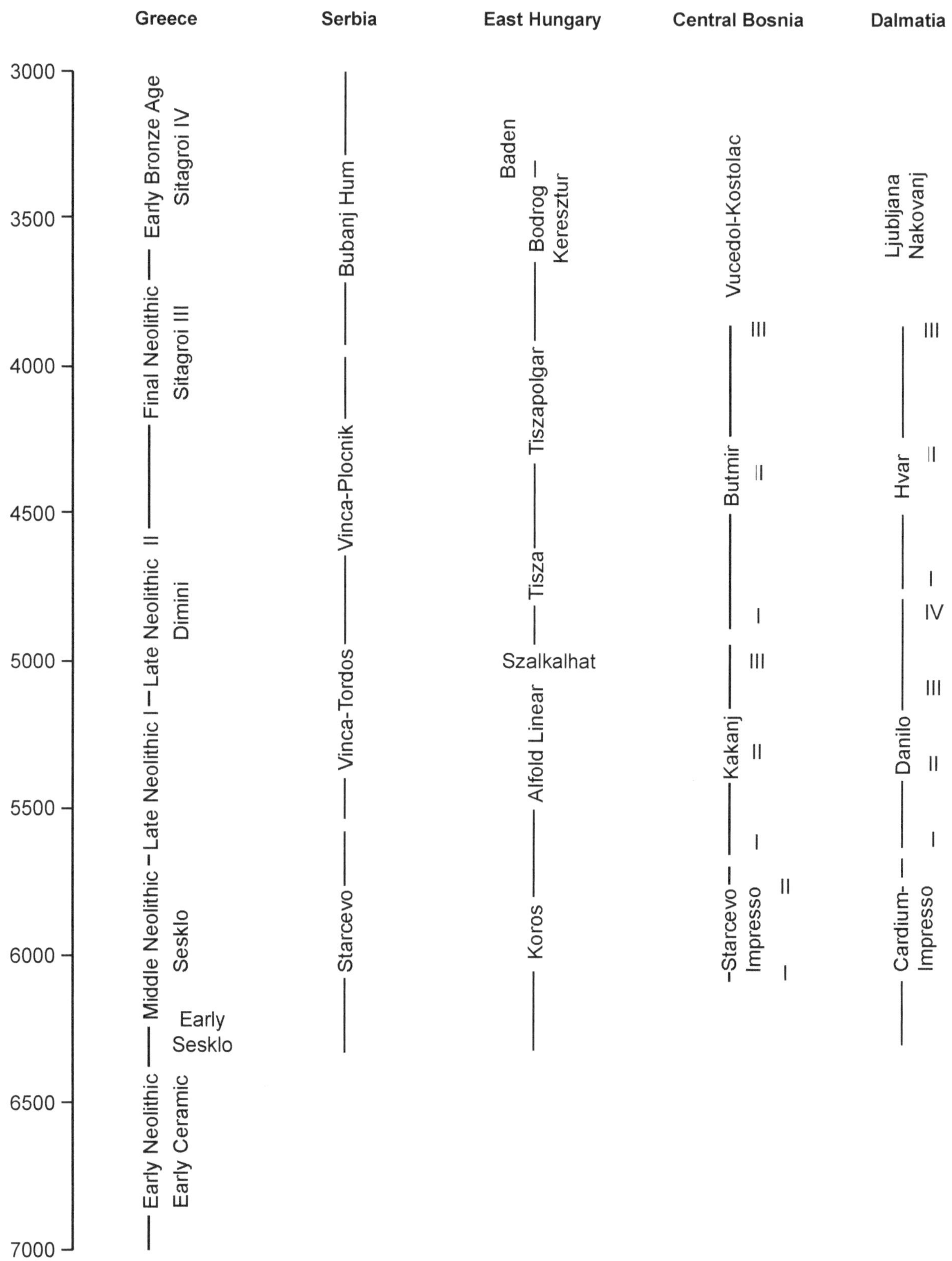

Figure 1.4. Simplified chronological chart of the Vinča culture and other contemporaneous cultures of Southeast Europe (modified after Whittle 1996 and Garašanin 1984).

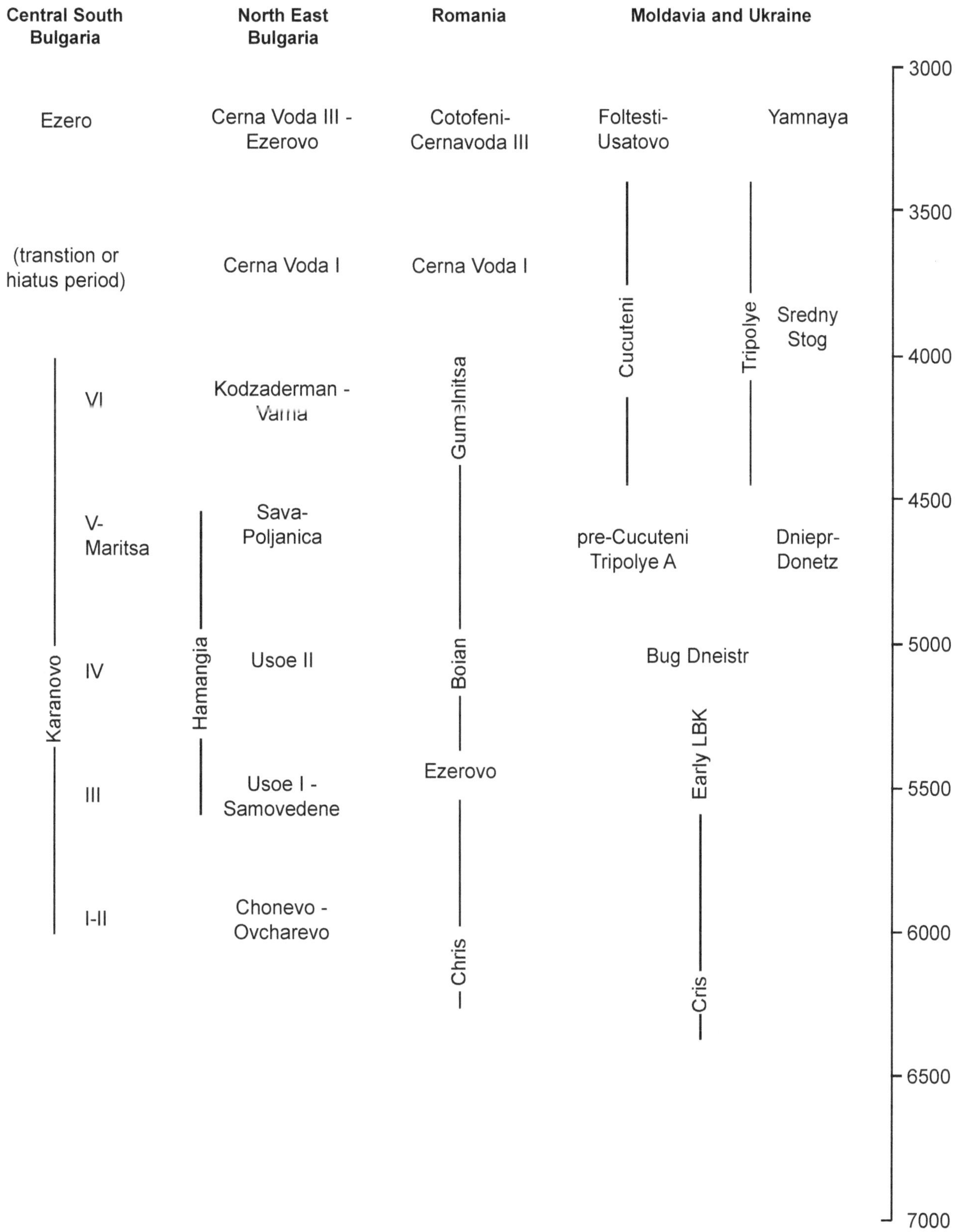
Central South Bulgaria
North East Bulgaria
Romania
Moldavia and Ukraine
3000
3500
4000
4500
5000
5500
6000
6500
7000
Ezero
Cerna Voda III - Ezerovo
Cotofeni- Cernavoda III
Foltesti- Usatovo
Yamnaya
(transtion or hiatus period)
Cerna Voda I
Cerna Voda I
Cucuteni
Tripolye
Sredny Stog
VI
Kodzaderman - Varna
Gumelnitsa
V- Maritsa
Sava- Poljanica
pre-Cucuteni Tripolye A
Dniepr- Donetz
Karanovo
IV
Hamangia
Usoe II
Boian
Bug Dneistr
III
Usoe I - Samovedene
Ezerovo
Early LBK
I-II
Chonevo - Ovcharevo
Chris
Cris

Figure 1.5. Remains of the burnt house built in wattle-and-daub at the Vinča site Banjica, Belgrade, 1979.

at the tell Uivar in the Romanian Banat, a complex system of ditched enclosures was discovered around a fortified settlement dated to the later phase of the Vinča period (Schier and Draşovan 2004).

Vinča houses were usually rectangular and built with a wattle-and-daub technique (Figure 1.5), although semi-subterranean dwellings were also recovered. The ground for house construction was usually well leveled. Most of the houses were densely built and almost parallel to each other, though some were grouped in clusters. Houses were spaced with only a few meters apart on tells, while in other cases there was more space between houses clusters. The houses were generally not large, typically around 8×4 m in size, indicating they were residential units for small families. However, a few larger houses are also known (e.g., 20 m in length). Internal division in the houses was also common (Brukner 1988). Some had domed ovens. In some of the houses, fragments of plaster with painted ornamentation were preserved. Most of the Vinča houses were burnt in the past, thus, large quantities of house rubble were preserved (Stevanović 1997a).

Human skeletal remains were excavated at the two cemeteries, Botoš and Gomolava. The deceased were buried in a flexed position and placed either on the left or right side. Both cemeteries are outside settlements, although in some cases individuals were buried inside a settlement. An individual grave was found at the type site Vinča-Belo Brdo, dated stratigraphically to the earlier phase Vinča-Tordoš II. On the basis of typological pottery analysis from the Gomolava graves, this cemetery is dated to the later phase, Vinča-Plocnik IIb (Vinča D2). Grave goods included mostly pottery, but also stone axes, limestone beads, shell bracelets, pendants, beads, and in the case of one Gomolava grave, a copper bracelet (Brukner 1988).

Figure 1.6. The Vinča culture ceramic objects (clockwise from upper left):

a) Zoomorphic lid (height 13.7 cm), Vinča-Belo Brdo (after Srejović 1967);
b) Black burnished pot (height 13.2 cm), Vinča-Belo Brdo (after Srejović 1967);
c) Fragmentary female figurine "Lady of Vinča" (height 15 cm), Vinča-Belo Brdo (after Srejović 1967);
d) Bear-like figurine (height 10 cm), Opovo (after Tringham et al. 1985).

Ceramics and Other Finds

Vinča sites are known for their wealth of pottery and other ceramic objects such as figurines, altars, zoomorphic lids, spheres, and conical weights that were found within and outside houses (Figures 1.6 and 1.7). Besides large storage jars, the most typical Vinča vessels are black burnished biconical bowls decorated by channeling or incising (Garašanin 1979). Technological and economic aspects of the Vinča ceramics were also investigated (Kaiser 1984). Anthropomorphic and zoomorphic figurines that vary in size and ornamentation through time and region have been studied in great detail (e.g., Gimbutas 1991; Srejović 1984). There are different interpretations of the meanings and the typological development of the figurine styles, but they were probably used in domestic cults because they were encountered in many domestic contexts, including pits. In several sites, bucrania were found, and in some in stances these houses are interpreted as shrines (Jovanović and Glišić 1960) or even as temples (Gimbutas 1991).

Various recurring incised signs were found on a number of Vinča ceramic vessels, figurines, seals, and on the few tablets. The signs are variously interpreted by various authors as representing either property marks, or ritual symbols, or the earliest form of writing, predating the Sumerian writing by two thousand years (see Gimbutas 1991). The signs have also been interpreted as the Vinča alphabet that is claimed to be ancestral to Etruscan (Pešić 2001).

Copper beads have been found on many sites from the Late Vinča period. Furthermore, a copper bracelet was found on an individual's hand in the Gomolava cemetery. Also, four hoards with copper and stone tools were found at the site of Pločnik and could be also assigned to this period. Copper was obtained from the ores in Rudna Glava in eastern Serbia, where finds of Vinča pottery testify to the existence of local metallurgy (Jovanović 1982). Imported goods include artifacts or ornaments made from spondylus shell, obsidian, and alabaster. Trade items also included pottery sherds from neighboring groups (Garašanin 1979). The chipped stone industry of the Vinča culture is a blade tradition, produced mainly from flint sources in the vicinity of the settlements. The raw materials for stone tools were mainly exploited from northern and eastern Serbia. The use wear analyses show high frequency of knives used for woodworking of small wooden artifacts. Harvesting tools and end-scrapers also appear, but the frequency of these tools changes through time on various sites and does not show a uniform trend (Kaczanowska and Kozłowski 1990).

Subsistence

The Vinča society is commonly perceived as comprising agricultural communities that grew cereals and raised mostly cattle and pigs, and to a lesser extent, sheep and goats. Cattle was the primary source of meat at many Vinča sites, including Gomolava, Divostin, Selevac, and

Figure 1.7. The Vinča bowl (height 5.2 cm; diameter 14.6 cm) with carbonized wheat (after Vasić 1936).

Vinča (Bökönyi 1990; Clason 1979; Greenfield 1986; Legge 1990). With the increasing prevalence of identified domesticated animal bones, it was assumed that hunting did not play an important role in the subsistence strategy of the Late Vinča period (Garašanin 1979:197). The bones of wild fauna are present in many Vinča assemblages, including red and roe deer, aurochs, wild pig, and some smaller animals (Bökönyi 1990; Clason 1979; Russell 1993). Bones and teeth belonging to a wild horse and a small horse-like animal were identified from the Vinča layers at Gomolava (Clason 1988). Discovery of fishing equipment and fish bones on some sites indicates that fish also formed part of the diet (Legge 1990; Russell 1993).

The most ubiquitous crops found in the Vinča culture were various types of wheat, particularly glumed wheats (see Table 2.5). Already, during the early excavations of the Vinča-Belo Brdo site in 1933, a pot that contained carbonized "grains of various species of wheat" was discovered along a wall of a house (Figure 1.7). The grains were identified by agronomist S. Lomejeko. The find provided evidence that Vinča inhabitants were growing and consuming wheat (Vasić 1936:170–173).

On many sites, emmer and einkorn were found together (Hopf 1974, Renfrew 1974, 1976). There is evidence for bread wheat. Barley was somewhat less common than emmer and einkorn, although it was recovered from several sites. Millet was identified only in the samples from Gomolava, dating to the Vinča period (van Zeist 1975, 1978, 2001/2002). Seeds of oats were recovered, although the distinction between wild or domestic species is generally unclear. Among pulses, the seeds of lentils, peas, and vetches were retrieved. In addition, flax seeds were found at Gomolava. Some reports mention wild fruits including the stones of Cornelian cherry and seeds of pears, apples, wild strawberries, dewberries, elderberries, and acorns. Among the weedy herbaceous plants, the seeds of black bindweed, corn cockle, goosefoot, rye, brome, and bristle grasses are mentioned. Available archaeobotanical data (see Table 2.5) implied that a similar spectrum of plants could be expected at the Opovo site.

The Opovo Site

The Opovo-Ugar Bajbuk site is located near the present-day village of Opovo in the Banat region in eastern Voyvodina, which lies within the southern part of the Pannonian Plain. The site of Opovo is situated ca. 4 km away from the present course of the river Tamis and less than 20 km away from the Tisa and the Danube rivers, respectively (Figure 1.8).

The Opovo Project

The archaeological excavations of the late Neolithic site Opovo-Ugar Bajbuk, Voyvodina, Serbia (1983–1989) were carried out by an international archaeological team lead by Professor Bogdan Brukner, University of Novi Sad, Serbia and Professor Ruth Tringham, University of California, Berkeley, USA.[3] The excavations revealed the remains of a small late Neolithic settlement of the Vinča culture dated to ca. 4700–4500 B.C. (Tringham et al. 1985, 1992).

The main research objectives of the Opovo project were to investigate: 1) change in the Neolithic economy north of the Danube; 2) degree of sedentism in the occupation of the Opovo site; and 3) emergence of the household as primary social unit and economic organization at the Opovo site (Tringham et al. 1985:427). Research at Opovo has focused on recovery and analyses of the following classes of data: architectural, ceramics, lithics, floral, and faunal remains. Within this general framework, my research has focused on the retrieval of floral remains and flotation during the excavations at the Opovo site and subsequent identification of plant macro remains. My main research of objectives were: 1) to infer prehistoric agricultural practices and wild plant procurement, and 2) to assess the suitability of the presented data for interpretation of land use and its consequences for social differentiation in the late Neolithic of the Balkans.

Setting and Chronology

The prehistoric, single component site is located on a well-drained hillock, covering an area of ca. 250×200 m that is surrounded by lower-lying, seasonally waterlogged land. The elevation of 78 m represents a degraded loess terrace following an old meander path of the Tamis River (Figure 1.8). According to pedological and geological investigations done by a team of Dutch soil scientists from the University of Wageningen, the site was located on a small island surrounded by waterlogged or marshland areas that were rich in wild game and plants and whose area of nearby agricultural soil was limited, but highly fertile. It is also supposed (but by no means proven) that the wooded areas around the site would have been limited so that the natural vegetation on the non-marshy area would have been steppe grassland (van der Plas et al. 1985). Such a site location, and such abundance of wild animal bones, is atypical of the Vinča settlements.

Excavations

Block I, measuring 16×20 m, was excavated to an average depth of 1.6 m below the modern soil surface. Within Block I, three Building Horizons (BH 1, BH 2, BH 3), or phases of building activity, were recognized (Figure 1.9). Each horizon represents an architectural phase, with BH 1 as the most recent. These three building horizons represent consecutive periods of construction and continuous occupation during approximately 200 years. Remains of three complete houses were excavated, in addition to the remains of three or four partially preserved houses and many pits. The houses were constructed using the traditional Vinča technique of wattle and daub and were interpreted as residential structures (Figure 1.10). The houses were smaller than typical Vinča houses and without clear internal divisions, except for House 4 which had a second story or mezzanine. A typical Vinča indoor oven was found only in House 1 (Feature 2), although remains of several outdoor ovens were discovered. At Opovo, houses were rebuilt on top of the previous ones and eventually were burnt in separate conflagrations. Most of the pits (e.g., Figure 1.11) were dated either to BH 2 or BH 3. The site of Opovo-Ugar Bajbuk is dated on the basis of pottery typology to the Vinča-Pločnik I and IIa (Vinča-Pločnik C) period (Table 1.1), which would place it ca. 4700–4500 B.C. One archeomagnetic date from a clay sample in House 1 provided a date of 3700 + 100 b.c. uncal. (Tringham et al. 1985, 1992).

Ceramics, Lithics, and Other Finds

A few Tisza culture pottery sherds were recovered from the site, but the pottery is predominantly of the Vinča type. Sherds of typical Vinča biconical bowls made of untemperd clay, followed by amphorae and casseroles were found in large number at the Opovo site (Figure 1.12). Most of the biconical bowls were made from locally available untempered clay. Pots tempered with agate and volcanic rock, were probably imported because these minerals were not locally available, and were most abundant in the BH 3 horizon. Grog was often used as a temper for open bowls and amphorae. Temperatures that were attained for firing pottery at the Opovo site are also typical for the Vinča culture, and ranged between 740–1050 °C. Functional analysis of the pots showed that the ratio between serving and cooking vessels versus storage jars varied among building horizons. In the earliest occupation of the site (BH 3), there were many more sherds of large pots made of coarse pottery, while in the latest horizon (BH 1), there were more

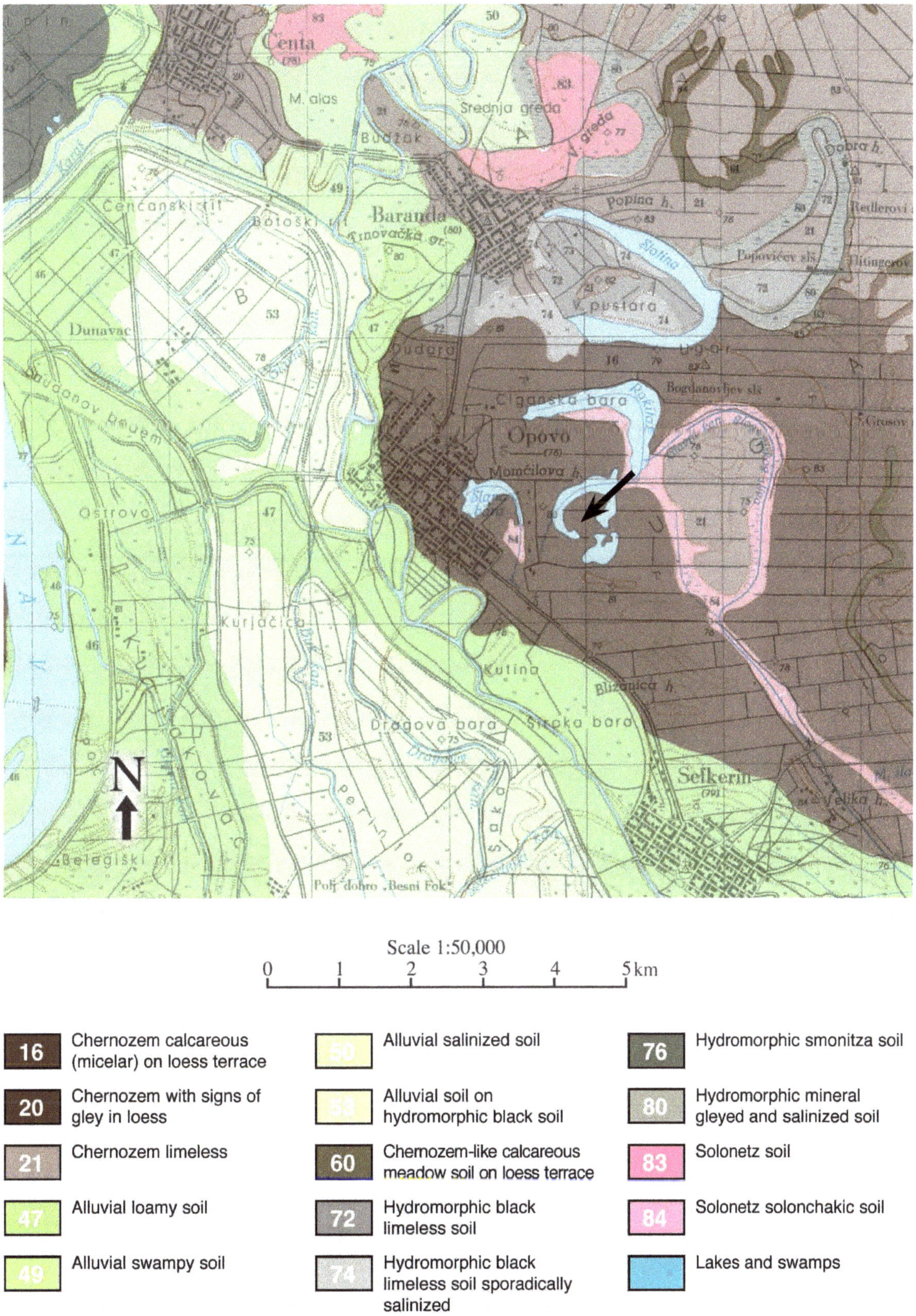

Figure 1.8. Map of soils around Opovo. Arrow indicates location of prehistoric site (adapted from *Pedološka Karta Vojvodine*, Nejgebauer et al. 1971).

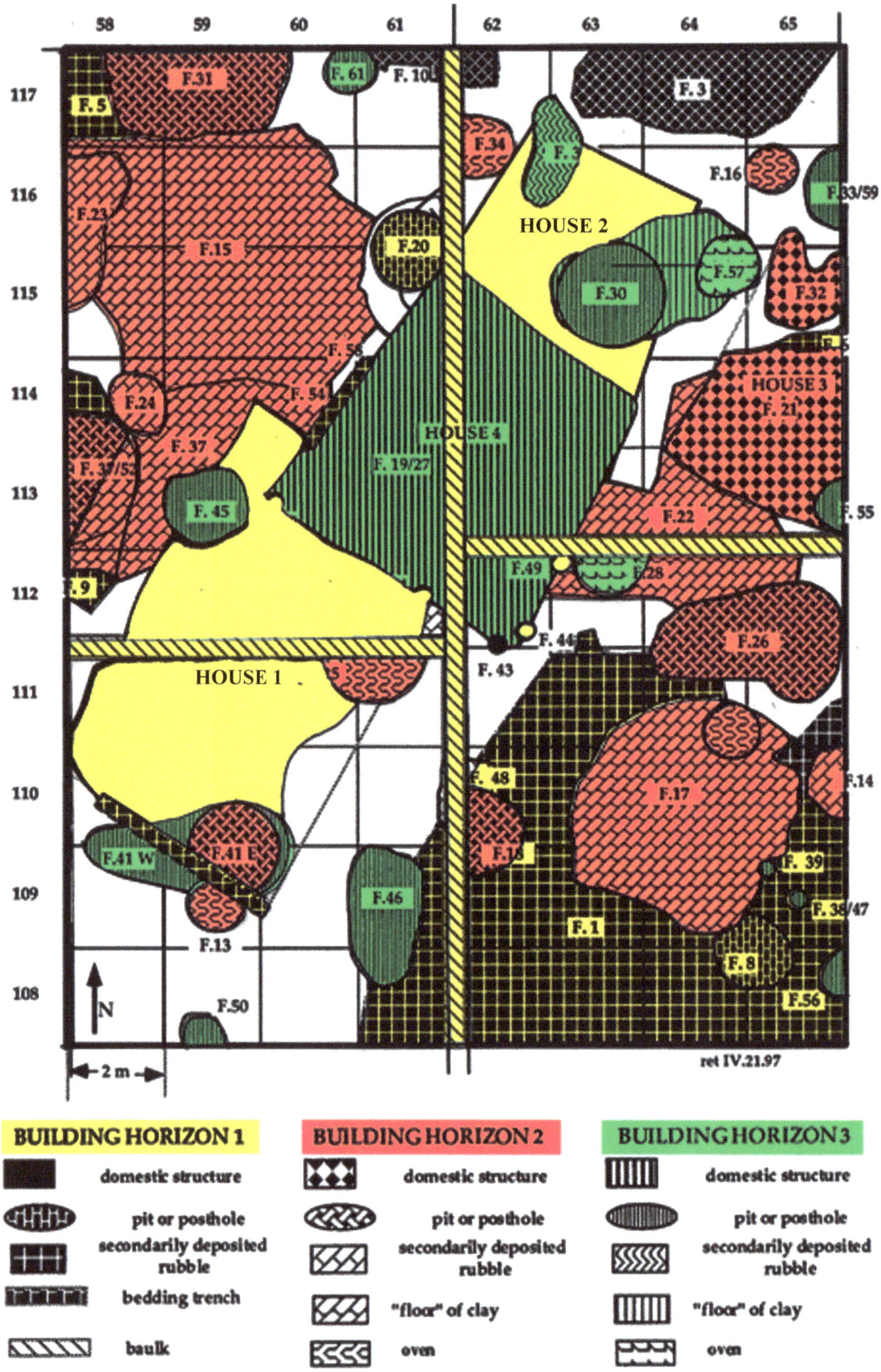

Figure 1.9. Building horizons at the Opovo site (modified from Stevanović 1997a ©Creative Commons License).[4]

Figure 1.10. View of the Opovo site looking north, excavations 1983.

sherds made of fine pottery used for cooking and serving (Kaiser in Tringham et al. 1992:376). Thus, it can be suggested that storage was more important in the older phase than in the younger, when the use of serving and cooking vessels became more prevalent. On the basis of pottery analysis, it seems that the "model of short-term residence and specialized food procurement is more appropriate than the model of sedentary, marginal Vinča hunters related to the backwaters of the middle Danube basin" (Tringham et al. 1992:377).

Relatively few ceramic figurines were found at the Opovo site in comparison with some other Vinča sites where hundreds of figurines were excavated. Stylistically, the figurines are rather simple and without ornamentation, and have a "bear-like" form (Figure 1.6d), that is more characteristic of the later phases of the Vinča culture than would be otherwise conjectured from the pottery typology.

Chipped stone artifacts, ground-edge stone tools, and grinding implements were recovered from the Opovo site (see Figures 5.5 and 5.6). Because there are no mineral sources in the vicinity of the site, it is believed that the source of most of the stone implements is the mountain Fruška Gora, 45 km away. Only a few stones had cortex, and there was an absence of debitage concentrations;

Figure 1.11. Pit (Feature 48) from the Building Horizon 2.

Figure 1.12. Floor with broken pots.

therefore, the preliminary core preparation of the chipped stone artifacts took place away from the settlement, perhaps at the source of rock procurement. Some of the tools were made from obsidian, and there is a higher percentage of obsidian tools in the later building horizons (9%) than in the earlier (3%). More obsidian tools were found in the Opovo site than in the southerly Vinča sites (e.g., Gomolava and Selevac). The closest available obsidian source are the mountains on the northeast border of Hungary and Slovakia, which are closer to the Opovo site than to the more southern Vinča sites. Use-wear analysis showed that many tools, including polished adzes and chipped stone artifacts, were used on wood, especially during the earliest occupation of the site (Voytek in Tringham et al. 1985; 1992). Also, the tools from the earliest building horizon seem to have been hafted in handles that were made of wood rather than of bone or antler. In the latest building horizon (BH 1), most chipped stone tools were used on softer materials such as cured skin or soft woods. The majority of grinding implements were recovered from this building horizon. From the wear pattern and choice of rock, it appears that the grinding implements were not used exclusively for cereal grinding. Blades and blade segments that had sickle gloss comprised 14% of the total blade assemblage in BH 1 (see Figure 5.5). In BH 3 and BH 2, a slightly smaller percentage of sickles was encountered (8.9% and 9.8–11.1%, respectively). Likewise, grinding stone fragments were less abundant in BH 3 (43.9 %) than in BH 2 and BH 1 (56.2–67.6%).

A fragment of woven linen textile 3.5×3.3 cm large (Figure 1.13), and a fragment of cord made from flax were discovered in House 4. Furthermore, two small fragments of copper oxide were discovered, implying links with the southeastern Vinča sites where copper ore was exploited (Tringham et al. 1992).

Subsistence

The Opovo site was exceptionally abundant in wild animal bones, compared with contemporaneous Vinča sites where bones of domestic animals, cattle, pigs, and occasionally sheep predominated. At the Opovo site, most abundant were the bones of roe deer, red deer, and wild boar, which are primarily forest species. In all three building horizons the percentage of wild taxa was above 50% based on the Diagnostic Zone measurements (DZ).[5] The greatest percentage of wild bones (70.9%) was in the latest building horizon; a smaller percentage (69.8%) was from the second building horizon; and the lowest (69.2%) from the earliest building horizon. However, if the percentages are computed on the basis of the Minimum Number of Individuals (MNI) a slightly different pattern occurs: the highest percentage (63.9%) was in the earliest building horizon with the smaller percentages in the second horizon (55%) and latest horizon (60.2%) (Russell in Tringham et al. 1992:367). Within the identified taxa, the number of red deer and cattle bones increased, while the number of roe deer and ovicarpids decreased over time. Wild cattle were almost completely absent from Opovo, while they were common in the Tisza culture settlements in the northern parts of the Pannonian Plain.

Figure 1.13. Textile fragment made of flax fibers from House 4 (Tringham ©Creative Commons License).

Previous studies of Vinča subsistence implied that at the Opovo site, a typical plant spectrum of emmer, einkorn, barley, and legumes should be expected, in addition to domestic animals (cattle, pig, sheep, and goats). However, studies from various Mesolithic and early Neolithic sites in the Mediterranean and Baltic (e.g., Donahue 1992; Halstead 1996; Zvelebil 1996) revealed that different subsistence patterns and pathways to agriculture are possible and should be considered in the interpretation of the subsistence data.

CHAPTER 2. OPOVO PLANT REMAINS

Recovery and analysis of ecofacts, including plant remains, were integral part of the Opovo excavations. Special attention was paid to the field recovery of samples during the excavations, flotation retrieval, and subsequent laboratory analysis of plant remains. I examined and identified macro plant remains, addressing the specific issues regarding the Opovo site: taphonomic processes and the mode of arrival of the identified plants within the various features and contexts, possible uses of plants, potential vegetation, permanency versus seasonality of the site occupation, differences in plant species composition among the features and the three building horizons, comparison between faunal and floral data from the same features, intra-site differentiation among households (features), and possible location of ancient fields.

Methods

Soil (sediment) samples were gathered in the field by the archaeological crew, and subsequently, I was in charge of flotation and the analysis of retrieved plant remains. In archaeobotanical samples gathered during excavations of the Opovo site, I identified more than twenty taxa (Tables 2.1–2.3). I divided the identified plant taxa from the Opovo samples into several large, arbitrary categories: Cereals, Legumes, Oil and Fiber Plants, Fruits, Water Chestnuts, Weedy Herbaceous Plants (Wild/Weed), Unknown Seeds, and Unidentifiable Fragments. Detailed accounts of field recovery and laboratory methods are presented here so that the results of the analysis of plant macro remains from the Opovo site could be compared with results from other sites. Within each large category of plant remains, each identified plant taxon from the Opovo archaeobotanical samples is fully described and compared with the finds from contemporaneous sites from the Pannonian Plain and central Balkan region.

Field Recovery of Plant Macro Remains

In 1985, a systematic screening and flotation sampling program was started. At first, stratified random sampling both for screening (mesh size of 1 cm) and flotation was carried out. The emphasis was placed on taking samples from and beneath features rather than around them. In subsequent years, below Level 3, the sampling strategy was changed to randomly selecting 1 m^2 out of every 2 m^2 quadrant (e.g., 58/108) throughout Block I (Figure 1.9) for screening thus giving a systematic 25% sample. Therefore, this sampling would ideally result in 20 routine screening samples and 20 flotation samples (ca. 10 liters each) from every randomly selected 1 m^2 taken from every level. Because differences between soil color and texture were almost invisible, excavation proceeded in arbitrary levels, except in the houses where the rubble was removed in natural layers.

Soil samples for flotation were taken from all features and often from various parts of the features. Not all designated areas were sampled, as some areas in a level consisted entirely of house rubble. The routine (random) sampling program was designed to provide a background sample to help balance the bias in the feature-oriented strategy, to explain the nature of the intrusive material and its changes across the site and between layers, and to increase the probability of finding more carbonized plant material (Tringham et al. 1985). Many samples were taken from features; therefore, many contained large quantities of burnt house daub. Daub was especially abundant in the heavy fractions of floated samples.

Samples for flotation were taken by archaeologists working on the site. Samples were tracked in the flotation sample field book in which provenience (including the locus number), the quantity of soil, and the flotation sample number were recorded. Each provenience unit was assigned a locus number which included the contextual description, recording quadrat, level, depth, and feature number. Locus numbers were the primary identifiers for the subsequent analyses.

In the upper five layers, only small samples were taken, weighing 3 kg of sediment per sample. Water flotation was carried out by hand in washtubs. Because the preliminary analysis showed that very few plant remains were recovered, the volume was enlarged to ca. 10 liters (the volume of a standard plastic wash-tub) weighing on average 13 kg. In addition, in cases where charred plant remains were observed in the field with the naked eye, more than one standard sample was taken in order to retrieve more diverse plant remains. Thus, in the lower layers, often an amount of sediment equivalent to 30 liters in volume was floated from the same locus. Further enlargement of the flotation samples did not yield any new plant taxa.

Because the site was located several kilometers away from a water source, bagged samples had to be transported by a jeep to the flotation spot. Flotation was carried out in a farmer's backyard in the open air. Since 1986, when soil samples were enlarged, the flotation was carried out with the aid of a flotation machine, a version of the "Ankara machine", which was locally made out of a metal gas tank (Figure 2.1). The water was supplied through a garden hose from the local water-pipe system. The sediment samples were poured into the inner large sieve, lined with a nylon mesh to hold the heavy fraction. The aperture of the mesh was ca. 1.0 mm. After the samples were poured into the inner large sieve, the flow of water was enhanced, and the sediment was agitated by hand. Floating light fractions were collected in a fine geological sieve with a mesh opening of 0.5 mm over which a fine mesh nylon cloth was placed with openings certainly smaller than 0.3 mm. The water was not recycled. After a few samples accumulated sludge was emptied through the drain based at the bottom of the tank. Dried samples of heavy and light fractions were packed separately in plastic bags labeled with their proveniences. I took all light fraction samples for further analysis, while heavy fraction samples were placed in boxes and deposited in the Pančevo Museum, in Voyvodina. Later, subsamples of some heavy fraction samples from various features were also sorted and prepared for analysis.

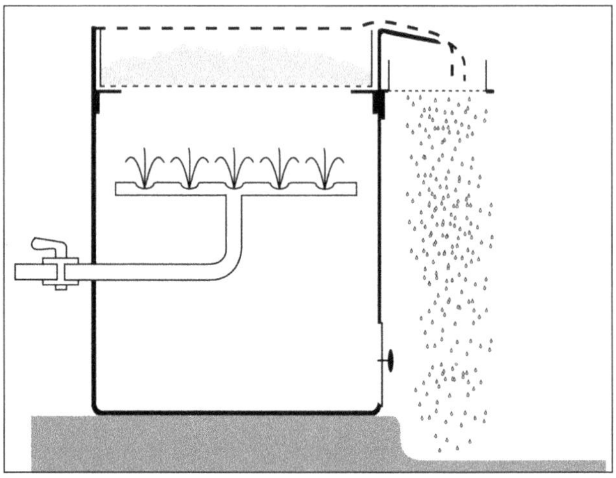

Figure 2.1. Diagram of the floatation set-up used in the Opovo project.

In summary, 523 sediment samples were taken for flotation during six years of excavation from 1984–1989. Flotation was initiated and carried out by J. Rossignol (then a graduate student at the University of New Mexico) from 1984 to 1986; subsequently, I performed most of the flotation with the help of a worker and two students. During 1984 and 1985, flotation was carried out by hand because the samples were rather small, weighing on the average 3 kg. Results of the analysis of these samples are not included in the final analysis because the samples were extremely poor in carbonized plant remains and did not contain any additional plant taxa. Beginning with the 1986 season, samples were enlarged to a volume of 10 liters, and the flotation machine described above was used. Some 3,000 liters of sediment were floated. Starting with the 1986 samples, all 267 light fraction samples were analyzed along with 40% of the heavy fraction samples (92).

Laboratory Analysis of Plant Macro Remains

I analyzed archaeobotanical samples primarily at the Paleoethnobotany Laboratory of the Department of Anthropology, Washington University in St. Louis, and later in the Paleoethnobotany Laboratory of the Department of Anthropology at University of Alabama at Birmingham (UAB). Sediment (soil) taken for flotation was generally not pre-sieved. Dry light-fraction samples were sieved through two U.S. graduated geological sieves (0.5 and 1.0 mm) to facilitate sorting. In later years, a fine geological sieve less than 0.3 mm was also used for separation. Only plant macro remains were sorted from stone/soil/house rubble/bone and snail; and uncarbonized, recent (modern) plant intrusion matrix. Hence, only carbonized or mineralized plant material further recorded and sorted for further analysis. All other material was left as residue. Carbonized wood was separated from identifiable seeds. Wood was not generally identified to the exact taxon, nor was it weighed if it was present only in a very small quantity (where a few particles, smaller than 2–3 mm were present). However, the presence or absence of wood was indicated for each sample and rated on the scale from 1–3, where 1 was equal to little, 2 was equal to a medium quantity, and 3 was equal to a large quantity of wood charcoal found in a sample. Larger wood charcoal fragments from the heavy fraction were also weighed. Wood charcoal specimens from six samples were identified by Neal Lopinot (personal communication, 1997).

The greatest importance was placed on the identification of the specific taxonomic category for seeds, stones, or other preserved parts such as spikelet forks, rachis fragments, water chestnut fragments, etc. In the cases when a specimen could not be identified to species, a new preliminary taxonomic category such as *Triticum monococcum/Triticum dicoccum* was introduced. This arbitrary category was established for plant parts which had features of both species but lacked diagnostic characteristics necessary for precise identification.

Identification of plant taxa was determined with the aid of low power (7–50×) binocular microscopes (Figure 2.2). In addition, scanning electron microscope photos were used for more precise identification. For seed identification, I used my own reference collection and that of the Paleoethnobotany Laboratory at the Department of Anthropology, Washington University in St. Louis. Various seed manuals and floras (Beijerinck 1976; Blamey and Blamey 1984; Dobrohotov 1961; Hanf 1990; Hubbard 1968; Katz et al. 1965; Martin 1987; Martin and Barkley 1973; Šarić 1989; Schoch et al.1988; Šilić 1988a, 1988b; Vračarić et al. 1977) were also consulted.

Scanning electron microscope photographs were taken by Michael Vieth, Department of Biology, Washington University in St. Louis and by Wendy Carden, Vista Engineering in Birmingham. I made all other photos of the Opovo plant remains. Dr. Helmut Kroll, University of Kiel in Germany helped with the botanical identification

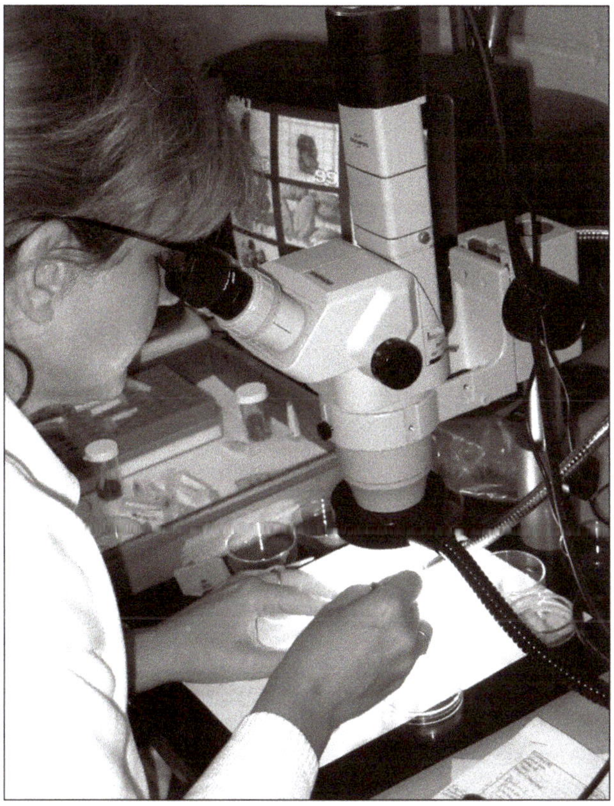

Figure 2.2. The author analyzing plant macro remains using stereomicroscope and digital camera at University of Alabama in Birmingham.

of the ambiguous specimens using SEM photographs. Dr. Gayle Fritz, Department of Anthropology, Washington University helped with unidentifiable specimens. Dr. Neal Lopinot, Center for Archaeological Research, Southwestern Missouri State University in Springfield identified wood charcoal from six samples.

For the scientific names of crops, including cereals, legumes, oil and fiber plants, and fruits, I followed the most recent nomenclature used by Zohary and Hopf (2000). For all other plants, I followed the nomenclature from *Flora SR Srbije* (Flora of Serbia, edited by Josifović 1970–1977 and Sarić 1986; Čanak et al. 1978), which is based on central European plant systematics. An entry for each taxon starts with a full scientific name, followed by an English common name, then an abbreviated scientific term in parenthesis, commonly used in archaeobotanical literature and used throughout this text. A plant family circumscription is indicated below a species name. The abbreviation "cf." (from the Latin word *confere*) following the genus name is used when the exact species identification is not positive. Also, the abbreviation "s.l." following the species name is used for the Latin term *sensu lato*. In case of heavy fragmentation when the preservation of seeds was incomplete, the seed fragments of cereal grains were either classified as *Triticum* sp. or Cerealia fragments, depending on the possibility of precise determination. Similarly, cereal spikelet forks/glume base fragments were classified as *Triticum monococcum/Triticum dicoccum*. In the heavy fraction samples, the preservation was so poor that I classified the spikelet forks/glume bases only as Triticum sp.

Preservation of Plant Macro Remains

The plant macro remains from the Opovo site were carbonized, except for 41 small seeds of various taxa that were mineralized and preserved their pale color. Such an exceptional preservation of mineralized seeds could perhaps be explained by the burnt house rubble that capped many of the pits and prevented disintegration of uncarbonized seeds, which instead became mineralized (silicified) through time. The majority of plant remains were recovered from pits, and very few were found within the house structures.

The floated samples from the top soil contained almost exclusively modern seeds and no carbonized ones. This pattern continued until a depth of 30–40 cm was reached. There were fewer modern seed intrusions in deeper layers. Many of the seeds present in flotation samples are typical of open waste ground and are often associated with modern agriculture. Most of the seeds of modern origin do not belong to the same species as the archaeological specimens. Only the seeds of *Polygonum* and "Cheno-Amaranth" species were present both in archaeological samples and modern weed assemblages. Polygonum and "Cheno-Amaranth" seeds found in archeological samples were carbonized, so confusion with uncarbonized ones could usually be avoided. Several mineralized seeds of Solanaceae and *Sambucus* species were identified from archaeological samples. Since seeds of these species were not found in the modern weed assemblage, they were most probably of the same antiquity as other archaeological specimens.

Generally, the plant remains from the Opovo archaeobotanical samples were poorly preserved, especially in the upper layers. The poor preservation may be due to frequent changes in the soil humidity that facilitated disintegration. Despite generally poor preservation of plant remains, the systematic flotation sampling strategy recovered representative samples of whatever plant macro remains were preserved at the site.

Quantitative Analysis

Each identifiable whole seed was counted as one seed, and two halves of an identifiable seed were counted also as one whole seed. The whole seeds (grains) were weighed for each sample. Measurements of length, thickness, and width were taken for well-preserved grains and seeds (Figure 2.3). Seed particles that could be identified but not counted due to fragmentation were weighed; the number was extrapolated on the basis of the count/weight ratio of the whole seeds. Also, unidentifiable seeds or seed fragments

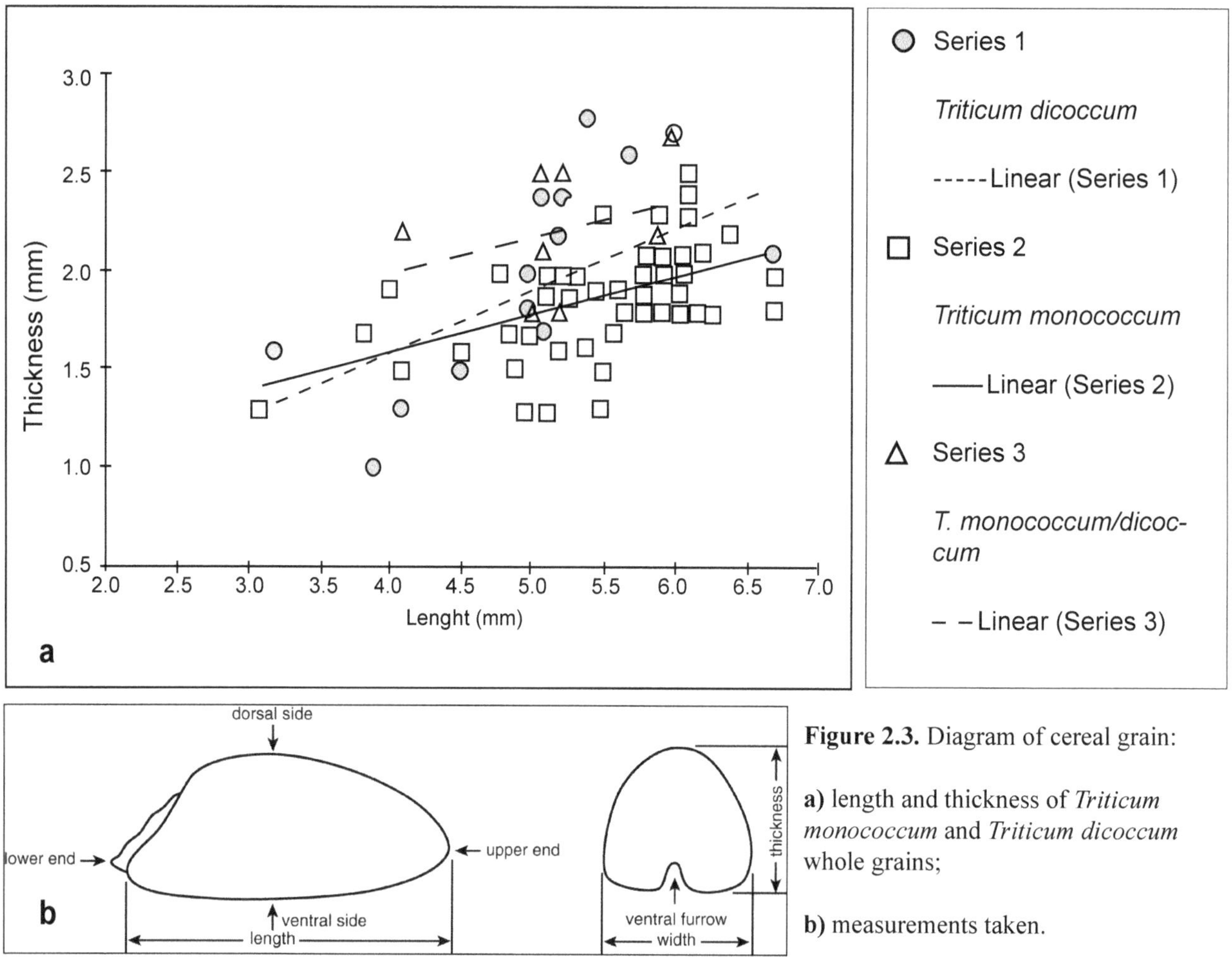

Figure 2.3. Diagram of cereal grain:

a) length and thickness of *Triticum monococcum* and *Triticum dicoccum* whole grains;

b) measurements taken.

were separately counted and indicated. Each spikelet fork (spikelet base) was counted as one, and two glume bases as one spikelet fork (base). The weight of each glume or even spikelet base was far below 0.01 g, and the weight was given only when large numbers of spikelet forks/glume bases were recovered. Many water chestnut fragments were retrieved, but not a single whole fruit was recovered from the Opovo samples. I was not able to extrapolate the number of whole fruits from the fragments present, and I presented instead the total count of *Trapa natans* fragments (see Tables 2.1–2.3).

Quantifying and comparing different classes of plant remains, e.g., weed seeds or ceral grains with water chestnut fruit fragments, remained problematic. In order to make the results comparable among layers, features, and horizons, I computed the density of taxa found (Pearsall 2000), expressed as the ratio of the number of specimens divided by the number of liters of floated sediment per feature and multiplied by 100 (liters) for 15 features and non-feature samples (see Table 2.2). To assess differences in density among features, I computed percentages of broad plant categories, i.e., cereal grains, spikelet bases (chaff), fruit seeds, wild/weed seeds, and water chestnut fragments per feature, regardless of the cause of uneven density (see Figure 3.11). To determine whether significant relationships can be delineated among various plant taxa, on the one hand, and between animal and other plant taxa on the other, I computed correlation coefficients *Pearson's r* or the *product-moment correlation coefficient* (see Chapter 3). I also calculated the ubiquity and percentage presence (see Table 2.4), that is the number of samples that a taxon is present in among the samples taken from the feature samples (N=137), non-feature samples (N=130), and the total number of examined light fraction samples (N=267), expressed as a percentage (%). For example, *Triticum dicoccum* is present in 66 samples (48.2%) from features, in 3.8% from non-features (random samples), and overall in 26.6% of all light fraction samples examined.

Opovo Plant Taxa

In an overview for each taxon identified from the Opovo samples, I briefly examine the morphology of the seeds, the modern plant and its distribution, plant use, archaeological distribution, and the Opovo findings. I include a section about the modern plants and their distributions, with special emphasis on the habitat for each species. It was necessary to describe plant habitats to try to reconstruct "plant behavior", similar to the way zoologists discuss animal behavior of recent species. I attempted to group recovered plant taxa as crops, weeds of winter or summer-sown crops (i.e. sagetals, ruderals), or as grassland and forest species.

I discuss at length plant use, including the use of plants in human and animal nutrition, medicine, and rituals. Whenever possible, I cite the ethnographic evidence of plant uses other than for food, and I concentrate on local ethnographic sources. By citing ethnographic and historic sources, I attempt to show ethnographic examples that may serve as archaeological analogies for plant uses in prehistory. There exists a rather recent trend of looking beyond subsistence in archaeobotany. Therefore, only as an example, I used the book *Rečnik srpskih narodnih verovanja o biljkama* (*Dictionary of Serbian Folk Beliefs about Plants*), edited by Vojislav Djurić (1985). This book was published posthumously from the manuscripts of the famous ethnographer Čajkanović, who investigated "the general religion of plants" in the late 1930s. As Čajkanović explains, rituals connected with plants are preserved in the folk traditions of many other European peoples. In many cases though, such rituals were preserved longer in almost unchanged forms among the Serbs. For example, the ritual cutting of an oak tree on Christmas Eve, comparable to the English yule-log tradition, is practiced among Serbs with similar precision as had been the case long ago. This tradition reflects an ancestral cult, and may date to pagan times (Radenković 1996). Furthermore, South Slavic people in the Balkans adopted Christianity rather late in comparison with other Mediterranean nations, and hence may have preserved many pre-Christian beliefs concerning plants. Once Christianized after the ninth and tenth centuries onwards, Serbs adopted Eastern Orthodoxy. In many instances, the Eastern Orthodox Church preserved Byzantine traditions that had succeeded ancient Greek tradition. Paradoxically, Serbs although Christianized, through Eastern Orthodoxy added to their belief systems rituals of the pagan Classical World. I might have used plant-related rituals and ceremonies among other European peoples as equally good examples for the folk, historic, and even possible prehistoric uses of plants. However, such a detailed and comprehensive analysis of rituals, myths, and folk traditions of various peoples was far beyond the scope of this book.

In the section about archaeological distribution, I list the archaeological sites where certain plant species were found. In this survey, I focus on the Neolithic period of the Balkans and South-central Europe because the Opovo site is a Late Neolithic site located in the south Pannonian Plain, which is on the border of central Europe and the Balkans. For the review, I mainly used publications of Kroll (1991a), Lisitzyna and Filipovitch (1980), and Renfrew (1979).

The final section is dedicated to the Opovo findings of a particular taxon. I include the number of finds and the features where each taxon occurred, density, and ubiquity, as well as dimensions of length, thickness, and width for whole seeds of particular taxon. I also note the number of uncarbonized specimens, if any. The plant remains found in the Opovo samples were almost all carbonized, or to be more precise, charred. However, some of the small fruit and weedy herbaceous seeds were mineralized. Different possibilities for preservation of both carbonized and uncarbonized specimens are discussed for each taxon separately. I cautiously tried to reconstruct the possible "mode of arrival" of the carbonized and uncarbonized seeds for each taxon found in the Opovo features, considering that in some samples only few plant remains were identified. In this reconstruction, I followed a taphonomic approach pioneered by Hillman which asserts that most of the seeds were likely brought to the site and later carbonized (charred) by human agency (Hillman 1981, 1984, 1991). I always tried to define the stage of plant processing for the Opovo plant assemblages and taphonomy. The taphonomic reconstruction for each feature allows me to compare like with like. Correspondingly, most seeds in archaeological assemblages, especially in agrarian sites, tend to represent plants from managed land and not plants from pristine wild communities. Indeed, I think that most of the small seeds present in Opovo pits were weedy plants of the crops or were gathered for a specific purpose. Some seeds may represent edaphic factors in certain fields, or they may indicate wild vegetation that preceded land management. Accordingly, the presence of the weed seeds can show under what field conditions certain crops were grown. *Trapa natans* or *Cornus* mas fruits did not arrive at the site as weedy intrusions. They were intentionally gathered in the wild and brought to the site, and they could be representative of parts of the Opovo environs.

Cereals

In the Opovo samples, cereals are the most abundant category. Altogether, 797 whole grains were recovered from the light fraction samples and the total weight of all cereal grains and fragments was 20.11 g (Tables 2.1–2.2). Additionally, 105 whole cereal grains, weighing 1.86 g, were recovered from the heavy fraction samples (Table 2.3). In this large cereal category, *Triticum monococcum* and *Triticum dicoccum* grains and spikelet forks/ glume bases were included. In total, 970 *Triticum* sp. spikelet forks were identified from the light fraction and 40 additional spikelet forks from the heavy fraction samples. In addition, this category includes wheat grain fragments that could not be identified more precisely than *Triticum* sp., *Hordeum vulgare* grains, and Poaceae seeds. In the cereal category, Cerealia fragments are also included.

Triticum sp. Fragments

In addition to the identified *Triticum monococcum* and *Triticum dicoccum*, some wheat grains were incomplete or too fragmented to be counted and identified more precisely. These grain fragments probably belong either to *Triticum monococcum* or *Triticum dicoccum* grains, but because identification was not possible, they were separated into a distinctive category. There were 4.66 g of carbonized *Triticum* sp. fragments found in the light fraction samples.

Triticum sp. was present in the same features where either *Triticum monococcum* or *Triticum dicoccum* were found.

Cerealia Fragments

Moreover, there were cereal grain fragments that were very poorly preserved and probably belong to one of the wheat species present, but could be attributed to *Hordeum* sp. or some other similar cereal species. Due to such incomplete preservation, these cereal fragments had to be separated into a distinct category, "Cerealia fragments". There were 8.52 g carbonized "Cerealia fragments" found in the light fraction and 0.31 g cereal fragments in the heavy fraction samples. The cereal fragments, in most cases, occured in the same features where identifiable cereal grain fragments were found; in the heavy fraction samples, however, a few specimens were found in additional features.

Poaceae

Besides the incompletely preserved cereal grains, there were eight smaller grains, weighing 0.06 g, that were different in shape from the regular wheat and barley grains. However, these carbonized grains had an overall shape with the distinctive embryo and hilum of grass seeds and were included in the Poaceae category (Table 2.1). These grass seeds may belong to any of the wild Poaceae species occurring in the region. The grass seeds could easily represent some weedy intrusions in the cereal crops that were brought to the settlement after harvests. Nevertheless, since these Poaceae seeds might have belonged to some other form of small cultivated cereals, I decided not to include them in the weed category. Poaceae seeds were identified from the following features: 42, 26, 41, 27, and 46; that is, in all the features where cereals were also identified. The ubiquity is 5.1% from feature samples (Table 2.4). One Poaceae seed from Feature 26 and one seed from Feature 41 were flat and resembled *Panicum*-type seeds. In addition, there were two small elongated Poaceae seeds. The one from Feature 46 had an embryo on the side, and one Poaceae seed from Feature 26 probably belonged to a long-seeded grass.

Triticum turgidum L. ssp. *dicoccum*
Emmer (*Triticum dicoccum*)
Poaceae (Gramineae)

The caryopses of *Triticum turgidum* ssp. *dicoccum*, often called grains or "seeds", are elongated asymmetrical grains that are broader than thick. The dorsal side is domed, and its highest point is above the embryo (or radical depression). The ventral side is usually flat, seldom convex, with a deep furrow (hilum). There is great variation in the form of the grains, and besides the described ones, others have wide tops and are called "drop shaped". There are plump grains called "clog-shaped", but also there are narrow grains that are called "spindle-shaped". The range of grain dimensions is ca. 7.2–9.0 mm in length×2.85–3.4 mm in width×2.6–3.1 mm in thickness, compared with wild emmer, *Triticum turgidum* ssp. *dicoccoides*, which are 9–12 mm×1.7–2.9 mm×1.85–2.7 mm, respectively (Percival 1921:191, cited in Renfrew 1973:51). In archaeobotanical samples, spikelet forks or glumes of emmer are found. There are also grains of *Triticum turgidum* ssp. *dicoccum* from the spikelets that contain only one grain instead of two. The one-seeded spikelets are typically from terminal parts of the ears; those grains have less flat ventral sides than do regular emmer grains. In general, the spikelets of *Triticum turgidum* ssp. *dicoccum* contain two grains, and therefore the spikelet forks are wide, i.e. "boat-shaped", usually lacking the distinctive keels (nerves) of einkorn glumes. The articulation scar, left by abscission of the spikelet attached above, is a result of threshing in domesticated wheats, and is also diagnostic. In *Triticum dicoccum*, the abscission scar is narrower and smaller relative to the whole spikelet width compared to *Triticum monococcum* spikelet forks.

The Modern Plant and Its Distribution

Hulled emmer wheat, *Triticum turgidum* ssp. *dicoccum* is a tetraploid wheat (2n=28) that is called *Triticum dicoccum*. Genetic and morphological evidence shows that hulled emmer wheat is closely related to wild emmer wheat (*Triticum dicoccoides*). It is the only wild stock in the genus *Triticum* that is cross-compatible and fully fertile with the cultivated *turgidum* wheats (Zohary and Hopf 1993:42).

Triticum dicoccum is a plant in which the products of threshing are individual spikelets. In emmer, as in einkorn (*Triticum monococcum*), threshing results in breaking the rachises of the ears at the weakest points below each spikelet. In *Triticum dicoccum*, the ears are dense, laterally compressed and narrow, and usually awned. The rachises are brittle, but less brittle than in *Triticum dicoccoides*. The tough glumes tightly enclose the grain at maturity, and the single prominent keel is extended into a beak. The palea is broad and does not split at maturity. Two grains develop in each spikelet. The caryopsis is broader and plumper than that in *Triticum dicoccoides*. The dorsal side is broad and flat in profile (Renfrew 1973:50).

From hulled emmer evolved the more advanced free-threshing tetraploids, like *Triticum durum*, *T. turgidum*, *T. polonicum*, and *T. carthlicum* wheats. Threshing of all these cultivated tetraploid types frees the naked grains. The rachis is usually uniformly tough (Zohary and Hopf 1993:39). Emmer is still sporadically grown in the Balkan countries, Anatolia, Iran, Caucasus and India, while in Ethiopia this wheat survives as a common and appreciated cereal (Zohary and Hopf 1993:39).

Plant Use

Triticum dicoccum and *Triticum monococcum* are wheats that can be used as food for humans and for animals.

Prehistorically, both wheats were used more for human consumption than for animals, although from Roman times onward, a considerable amount of einkorn was fed to animals. Various kinds of porridges and gruels can be made both from einkorn and emmer. Also, both wheats can be ground into flour, and various sorts of breads, cakes, and other products can be made from emmer and einkorn. Both wheats are nutritious but the rising quality of flour is poor, especially for einkorn. Because both einkorn and emmer have hulled grains, typically they have to be parched before they are dehusked. The parching process usually destroys glutens, necessary for rising, hence leavened bread is difficult to make. However, various pita breads can be made from these wheats. In some parts of Africa, and in experiments at the Butser farm, emmer was dehusked in a mortar with an iron-capped pestle without being parched. Bread made from this flour rises better than does flour ground from the parched grain (Braun 1995:35). Also, the grains can be fermented and used to make various alcoholic drinks. The grains can be also popped and roasted like maize, and immediately consumed. Furthermore, straw and chaff can be used as temper in clay and house daub, and the harvested stalks used for thatching roofs and for fodder.

In the former Yugoslavia, before World War II, *Triticum dicoccum* was grown in some mountain regions as a cereal crop used for human consumption, but also in mixture with *Triticum monococcum*, oats, and some other grasses, and was also used as food for domestic animals. After World War II, emmer was rarely grown for human food, but there are still some regions where it is grown and used as fodder, e.g., Montenegro and Banija (Slavko Borojević, personal communication 1997; Ljubo Pavičević, personal communication 1997, see Figure 5.1).

There are many rituals connected with wheat in general, which are documented in historical records as well as in folklore of many peoples. Wheat is also present in Egyptian, Greek, Roman, and Christian religions as being a sacrificial food and a symbol of resurrection. Probably many of the beliefs and ceremonies connected with naked wheat (*Triticum aestivum*) were passed on from the rituals and traditions once characteristic of hulled wheats (emmer and einkorn).

Archaeological Distribution

Triticum dicoccum was the principal wheat of Old World agriculture in the Neolithic and early Bronze Ages. The grains were used for food. This wheat was gradually replaced by more advanced free-threshing tetraploid and hexaploid types of the genus *Triticum*. It should be pointed out that among the three main Neolithic grain crops (einkorn, emmer, and barley), the wild progenitor of emmer has the most limited distribution, it is confined to the "arc" of the Near East. Thus, the numerous finds of cultivated emmer retrieved from settlements outside the Near East serve as evidence for the initiation of Neolithic agriculture in the Near East and its subsequent spread from this core area to the Aegean and the Balkans. Emmer was also the principal crop in the LBK culture (Central Europe) and continued throughout the Neolithic and Bronze Ages. A similar situation is found in the Mediterranean at the Impresso culture sites (the first half of the sixth millennium B.C.), while at the Cardium Ware culture sites (the second half of the sixth millennium B.C.), there was already evidence of naked wheats (Zohary and Hopf 2000). *Triticum dicoccum* was also the most ubiquitous component of the crop assemblage in the Balkans during Neolithic and Bronze Ages (Kroll 1991a). Emmer was identified in most of the Neolithic sites in the former Yugoslavia and in all Vinča sites where analyses were performed.

Opovo Findings

A total of 732 carbonized *Triticum dicoccum* grains was retrieved from the Opovo samples, out of which 636 (5.99 g) grains were recovered from light fraction samples from three building horizons (Table 2.1). The percentage presence (ubiquity) in features samples was 48.2 % and 26.6% in all samples (Table 2.4). The highest density was from Feature 41 (Table 2.2). In addition, from the same Feature, 96 grains were recovered from the heavy fraction samples (Table 2.3). Only 32 carbonized *Triticum dicoccum* spikelet forks were obtained from both the light and heavy fractions of the Opovo samples (Tables 2.1–2.3). The percentage presence (ubiquity) of *Triticum dicoccum* spikelet forks was also low, only 7.3% from feature samples and 3.7% from all samples (Table 2.4). In Feature 26, 13 spikelet forks were identified, representing the highest number and density from the Opovo samples (Table 2.2). *Triticum dicoccum* was found in the pits and in the floor of House 4. Many of the grains came from floating a concentration of house rubble, and some were picked out of the heavy fraction. Some of the grains might have been deposited in the pits after they were burned during parching or cooking.

It was extremely difficult to identify the exact taxon of *Triticum* grains from Opovo samples. Many of the grains were exploded and could not be identified further than Cerealia fragments or *Triticum* species fragments. Furthermore, most of the grains could not be categorized as *Triticum monococcum* or *Triticum dicoccum*, and in many cases did not have pronounced dorsal ridges characteristic of *Triticum monococcum* grains (Figures 2.4 and 2.5). Because not a single spikelet of *Triticum spelta* was found in the samples, the possibility that some *Triticum dicoccum* grains could actually be *Triticum spelta* was excluded. Some of the grains were shorter and resembled *Triticum compactum* (e.g., two grains from Feature 31), a form of *Triticum aestivum*; a type of grains now more commonly classified in a new encompassing category of *Triticum aestivum/durum/turgidum.* *Triticum dicoccum* grains from Feature 41 were wide and rounded at the top, and could be categorized by their form as "drop shaped" (Figure 2.4).

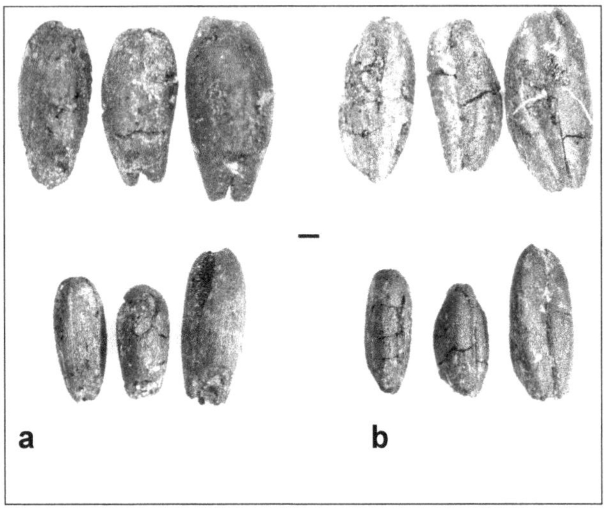

Figure 2.4. *Triticum dicoccum* carbonized grains from Feature 41: **a)** dorsal side **b)** ventral side (scale 1 mm).

Figure 2.5. *Triticum monococcum* cf. *dicoccum* carbonized grains from Feature 41, lacking dorsal ridge (scale 1 mm).

The dimensions of 68 whole *Triticum dicoccum* grains (Figure 2.3) from the Opovo samples were on the average 5.50 mm long (3.10–6.70 mm)×2.38 mm wide (1.50–3.10 mm)×1.88 mm thick (1.30–2.50 mm). The indexes were also computed: length/breath, L/B =2.35 (1.73–3.88); length /thickness, L/T= 2.95 (2.11–4.23); and breadth/thickness, B/T 1.28 (0.70–1.69). Although many more grains were identified they were too poorly preserved to be measured. The measured grains were shorter on average, including their maximum length, compared to *Triticum dicoccum* grains from other archaeological sites. There are at least two possible explanations. These grains may represent previously processed and sieved grains that were not the prime grains intended for immediate consumption but by-product of coarse sieving. However, the grains were not exceptionally thin, which would allow them to pass through fine sieves. It is also possible that these grains represent single-seeded terminal florets of *Triticum dicoccum* that were sieved out. Certainly, among the *Triticum dicoccum* grains some of the two-seeded einkorns could have been included. In the archaeobotanical literature, there are discussions about how to classify these grains because about 2% of einkorn grains do not show the typical einkorn form (Kroll 1992:181–185). It is possible that some of the grains with lower values of the length/thickness index were actually two-seeded einkorns. Yet, another possibility is that the grains due to poor growing conditions in the Opovo fields could not develop fully. However, the grains from Opovo do not have sunken sides (cheeks) that would indicate that they were harvested not fully ripe or that they did not have a chance to develop fully in the field.

Triticum monococcum* L. spp. *monococcum
Einkorn (*Triticum monococcum*)
Poaceae (Gramineae)

The caryopses of *Triticum monococcum*, often called grains or "seeds", are elongated, thin, and pointed at both ends. The dorsal side is highly ridged, and the highest point is in the middle of the grain. The ventral side is narrow, slightly domed, and the furrow (hilum) is very narrow. The grains are thicker than wide. Usually there is one grain per spikelet, although there are forms with two-seeded (two-grained) spikelets, and in that case the grains are more emmer-like than typical einkorn grains. The embryo is relatively large. In two-grained spikelets the caryopses are usually smaller and narrower but otherwise similar to those of emmer. The distinguishing factor between two-grain einkorn and emmer caryopses is their ratio of breath to thickness (B/T index); in einkorn, the index ranges between 100–179 and in emmer 75–95 (Renfrew 1973:48). Two-seeded emmer grains have a depression below the top of the grain (viewed laterally), shaping points of the grains in a different manner from those of typical einkorn grains (Kroll 1992:181-185). In addition to 2% of these emmer-type einkorns, some grains of two-seeded (with two grains) einkorn retain their typical einkorn form and hence are properly included among *Triticum monococcum* grains. The only distinguishable characteristic is that they are flatter on the ventral side. The dimensions of single-seeded einkorns range between 7.0–8.5 mm in length×1.8–3.0 mm in width×3.0–3.5 mm in thickness (Percival 1921:174, cited in Renfrew 1973:49). Another point that may be difficult to establish is the difference between the single-grained einkorn caryopsis and the grain that may develop in the terminal spikelet of emmer (Schiemann 1951:316, cited in Renfrew 1973:48). Because einkorn is hulled wheat, its spikelet forks are often preserved in archaeological samples. The spikelets are narrow with prominent glume nerves. The abscission scar is long in comparison to the width of the whole spikelet. Also, the bases of the spikelet forks are wider than in emmer.

The Modern Plant and Its Distribution

Einkorn is a small plant, rarely more than 70 cm high with rather low yield, but it can be grown on poor soils where other wheat types fail. *Triticum monococcum* is a diploid species (2n=14 chromosomes, AA) with hulled grains and delicate ears and spikelets. Most cultivated einkorn varieties produce one grain per spikelet, hence its name, but some varieties have two grains. In the past, einkorn cultivation was much more extensive, but it is still sporadically grown in western Turkey, the Balkan countries, Germany, Switzerland, Spain, and the Caucasus. This wheat was one of the founder grain crops of Neolithic agriculture in the Near East and in Europe. Since the Bronze Age, its importance has gradually declined in competition with free-threshing wheats (Zohary and Hopf 1993:32).

Cultivated *Triticum monococcum* is closely related to a group of wild and weedy einkorn, *T. boeticum*, which spread over the Near East. These species are morphologically similar; hybrids between them are fully fertile and have regular chromosome pairings. The main distinguishing trait between wild einkorn and cultivated einkorn is the mode of seed dispersal. Wild einkorn has brittle ears, and individual spikelets disarticulate at maturity to disperse the seed. In cultivated einkorn, the mature ear remains intact and breaks into individual spikelets only upon pressure or threshing (Zohary and Hopf 1993:32).

Plant Use

Einkorn has nutritious grains that can be ground into fine yellow flour. But due to its poor leavening quality, einkorn has been consumed primarily as porridge or as cooked whole grains. Since Roman times, a considerable part of the yield has been fed to animals (Schieman 1948, Harlan 1981, cited in Zohary and Hopf 1993:32). The chaff of wheats together with straw can be used for tempering house daub and pottery. And, indeed, in the Opovo house rubble many impressions of hulled wheat were found.

Archaeological Distribution

Triticum monococcum, like *Triticum dicoccum*, is one of the principal wheats of Old World agriculture. *Triticum monococcum* is found on the majority of archaeological sites starting from the very early Neolithic period. Even in the later periods when free-threshing wheat *Triticum aestivum* becomes more common, both *Triticum dicoccum* and *Triticum monococcum* still appear on many sites. They are the most ubiquitous crops identified on the Balkan Neolithic sites (Kroll 1991a). *Triticum monococcum* grains and/or spikelet bases/forks were identified at all the early Neolithic sites in the former Yugoslavia, and at all the Vinča sites except Predionica and Valač in Serbia (Renfrew 1979). Einkorn seems to be of far greater importance than emmer at the multi-layer site Gomolava, (van Zeist 2001/2002), also at the early Neolithic site Vršnik (Hopf 1961), and the late Neolithic site(s) Staro Selo/Selevac and Selevac (Hopf 1974; McLaren and Hubbard 1990).

Opovo Findings

A total of 127 (1.01 g) carbonized grains of *Triticum monococcum* has been found in the light fraction samples and 9 grains (0.09 g) in the heavy fraction samples in all three building horizons (Tables 2.1–2.3). Overall percentage presence (ubiquity) of *Triticum monococcum* is 14.6% of all light fractions samples and 27% from feature samples (Table 2.4). The highest density of *Triticum monococcum* grains is in the pits: Feature 52 and Feature 41 (Table 2.2). Also, five *Triticum monococcum* grains were found on the floor of House 4. In general, very few grains were typical einkorn specimens in the Opovo samples. Many grains did not have the highest point of the dorsal ridge in the middle of the grain, but the highest point was placed slightly towards the embryo. Furthermore, the ventral side was not often strongly convex as in the typical einkorn grains. In some cases, the grains more closely resembled the single-seeded emmer than the typical einkorn grains. These grains were, however, included among possible *Triticum monococcum* grains. Altogether, 29 whole *Triticum monococcum* and 8 *Triticum monococcum* cf. *dicoccum* grains were measured (Figure 2.5) and their indexes were computed. The dimensions for these *Triticum monococcum* are, on the average, 5.23 mm (3.20–6.70 mm) long×1.75 mm wide (1.10–2.80 mm)×1.99 mm thick (1.00–2.80 mm). The indexes are: L/B= 3.11 (1.82–4.50); L/T=2.72 (1.93–3.90); B/T=0.90 (0.43–1.20). The dimensions for *Triticum monococcum* cf. *dicoccum* are on average 5.20 mm (4.10–6.00 mm) long×1.90 mm wide (1.60–2.20 mm)×2.23 mm thick (1.80–2.70 mm). The indexes are: L/B=2.77 (1.86–3.28); L/T=2.37 (1.86–2.89); B/T=0.86 (0.68–1.00). When length was plotted against thickness (Figure 2.3), the values for *Triticum monococcum* cf. *dicoccum* grains fall more in the range of *Triticum monococcum*, and hence were included among *Triticum monococcum*. The Opovo einkorn grains were generally smaller, shorter and flatter than grains found in other sites. This trend was also apparent with *Triticum dicoccum* grains, which have been found in much larger numbers. Both *Triticum monococcum* and *Triticum dicoccum* grains have obviously shorter grains, although they are not narrower than the grains from contemporaneous sites. If the grains were apparently slimmer than usual, it could be suggested that they were not prime grains of the crop, but rather the byproducts of coarse sieving. Probably, due to intensive heat, carbonization distorted and expanded the grains.

In contrast to a very few *Triticum monococcum* grains found in the Opovo samples, many spikelet forks and glume bases were identified. A total of 947 spikelet forks were identified from light fraction samples (Tables 2.1). While the weight of one spikelet fork was far below 0.01 g, 589 spikelet forks had a weight of 0.4 g. Most of spikelet forks were from Feature 52, followed by Feature 59. Because of the high concentration of spikelet forks in just a few features, the percentage presence (ubiquity) of *Triticum monococcum* spikelet bases is only 15.3% from feature samples and 8.2% from all light fraction samples (Table 2.4).

The identification of spikelet forks and glume bases was even more ambiguous (Figures 2.6–2.9). The *Triticum* spikelet forks found in the Opovo samples were typically narrow, and had quite broad abscission scars, but only the first nerves of the spikelet were prominent, the ones from the ventral sides, not the second ones from the dorsal sides. Looking only at the cross-section of the glumes from the top, one could assign many of the *Triticum* cf. *monococcum* spikelets to *Triticum dicoccum*. However, the overall size and shape of the spikelet forks suggests that the larg-

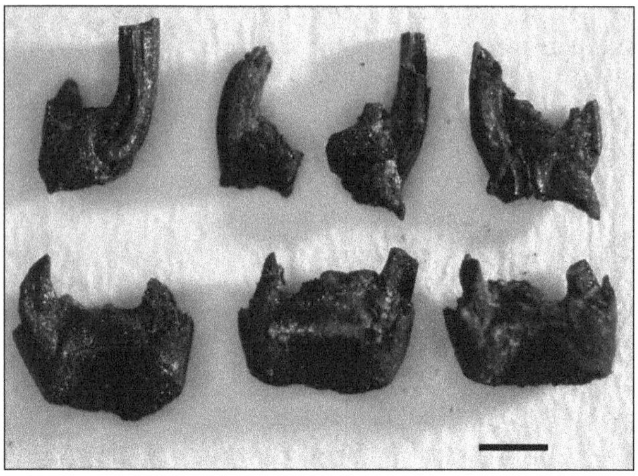

Figure 2.6. *Triticum monococcum* and *T. dicoccum* carbonized spikelet forks and glume bases from various features (scale 1 mm).

est number of wheat spikelet forks could be attributed to *Triticum monococcum*. Such identification causes further irregularities because far more grains of *Triticum dicoccum* were recovered than *Triticum monococcum*. On the other hand, far more *Triticum* cf. *monococcum* spikelet forks were identified than *Triticum dicoccum*. Because both species have hulled grains, there should be half as many spikelet forks of *Triticum dicoccu*m as grains, and the number of *Triticum monococcum* grains should equal the number of spikelet forks. This discrepancy can partially be explained by the fact that *Triticum monococcum* spikelet forks are tougher, and hence less susceptible to destruction, than the spikelet forks of *Triticum dicoccum*. However, charring experiments have shown that among the glumed wheats eikorn glumes were the first to turn to ashes among the glumed wheats (Boardman and Jones 1990). It is possible that a number of *Triticum monococcum* grains could be *Triticum dicoccum* (Figure 2.5), and vice versa, and the same holds true for spikelet forks (Figure 2.6). Nevertheless, the discrepancy is too great to be explained solely by mistakes in identification. Furthermore, in the pits where the largest number of spikelet forks was found, proportionally few grains of either *Triticum monococcum* or *Triticum dicoccum* were found. It seems that different Opovo pits served different functions, which resulted in different depositional processes of plant remains.

Yet another possibility is that some *Triticum dicoccu*m grains are actually some of the free-threshing wheat grains such as *Triticum aestivum* or *Triticum durum*, hence the lack of spikelet forks. Both possibilities seem rather unlikely, however. On the one hand, not a single rachis of a free-threshing wheat was recovered from the Opovo samples, and large quantities of *Triticum aestivum* grains from the Neolithic period are not commonly found in the Balkans. On the other hand, another naked wheat, *Triticum durum* is usually limited to the Mediterranean climate because it is susceptible to harsh winters. The spikelet bases from the Opovo samples did not generally look like the "new type of glume wheat" (Glynis Jones and Helmut Kroll, personal communications 2004) that were recently documented by Jones et al. (2000). However, under closer examination perahps some of the *Triticum dicoccum* spikelet forks may resemble the "new type of glume wheat" (Figure 2.9) that was recognized from the Vinča layers at the tell site Uivar in Romania, and identified as *Triticum* cf. *timopheevi* (Elske Fischer, personal communication 2005).

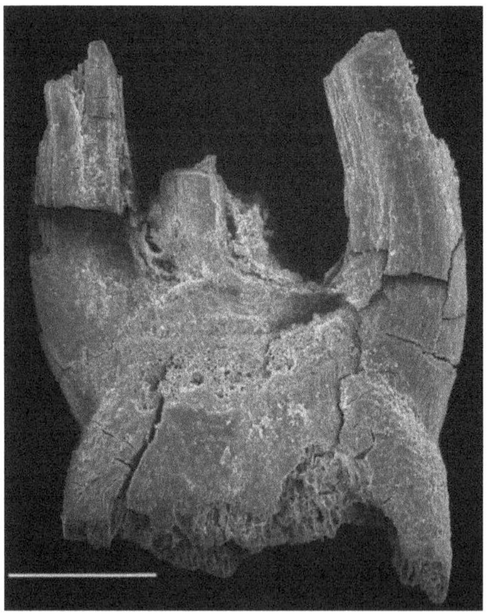

Figure 2.7. *Triticum monococcum* carbonized spikelet fork from Feature 52 (scale 1 mm).

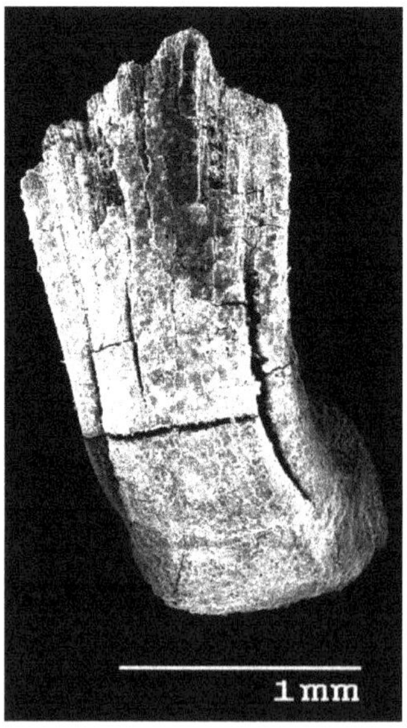

Figure 2.8. *Triticum monococcum* cf. *dicoccum* carbonized glume from Feature 41 (scale 1 mm).

Figure 2.9. *Triticum* sp. carbonized glume base, possibly the "new type of glume wheat" from Feature 52 (scale 1 mm).

Hordeum vulgare* L. *vulgare
Barley (*Hordeum vulgare*)
Poaceae (Gramineae)
The caryopses, so called grains or "seeds", of *Hordeum vulgare* are spindle shaped and slightly angular as a consequence of closely enclosed glumes. Both the top and the basal parts are truncated. The ventral and the dorsal sides are slightly domed, and almost symmetrical in cross-section. The greatest width is in the middle of the grain. The ventral furrow is prominent and "V" shaped. Dimensions of hulled barley forms range ca. 3.3–7.5 mm in length×1.7–4.0 mm in width×1.3–3.1 mm in thickness. Naked barley grains are usually smaller (Schoch et al. 1988:83). In two-rowed barleys, all kernels are straight and symmetrical, while in six-rowed barleys, the lateral grains are often slightly bent and asymmetrical. Kernels of naked barley (var. *nudum*) can frequently be recognized by their shriveled skin and a furrow that stays narrow near the apex, in contrast to the "V" shaped furrow in hulled grains (Zohary and Hopf 1993:58–60).

The Modern Plant and Its Distribution
According to the morphology of the spikelets, barley under domestication can be divided into two principal types. Two-rowed forms, called *Hordeum distichum*, and those in which the middle spikelet in each triplet is fertile and usually armed with prominent awns. The two lateral spikelets are reduced and are grainless and awnless. Each ear thus contains only two rows of fertile spikelets, and the grains are more asymmetrical. On the other hand, six-rowed forms, called *Hordeum hexastichum* or sometimes *tetrastichum*, are those in which there are three spikelets in each triplet bearing a grain and usually are all awned. The grains of this form are more asymmetrical because the lateral grains and the middle grains develop differently. Two-rowed forms are found in the wild progenitor of the crop, whereas six-rowed types were derived under domestication (Zohary and Hopf 1993:55).

Barley grows best where the ripening season is comparatively long and cool, and with moderate rainfall. Some spring barleys may mature in 60–70 days after sowing, which is earlier than any other spring cereal. On the other hand, barley penetrates warm steppe and desert areas of the Near East more successfully than the wild wheats (Renfrew 1973:80). Barley is a very tolerant plant in saline and alkaline soils, but it is sensitive to soil acidity. It can grow on thin limestone soil as encountered in Greece and many other parts of the Mediterranean World (Leonard and Martin 1963:500, cited in Renfrew 1973:81).

Plant Use
Wild barley as well as the majority of cultivated forms have hulled grains, i.e. the palea are fused with the kernels and still cover them after threshing. In naked barleys, the ripe grains are free and are released by threshing without additional processing. In some farming communities, naked barleys are favored for food preparation, while hulled forms are preferred for brewing beer and feeding animals (Zohary and Hopf 2000).

Barley was the main staple of ancient Greece, where it was ground into flour and kneaded into barley-cakes or "maza", a standard form of flat bread baked by the Greeks. Braun (1995:25-38) describes how ancient maza was made from roasted barley kneaded with water, milk, or oil, and was then eaten without baking. Roasted barley grains are a common food cited in the Old Testament and were also used by the ancient Greeks. The Greeks also used barley grains in sacrifices, a practice that may point to the antiquity of barley consumption. Similarly, the Egyptians preferred barley to wheat (Währen 1963:16, cited in Braun 1995:25). During Roman times and in most of the Near East today, barley was considered an inferior food and more commonly used to feed animals. Egypt became one of the main suppliers of wheat for Rome. However, Romans used barley grains medicinally and for making porridges.

Archaeological Distribution
Cultivated barley, *Hordeum vulgare*, is one of the main cereals in Mediterranean agriculture and a founder crop of Old World Neolithic food production. Carbonized grains are the most common barley remains found in archaeological excavations. Occasionally whole triplets are retrieved, permitting the identification of the cereal to the two-rowed or six-rowed type. In the Balkans, barley was ubiquitously present in the Early, Middle, and Late Neolithic sites. In many of the Bulgarian sites, *Hordeum nudum* is present in large quantities (Kroll 1991a; Popova 1995:193–209). In the former Yugoslavia, *Hordeum hexastichum* was identified in the sites of the Early and Late Neolithic periods. Barley was present in most of the late Neolithic sites in Bosnia and Macedonia, but it was not identified in several Vinča sites from Serbia (Hopf 1974:1-11), except for small quantities at Gomolava and Selevac (McLaren and Hubbard 1990:247-254; van Zeist 1975, 2001/2002).

Opovo Findings
A total of 26 (0.21 g) carbonized grains of *Hordeum vulgare* was identified in light fraction samples from Feature 41 only (Table 2.1), and the percentage presence (ubiquity) is also low (6.6%), reflecting the large number of samples floated from this feature (Table 2.4). As in the case of other cereals, carbonized *Hordeum* grains were badly distorted and identification to *distichum* or *hexastichum* form was not possible (Figure 2.10).

Only nine grains could be measured. The grains are on average 4.98 mm long (4.20–6.00 mm)×2.82 mm wide (2.20–3.30 mm)×1.97 mm thick (1.80–2.30 mm). The indexes are L/B=1.78 (1.50–2.27); L/T=2.54 (2.17–3.00); B/T=1.43 (1.22–1.56). The grains fall in the range of smaller barley grains, especially considering the thickness. The smaller dimensions of barley grains coincided with

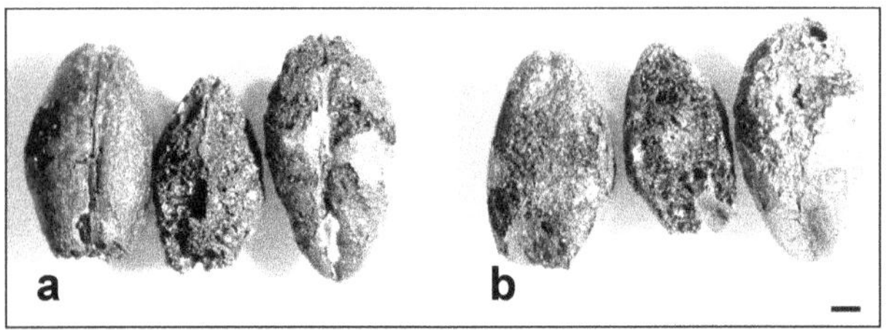

Figure 2.10. *Hordeum vulgare* carbonized grains from Feature 41: **a)** ventral side; **b)** dorsal side (scale 1 mm).

the dimensions of other cereal grains, which are generally smaller than the typical cereal grains from contemporaneous sites. However, the grains from the Opovo site were not necessarily always slimmer. The dimensions of cereal grains were smaller because all the cereal remains were from refuse pits, and no storage with prime grains was recovered from the Opovo site.

Legumes

The legume category contains 49 (0.33 g) grains in the light fraction samples and two additional grains from the heavy fraction (Tables 2.1–2.3). Legumes are identified almost exclusively from feature samples, with the percentage presence (ubiquity) of 10.9% for feature samples (Table 2.4). The most numerous legume seeds are *Lens culinaris*. Additionally, there is a Fabaceae seed category in which mostly *Vicia*-type seeds were included. Several of the *Vicia* seeds could easily have been weedy intrusions of cereal crops, but some legumes could also have been collected or cultivated intentionally.

Lens culinaris Medik. ssp. *microsperma*
Lentils (*Lens culinaris*)
Fabaceae

Lens culinaris seeds are circular and greatly flattened (lens-shaped) with sharp or blunt margins. The hilum is about 1 mm long and very narrow. The surface of a seed is smooth, rarely slightly pitted, and is often distended by carbonization. Modern cultivated varieties have larger seeds. Conventionally, *Lens culinaris* is grouped according to the seed size in two subspecies: *Lens culinaris* ssp. *microsperma*, with small pods and seeds that range between 3–6 mm in diameter; and *Lens culinaris* ssp. *macrosperma*, with large seeds attaining more than 6 mm in diameter. The *macrosperma* subspecies is more recently bred subspecies; it began to appear in archaeological sites in the first millennium b.c. (Zohary and Hopf 2000).

The Modern Plant and Its Distribution

There are several wild species of *Lens* occurring in warmer regions of the Mediterranean area, including the Balkans, e.g., *Lens nigracans* and *Lens ervoidens* (Polunin 1980:296). However, only *Lens culinaris* is considered here. *Lens culinaris* ssp. *microsperma* is an annual upright plant that reaches a height between 13–35 cm. Its fruit is a pod that usually contains 1–3 seeds. *Lens culinaris* plants are susceptible to frost and therefore usually sown in spring, with a few exceptions of winter-sown varieties. *Lens culinaris* thrives best in chernozem and alluvial soils. It cannot successfully grow on salty soils, nor where underground water is high. Fruits ripen unevenly in summer, and the pods dehisce irregularly. Pods must be harvested at the right time; if harvested not fully ripened, they must be left for a few days to dry in the fields. In Mediterranean agriculture, *Lens culinaris* is a common crop, companion to wheat and barley. Although lentil yields (ca. 500–1500 kg/ha) are considerably smaller than those of cereals, it is highly valued for the protein content (25%) of its seeds (Zohary and Hopf 2000).

Plant Use

Lens culinaris is cultivated in southern Europe and consumed as soup or paste, and mixed with other cereals. Similarly, in Asia it is an important substitute for meat. In the Balkans, it is most commonly consumed as a soup or in vegetable stew. Also it can be ground and the flour used to make special small breads. Only ground seeds are used for fodder because whole seeds may cause problems in animal digestive systems. However, the green plants are very nutritious and can be given to animals either fresh or dried as hay. In Serbian folklore, *Lens culinaris* was used in healing rituals and was considered food for the dead (Čajkanović 1985:311).

Archaeological Distribution

Lens culinaris is usually found in archaeological samples along with emmer, einkorn, and barley. In Kroll's bibliographical review (1991a:161-177), *Lens* species are among the most ubiquitous seeds found in Pre-ceramic, Early, Middle, and Late Neolithic sites in south-east Europe. *Lens* sp. seeds were found in the Pre-Neolithic levels of Franchthi Cave, and in Early Neolithic sites in Greece such as Ghediki and Argissa-Magula. More precisely identified, cultivated *Lens culinaris* seeds are present in many Neolithic settlements in Greece and in Bulgaria. It was also found at Anza, Obre, Gomolava, and Selevac, the Neolthic sites in the former Yugoslavia (Renfrew 1979:243–265; Zohary and Hopf 1993:93; van Zeist 2001/2002).

Opovo Findings

A total of 26 carbonized *Lens* seeds was recovered from the Opovo samples, weighing only 0.18 g (Tables 2.1). The seeds have the distinctive flat shape, although other characteristics are not distinctive enough for positive identification as *Lens culinaris* species. The wild *Lens* species are not common in the area, and if we assume that the situation was similar at the time of the Opovo settlement, then the archaeological seeds most probably belong to *Lens culinaris*. One possible *Lens culinaris* seed, from Feature 42, whose identification was even less certain, was included among Fabaceae. *Lens culinaris* (Table 2.4) was identified

Figure 2.11 *Lens culinaris* carbonized seeds from Feature 46 (scale 1 mm).

primarily from features (percentage presence 10.9%) and rarely from non–features samples (percentage presence 1.5%). It is present in the light fraction of samples from Features: 26, 41, 52, 30, 46, and 59.

It is unceratain whether *Lens culinaris* was cultivated as a separate garden or field crop in the Opovo settlement. Certainly, the seeds could have been mixed together with other cereals for culinary purposes, or seeds could have been mixed later in pits. It is interesting to note that the highest number of *Lens culinaris* seeds (13) occurs in Feature 46 (Figure 2.11), a pit where the highest concentration of *Trapa natans* fragments was also found. A similar practice is cited by Brothwell and Brothwell (1969:107) where instead of water chestnuts (*Trapa natans*), chestnuts (*Castanea sativa*) were cooked together with lentils.

Vicia L.
Vetch-type (*Vicia*-type)
Fabaceae

Vicia seeds are mostly included in the Fabaceae family with other legume seeds that could not be identified more precisely. Some of the seeds from the Opovo samples are similar to *Lens culinaris*, but are less flat and more rounded. Most of the seeds are rounded and resemble some kind of *Vicia*-type seeds. The vetches are legumes and their fruits are pods with few seeds. The seeds are either angular, as in the case of domesticated *Vicia ervilia*, or rounded, as in the case of the *Vicia sativa* and many other species. The seeds identified in the Opovo samples are rounded and therefore probably do not belong to the species *Vicia ervilia*, which has angular seeds. I was not able to determine whether these seeds belong to the domesticated species *Vicia sativa* or to some of the wild species such as *Vicia sativa* ssp. *angustifolia*, which are also included in the same *Vicia sativa* complex. *Vicia sativa* seeds are smooth, ranging from 4.5–7.0 mm in diameter. The hilum is rather long and narrow, comprising ca. 1/5 of the seed circumference (Kroll 1983:95–96 ; Zohary and Hopf 2000).

The Modern Plant and Its Distribution

There are about 42 *Vicia* species found in the Balkans. Several species such as *Vicia cracca* and *Vicia sepium* are common plants in meadows; the latter one thriving also in birch forests (Polunin 1980:293-296). Many of the species such as *Vicia pannonica* and others are widespread weeds in crops. Some of the species are perennial, e.g., *Vicia cracca*, while others are annual, e.g., *Vicia sativa*, which is found in the Mediterranean region.

Plant Use

Many of the *Vicia* species are excellent fodder. *Vicia* seeds are occasionally used for human food in times of need, but they are not considered first-choice food because the seeds are toxic and must be soaked in water to be detoxified before eating (van Zeist 1988). *Vicia cracca* seeds have medicinal properties, however.

Archaeological Distribution

In Europe, the seeds of *Vicia ervilia* were present in the Pre-ceramic levels of Franchthi Cave in Greece. They were also recovered from the Early Neolithic site of Nea Nikomedeia (van Zeist and Bottema 1971) and other Early, Middle, and Late Neolithic sites in Greece and Bulgaria (Kroll 1991a:161-177). The seeds are almost ubiquitously present in Bulgarian Neolithic sites (Popova 1995:193-203). However, at the Neolithic sites of the former Yugolsavia, they were recovered only from the tell site Gomolava from the late Vinča layers, but *Vicia faba* var. minor was found from the Iron Age layers at the same site (van Zeist 1975:315–325, 2001/2002:87–115). *Vicia sativa* is another species that is present in Mediterranean agriculture. The seeds are found both in the Near East and in Europe, and were identified in a few Bronze Age sites in Bulgaria (Lisitzyna and Filipovitch 1980:5-90). Often the seeds recovered are not identified to the exact taxon and instead are classified in the *Vicia/Lathyrus* category (Kroll 1991a: 161-177).

Opovo Findings

There are 25 carbonized Fabaceae seeds recovered from the Opovo features, both from light and heavy fractions. Fabaceae seeds were found in all three building horizons. from Features 42, 26, 41, 48, 52, 46, and 59 (Table 2.1–2.4). Sixteen seeds were recovered in Feature 41. Because only a few *Vicia*-type seeds were found from Feature 41, they are included in the more general Fabaceae category. The Opovo seed dimensions range between 0.01–0.15 mm. Some seeds were not typical, but are included *Vicia*-type seeds, e.g., one seed from Feature 59 was bigger than the other *Vicia* seeds. Also, one seed from Feature 26 and one seed from Feature 41 resemble *Pisum sativum* seeds. These seeds certainly belong to the Fabaceae family, but could belong to genera other than *Vicia*. None of the seeds identified is angular in shape, so *Vicia ervilia* and most *Lathyrus* species seeds can be excluded. The seeds from the Opovo samples are considerably smaller than *Vicia faba* and different in shape. Because no greater concentration of *Vicia* seeds was found in the Opovo settlement, these Fabaceae seeds may belong to wild species and were intrusions in crops. Kroll (1983) has identified *Vicia* seeds in the Bronze Age layer of the settlement Kastanas in Greece and has classified them as weeds in einkorn crops. The *Vicia* seeds from Opovo could also be considered weeds in the einkorn crop because einkorn grains were found in the same features.

Oil and Fiber Plants

***Linum usitatissimum* L.**
Fiber flax (*Linum usitatissimum*)
Linaceae
Only flax seeds were found in this category. *Linum* has a globular fruit capsule, which is usually composed of ten segments, each containing a seed. The seeds are oval in outline with enclosed points and with sharp margins. The seed surface has shallow rounded pitted cells. The dimensions range from ca 2.7–4.8 mm in length×1.3–3.0 mm in width×0.5–1.6 mm in thickness. Archaeological seeds are distorted by carbonization, shorter, and have clearly interrupted tips. The seeds often burst open and are fragmented as a result of pressing for oil (Schoch et al. 1988:119).

The Modern Plant and Its Distribution
Flax is an annual plant that has characteristically slender, strong stems and rounded capsules. Many *Linum* species such as *Linum catharticum* are present in the local flora, which grow in meadows and forests; others such as *Linum liburnicum* and *Linum hirsutum* are ruderal and segetal weeds. The genus *Linum* includes some 200 species spread over the Mediterranean and steppe belts in the northern hemisphere. *Linum bienne* is considered the wild progenitor of *Linum usitatissimum* (Zohary and Hopf 2000). Some wild forms are biennial or perennial; others are annual and predominantly self-pollinated, although hybridization between wild and domesticated forms occurs sporadically. *Linum bienne* grows mainly in moist grassy areas, springs, moist clay soils, and marshy lands. *Linum usitatissimum* as a crop has high nutrient and soil humidity requirements because of its undeveloped root system and short vegetation period. Occasionally, wild *Linum* plants are found as weeds in flax crops or some other crops. At least two wild *Linum* species, *Linum perenne* and *Linum austriacum*, can be found in the Voyvodina region. *Linum perenne* was found among the seeds of modern origin collected in the vicinity of the Opovo site. *Linum austriacum* thrives in dry grassy areas.

Plant Use
Linum usitatissimum was principally an oil and fiber source in the Old World, and probably the earliest cultivated plant used for weaving clothing. The seeds contain ca. 40% oil and were used as a source of edible oil and lighting oil. Two specialized types of cultivated *Linum* are apparent: oil varieties which are relatively short (30–70cm), and fiber varieties, which are taller with fewer little branches and smaller seeds. Transitional forms also occur, which are cultivated both for fiber and oil. Accordingly, different processes are used for extraction of oil or fibers. Cooking oil is produced by the cold pressing of linseeds; the oil used for varnishes is produced by a hot treatment of the seeds before pressing. Stalks are used for producing linen fibers. The flax stems are harvested before seed maturation. Usually after harvesting, the stems are dried then immersed in water for decomposition of pectins which connect fibers with other cells. After this retting process, the stems are dried again, and the fibers, on the average 4 cm long, are separated by pounding and combing (Zohary and Hopf 2000).

In Serbian and German folklore, flax is associated with female divinities. Furthermore, among both Indo-European and Semitic people, flax tissue is a sign of purity. In many old religions, priests had to dress in linen clothes. In folk medicine, cooked flax seeds are used for constipation and urinary problems. Porridge made from flax seeds is used in compresses for various sorts of inflammations and wounds. In more recent periods, flax was replaced by hemp and most of the rituals and beliefs connected with flax were transferred to hemp (Čajkanović 1985:156–158).

Archaeological Evidence
In archaeological sites, both flax seeds and capsules were found. The oldest linseeds were remains retrieved from sites in the Near East and Europe. These seeds have been identified as wild *Linum bienne*. Fragments of a capsule from Pre-Pottery Neolithic B Jericho may represent a cultivated flax. In addition, a complete capsule was retrieved from Bronze Age. Available archaeological evidence clearly suggests that flax belongs to the first group of crops domesticated in the Near East (Zohary and Hopf 2000). In Kroll's review (1991a:161–177) of archaeobotanical finds in southeast Europe, the author notes that *Linum* sp. was identified in one Pre-ceramic site in Cyprus, in three of 20 Early Neolithic sites, in one of 12 Middle Neolithic sites, and in only two of 42 Late Neolithic sites. Van Zeist (2001/2002) identified *Linum usitatissimum* seeds in a few samples from the Vinča layers at Gomolava (one sample contained nine seeds). The author suggests that flax was grown primarily for its seeds and oil and possibly for fibers.

Opovo Findings
Only five carbonized seeds of *Linum* cf. *usitatissimum* were identified in Features 26, 41, and 52 (Table 2.1). The dimensions of one whole preserved *Linum* seed from Feature 52 are 2.1 mm long×0.9 mm wide×0.7 mm thick (Figure 2.12). It uncertain whether the finds of flax seeds

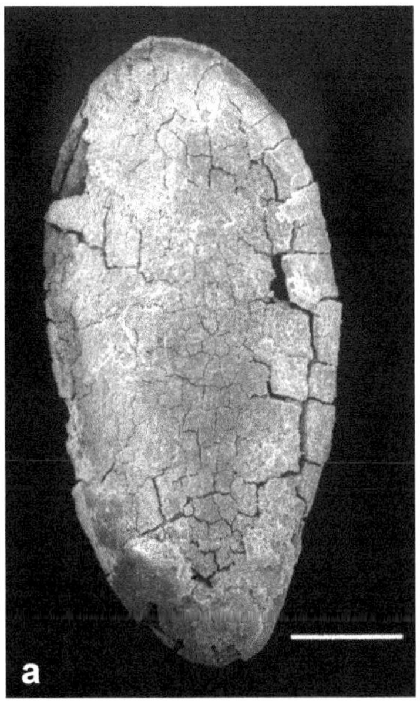

Figure 2.12a. *Linum usitatissimum* from Feature 52, carbonized (SEM, scale 500 μm).

Figure 2.12b. *Linum usitatissimum* from Feature 52. Detail showing characteristic longitudinally arranged cells (SEM, scale 50 μm).

in Opovo samples were part of a crop or were merely weed intrusions from some other crops. Of particular interest is the discovery of both *Linum* and *Lolium* seeds in Feature 41. If the *Linum* sp. seed is positively *Linum usitatissimum* and if the *Lolium* sp. seeds are positively *Lolium remotum*, then their occurance together could indicate the cultivation of flax. (*Lolium remotum* is a common weed in flax crops.) The *Linum* seeds may belong to one of the wild species.

In archaeobotanical literature, *Linum* seeds are commonly classified among oily plants, particularly when they are found in Neolithic sites (e.g., in Gomolava, van Zeist 1975:322-323). If *Linum* seeds were used to produce oil, one might expect to find larger numbers of crushed *Linum* seeds, such as the cache of *Linum usitatissimum* seeds found in a Bronze Age layer at Kastanas, in Greece (Kroll 1983:57). In Opovo, in House 4, a fragment of linen cloth (Figure 1.13) and a fragment of linen cord were discovered (identification by Hundt, personal communication, cited in Tringham et al. 1992:378). These fragments from Opovo show that even in the Late Neolithic, flax was cultivated as a crop and used for textiles and cordage. Perhaps the flax was grown and processed locally in Opovo, but we cannot exclude a possibly that finished linen products might have been brought from elsewhere.

Fruits

A total of 137 fruit seeds was recovered from the light fraction samples, and six additional fruit seeds were found in the heavy fraction samples (Tables 2.1–2.3). The majority of the fruit seeds were carbonized, although 12 small seeds were mineralized. This category, similarly to the weed category, is rather ambiguous. Although most of the fruits like *Cornus mas* were gathered from the wild, others could have been weeds in the field margins and classified as weeds. Another possibility is that these tree-shrubs were managed, encouraged, or cultivated. Because all of these plants, except *Sambucus ebulus*, have palatable fruits, I included them in the fruit category. I decided to include the *Sambucus ebulus* seeds among the fruits because the distinction among the *Sambucus* seeds is not always absolutely clear, and because the *Sambucus ebulus* plants can be used medicinally and for making dyes.

Further complications arose when calculating the number of achenes, stones, and seeds of fruit berries or stones recovered from the samples. *Cornus mas* stones were found whole as well as fragmented; thus, it was possible to extrapolate the number of whole stones present in the Opovo samples. Each stone represented one drupe of cornelian cherry. On the other hand, one strawberry has many seeds (achenes) and many of the strawberry achenes could have been part of only one fruit. Because of these ambiguitites, I tried not to overemphasize the quantities of these composite fruits.

***Cornus mas* L.**
Cornelian Cherry (*Cornus mas*)
Cornaceae
The fruit of *Cornus mas* is a drupe that has an elongated, narrow, barrel-shaped stone. The hilum is a small oval depression found on the tip of the stone. The stone surface is smooth, but has four narrow lateral ridges with two opposite, short, shallow furrows of unequal length radiating from the hilum. In a cross-section, the stone is almost round. It has two distinctive inner loculi, containing seeds (when fresh) surrounded by a thick wall of stony endo-

carp with numerous spherical resinous cavities (Renfrew 1973:142). Dimensions of the stones according to Schoch et al. (1988:63) are 10.0–16.0 mm (length)×4.0–7.4 mm (width)×4.0–7.0 mm (thickness).

The Modern Plant and Its Distribution

Cornus mas is a bush or small tree that grows in hedges and forest margins, frequently in oak forests and open woodlands and rarely in alluvial forests. It may grow up to 1300 m above sea level. The best growth is on carbonate, dry, warm soil which is not exposed to late frost. Plants flower in February and March, and the fruits ripen from late August throughtout September. *Cornus mas* is a plant that likes light and warmth. It is a plant that prefers dry habitats, unlike *Cornus sanguinea,* which can thrive in moister areas. *Cornus mas* grows very slowly, thought it may thrive more than 200 years. The modern distribution of *Cornus mas* is mainly in central and southern Europe, Asia Minor, the Crimea, and the Caucasus (Šilić 1988b:140).

Plant Use

Cornus mas fruits are collected from the wild and utilized in several ways. Fruits have a bitter taste if eaten unripe, however they can be eaten pickled in brine (Tibbles 1912:661, cited in Renfrew 1973:142). Ripe fruits have sweet pulp and are eaten fresh. Fresh fruits have a very high vitamin C content; 100 g of fruit contains 92 mg of vitamin C (Vračarić et al. 1977:324). In the Balkans, juice and jam are made from the fresh fruits. In France, Cornelian cherries are sometimes used to make wine, *vin de cornuille*, and jelly (Blamey and Blamey 1984:146). Fruits are also preserved in honey or sugar (*Larousse Gastronomique* 1965:307, cited in Renfrew 1973:142). In Russia, pancakes are prepared with sun-dried fruit (Zukovskij 1962:41, cited in Renfrew 1973:142), and in northern Greece, fruits are used to make a type of cherry brandy (Renfrew 1973:142). Besides its use as a food, *Cornus mas* played a very important ritual and magical role in Serbian folklore, and probably in the folklore of other Slavic people, because the word *dren* is of pre-Slavic origin. In the symbolism of southern Slavs, the Cornelian cherry has a medial position between the wild and cultivated world because it grows on the edge of forests and fields; it is neither a big tree nor a typical bush, and it flowers early and fruits late (Radenković 1996). The Cornelian cherry tree was a symbol of health, known for its healthy, hard, elastic wood. It is also known as an *anthropoion*, with magic properties against witches, werewolves, and other creatures. In folk medicine, *Cornus mas* was used for diarrhea, inflated spleens, skin rashes, and hemorrhoids. Also, the fruits are still given to pregnant women to prevent nausea, and after delivery of a baby to aid recovery of the mother (Čajkanović 1985:90–93).

Archaeological Distribution

Stones of the *Cornus mas* fruits have been found in many European Neolithic sites, including lake-side dwellings in

Figure 2.13. *Cornus mas* carbonized fruit stone and stone fragments from various features (scale 1 mm).

northern Italy, sites in Switzerland, Germany, and Hungary (Renfrew 1973:142). They were also found at lake site dwellings in Bosnia, but from the Iron Age. In Bulgaria, Greece, and the former Yugoslavia, *Cornus mas* was identified in eight of 20 Early Neolithic sites and in two of 12 Middle Neolithic sites. Among 42 Late Neolithic sites, besides Opovo, *Cornus mas* was identified only in Gomolava and Vinča (Kroll 1991a:161-177). At Gomolava, *Cornus mas* is well represented in the samples from Vinča layers (van Zeist 2001/2002).

Opovo Findings

A total of 119 carbonized fragments of *Cornus mas*, including several whole stones, weighing 3.92 g were identified from the Opovo samples. *Cornus mas* was found in light fraction and some heavy fraction samples from the majority of the features and from several non-feature samples (Tables 2.1–2.4).

In many Opovo samples, carbonized fragments and occasionally whole stones of *Cornus mas* were found, reflected in the rather high percentage presence (ubiquity) of 37.2% in feature samples and 6.2% in non-feature samples (Table 2.4). Because of its distinctive inner structure, even the small fragments of carbonized stone were easily identifiable (Figure 2.13).

The number of whole stones was extrapolated as the weight of fragments divided by 0.08 g, an average obtained by weighing several whole stones. Whole *Cornus mas* stones were counted as one fragment each. Although it seems that there were many *Cornus mas* fragments in the Opovo samples, the extrapolated total is 34 whole stones. The maximum density is 10.77 fruit stone fragments (per 100 liters) in Feature 31 (Tables 2.1 and 2.2). If the fragments are extrapolated to the number of whole stones, this Feature contained about eight whole stones. Another stone of a drupe resembling a *Cornus mas* stone was identified in Feature 41, but its outer surface was considerably more wrinkled and was therefore included in the "Unknown

Seed" category. The dimensions of one carbonized whole stone from Feature 41 are 11.2 mm in length and 4.7 mm in width. Three whole stones (not included in the flotation tables) were picked out by hand during excavation of Feature 30.[1] They measure 11.8–12.5 mm in length×4.0–5.8 mm in width. From Feature 31, two whole stones were identified, measuring 10.0–10.2 mm in length×4.8 mm in width. Because few whole stones of drupes were preserved, it is difficult to draw conclusions about the dimension from these three features. Nevertheless, if we assume that the dimensions of whole stones might be representative of the other stones found in fragments in these features, it seems that the largest drupes of *Cornus mas* were in Feature 30 and the smallest in Feature 31. Feature 30 is assigned to BH 3, Feature 31 to BH 2/1, and Feature 41 to BH 2/3. If these differences in sizes of fruit stones are not merely a reflection of the variability of fruits from one plant, then the difference in fruit stones from various features might indicate that the fruits were gathered in various years and probably from different plants. In addition, *Cornus* wood charcoal was identified from two of five charcoal samples examined (Table 4.2).

The presence of *Cornus mas* stones in the Opovo samples is indicative of its use since stone fragments were found both in pits and on the floor of House 4 (Tables 2.1–2.3). The presence of stones in both types of context implies that the fruits were intentionally gathered from the wild (or perhaps encouraged trees) in all three building horizons. However, it is difficult to comprehend how they were used at Opovo. Because all the finds are carbonized, I assume that the stones were charred after the fleshy fruit was eaten or extracted for processing. It is possible that the stones were spit out onto the house floors after the fleshy part was eaten, and then were carbonized when the house burned down. Another possibility is that after the fleshy part of the fruit was separated from the stones in making juice or jam, the stones were thrown into the hearth and the ash was taken and dumped into pits. As a result, charred fragments of stones were found in pits. The presence of *Cornus mas* plants in the vicinity of the settlement is important for reconstructing Opovo's environment. The presence of both wood charcoal and fruit stones of *Cornus mas* implies that there were places dry places, since the natural habitat of modern *Cornus mas* plants is dry and sunny.

While *Cornus mas* has been identified from many Early Neolithic sites in the former Yugoslavia, Bulgaria, and Greece, its ubiquity from the Late Neolithic period onwards rapidly dicreases. The drupe stones are present in only a few sites from the Bronze Age and it is almost completely absent from later periods (Kroll 1991a:161-177). Such a diminishing trend of *Cornus mas* stones from the sites of later periods is difficult to explain, considering the evidence of its use until modern times. It cannot be explained as a consequence of methodological mistakes because stones of *Cornus mas* are rather big and easily observable in deposits. Furthermore, many of the sites from early and late periods were analyzed by the same archaeobotanists. Also, the context of the archaeobotanical remains from various periods is similar: pits and house floors. I assume that the processing of *Cornus mas* drupes including the final discarding of stones might have changed since Late Neolithic-Bronze Age times. Therefore, in the later periods, *Cornus mas* stones or drupes might not have been frequently exposed to fire and hence not preserved in archaeological contexts. All the *Cornus mas* finds from Opovo and other sites (e.g., Kastanas in Greece) were carbonized. Citing classical sources (*Odyssey*), Kroll (1983:73) explains the paucity of the *Cornus mas* finds at Kastanas (Bronze-Iron Age) is attributed to its modest food value (it was often mixed with acorns and offered to pigs). This practice could certainly diminish the presence of carbonized stone fragments of *Cornus mas* in the archaeological record. However, the practice would not explain what happened to the stones of fruits that were used for human consumption, unless they were also thrown out and eaten by pigs. It is difficult to discern whether the human use of *Cornus mas* as fruits gathered from the wild diminished in the later periods, or whether its disappearance was the consequence of different taphonomic processes.

***Sambucus nigra* L.**
Elder (*Sambucus nigra*)
Caprifoliaceae

The fruit of *Sambucus nigra* is a berry fruit that usually contains three seeds. The oval seed is thin and elongated. The dorsal surface is weakly domed and the ventral surface is slightly "V"-shaped. The hilum is circular to oval and found at the pointed end of a seed. The surface of the seed is covered with coarse transverse wrinkles. Some authors note that it is difficult to distinguish seeds of *Sambucus nigra* from the seeds of *Sambucus ebulus*, except for the degree to which the hilum extends downward over the top of the ventral side (McLaren and Hubbard 1990:250). *Sambucus nigra* seeds are more elongated, while *Sambucus ebulus* seeds are more rounded with more pronounced ridges on the ventral sides. Dimensions of seeds range from ca. 3.0–5.0 mm in length×1.5–2.0 mm in width×0.7–1.2 mm in thickness (Schoch et al.1988:35).

The Modern Plant and Its Distribution

Sambucus nigra is a bush that grows on wet and humus soil. Bushes can be found at forest edges, in wasteland, and near roadsides. They can grow up to 10 m in height. This mesophytic species can be found up to 1200 m above sea level (Šilić 1988b:173). The plant develops leaves first and then flowers in May and June; its fruits ripen in autumn. The whole plant has a distinctive smell, with especially fragrant flowers. They grow throughout Europe, except in Scandinavia, but also in Asia Minor, the Caucasus, West Siberia, and North Africa (Šilić 1988b:173).

Plant Use

The tiny gooseberry-like fruits are shiny black berries which are often gathered in the wild. For centuries they have been used for making delicious red wines (Blamey and Blamey 1984:198-199). The berries are also cooked and used for making preserves, jams, and juices. Such preparations are still produced in the Balkans. From flowers, sweet syrups are made. Flowers and leaves are also used medicinally. Tea made from flowers is still used for respiratory difficulties, and tea made from leaves is used for headaches. Fresh leaves are used for healing wounds. Seeds mixed in white wine are used for constipation. Drinks that facilitate child birth, and lotions for skin rashes are made from the heart of the wood. Besides its medicinal use, *Sambucus nigra* symbolizes demonic trees in Serbian folklore. Fairies reside in trees according to legend, and if someone cuts them, the perpetrator may suffer death or paralysis. *Sambucus nigra* played a very important role in rituals for domestic animals. For example, it was mixed with cereals and given to cows to produce more milk, and was used in various cults in spring festivities connected with livestock (Čajkanović 1985:102).

Archaeological Distribution

Sambucus nigra has been used since the Neolithic period in many settlements in Europe. Uncarbonized fruits were identified from a number of prehistoric sites, mainly in the Alpine region, but also in other parts of Europe including Germany and Britain (Renfrew 1973:150). Only a few seeds were found on the Neolithic site of Selevac (McLaren and Hubbard 1990:252), and in the Iron Age samples from Gomolava (van Zeist 2001/2002). *Sambucus nigra* seeds were identified in the Bronze Age site of Feudvar in Voyvodina (Kroll 1991a:161–177).

Opovo Findings

Only five seeds of *Sambucus nigra* were identified in the light fraction from Features 26 and 41, and from a non-feature sample (Table 2.1), and thus the percentage presence (ubiquity) is only 1.1% overall (Table 2.4). Two seeds were carbonized and three uncarbonized. The seeds are slightly different in shape and longer than *Sambucus ebulus* seeds.

Dimensions of two carbonized seeds of *Sambucus* sp. from Feature 41 are: 2.9 mm in length×1.7 mm in width×1.0 mm in thickness. In sum, only a few seeds were found in the Opovo samples. Considering the mode of arrival, one can suggest several ways for seeds of *Sambucus* sp. to end up in the pits of BH 2 and BH 2/3. One possibility for uncarbonized seeds is that they were contained in paleofecal remains, although paleofeces were not observed as such on the site. Some of the berries could have charred been while being cooked or dried, or else were spilled on the floor and swept into house hearths and emptied into pits. Certainly, there are other explanations for the preservation of these seeds in pits, but the two cited seem plausible in the case of the Opovo samples.

***Sambucus ebulus* L.**
Dwarf Elder (*Sambucus ebulus*)
Caprifoliaceae

Sambucus ebulus is a berry that usually has two or three seeds. The seed is a truncated, ovate stone. The dorsal side is domed and the ventral side is "V"-shaped. The hilum is located at the narrow end of the seed. The seed surface has marked lateral wrinkling. *Sambucus ebulus* seeds are distinguished from other forms of *Sambucus* species seeds by their short form and orange-yellow color when fresh (Kroll 1983:74). Dimensions of the seeds range from ca. 2.0–3.5 mm in length×1.5–2.3 mm in width×1.0–1.5 mm in thickness (Schoch et al. 1988:35).

The Modern Plant and Its Distribution

Sambucus ebulus is a small, herbaceous plant (rather than a bush) in contrast to *Sambucus nigra*. *Sambucus ebulus* grows on fertile humus soils, in forest clearings, and near roadsides. The whole plant has a very offensive odor (Šilić 1988b:172). The fruit ripens in autumn. It grows in all of Europe, particularly in warmer regions with temperate climates (Schoch et al. 1988:34).

Plant Use

Sambucus ebulus is a poisonous plant with medicinal properties. The black berries were used as a source of blue dye. In Serbian folklore, *Sambucus ebulus* had magical powers. It was believed that when a person lies down near the plant, he or she would be protected from snake bites. Probably such a belief emerged as a consequence of the unpleasant odor that *Sambucus ebulus* emits. Similarly, the roots, stems or leaves are used for snake or scorpion bites. In addition, in folk medicine, *Sambucus ebulus* relieves diarrhea, back and leg pains, and ulcers. It is also used to treat fever in magical rituals.

Archaeological Distribution

In Europe, *Sambucus ebulus* seeds were found from the Neolithic period. In the Balkans, the finds were reported from Neolithic sites in Bulgaria (Renfrew 1979:257; Lisitzyna and Filipovitch 1980:69) and from the Neolithic settlement of Selevac (McLaren and Hubbard 1990:252). Kroll (1983:74) explains the occurrence of numerous uncarbonized seeds of *Sambucus ebulus* in the Iron Age layers in Kastanas as the consequence of using fruits to make fiber and food dyes rather than food or medicine. *Sambucus ebulus* were identified at Gomolava, including the Vinča layers (van Zeist 2001/2002).

Opovo Findings

Only five *Sambucus ebulus* seeds (four carbonized and one uncarbonized) were identified in the light fractions of Feature 30 and 41, and one in a non-feature sample (Table 2.1). The percentage presence (ubiquity) is 1.9% from all samples (Table 2.4). The dimensions of one whole carbonized *Sambucus ebulus* seed from Feature 30 are 1.9 mm in length×1.8 mm in width×1.0 mm in thickness. Kroll

Figure 2.14a. *Sambucus ebulus* from Feature 30, carbonized. (SEM, scale 500 μm).

Figure 2.14b. *Sambucus ebulus* from Feature 30. Detail showing hilum. (SEM, scale 500 μm).

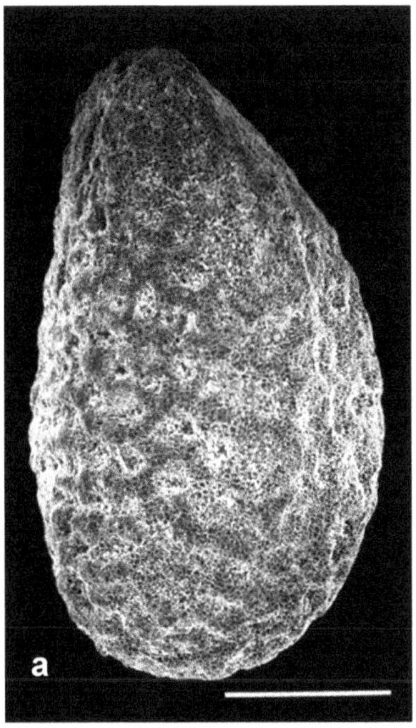

Figure 2.15a. *Sambucus* cf. *ebulus* long-seeded type from Feature 41, carbonized. (SEM, scale 500 μm).

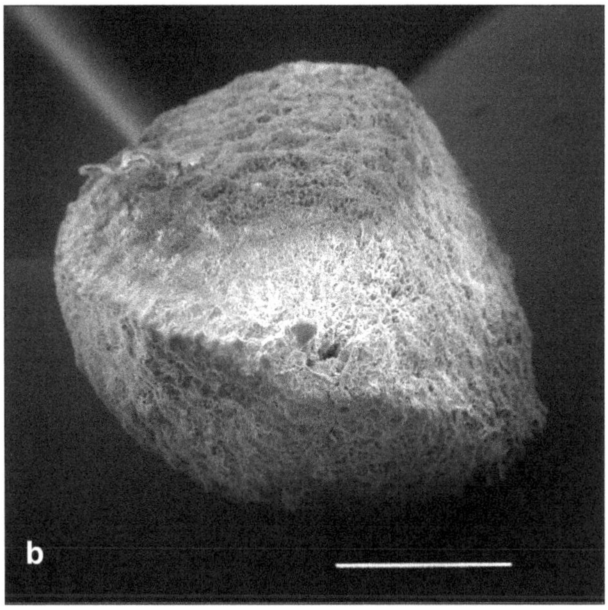

Figure 2.15b. *Sambucus* cf. *ebulus* long-seeded type from Feature 41. Detail showing hilum (SEM, scale 500 μm).

(1983:75) reports similar dimensions for 25 *Sambucus ebulus* seeds from an Iron Age layer at the site of Kastanas: 1.7–3.4 mm (average 2.77 mm) in length×1.3–2.1 mm (average 1.61 mm) in width×0.6–1.3 mm (average 1.06 mm) in thickness.

It is difficult to explain the presence of a few seeds of *Sambucus ebulus* in both carbonized (Figures 2.14–2.15) and uncarbonized forms in Opovo samples. It is also uncertain whether the berries were intentionally gathered for producing dyes or medicines, or whether the berries fell off a plant that was present on wasteland near the settlement and somehow ended up in pits.

Vitis vinifera* L. ssp. *sylvestris
Grape (*Vitis vinifera*)
Vitaceae
Grapes are fruits that usually have two-to-four pips per berry. Pips are pear-shaped seeds that vary in form and size. The dorsal side of the seed is domed and often indented

on top by narrow notches, with a central depression. The ventral side is flattened, with clear elongated edges and two elongated depressions turning parallel to them. Both wild and cultivated grapes belong to the same species *Vitis vinifera*, but to two different subspecies. Wild forms are classified as *Vitis vinifera* ssp. *sylvestris* while cultivated forms are classified as *Vitis vinifera* ssp. *vinifera*. *Vitis vinifera* ssp. *sylvestris* has rounded thick pips with rather short stalks. On the other hand, cultivated *Vitis vinifera* ssp. *vinifera* has elongated, slender, and often fragmented seed stalks. Intermediate forms of seeds also exist. Dimensions of *Vitis vinifera* ssp. *vinifera* pips are between 3.5–6.0 mm in length×3.0–4.1 mm in width×2.2–3.0 mm in thickness (Schoch et al. 1988:196).

The Modern Plant and Its Distribution

Vitis is a perennial woody climber that thrives in areas with a relatively mild climate. The cultivated *Vitis vinifera* ssp. *vinifera* is closely related to both wild and feral forms. The main difference between a wild grape (*sylvestris*) and a cultivated grape (*vinifera*) lies in the sexual distribution of the flowers. According to Kroll (1983:62), wild grapes are monosexual and a plant is either male or female. Zohary and Hopf (1993:147) state that *sylvestris* grapes are dioecious and that their populations comprise an equal proportion of male and female individuals; hence they can cross-pollinate while *vinifera* is self-pollinated. *Sylvestris* and *vinifera* may be crossed when they are grown in close proximity, and F_1 hybrids are fully fertile, which blurs the differences between wild and cultivated forms (Zohary and Hopf 1993:144). The grape fruits ripen in autumn. *Vitis sylvestris* is widely distributed over Europe and western Asia. Plants are primarily forest climbers and seem to be native in the humid mild climates south of the Caspian Sea and the Black Sea. Grapes also thrive in the more mesic and cooler scelorphyllous vegetation belt spanning from Turkey and Crimea through Greece and the former Yugoslavia to Italy, Spain, France, and northwest Africa. Wild grapes also penetrated more deeply in Europe along the Danube and the Rhine and are found in alluvial forests. According to Zohary and Hopf (1993:144), there is little doubt that *Vitis vinifera* ssp. *sylvestris* vines are indigenous in southern Europe, the Near East, and the southern Caspian Belt.

Plant Use

Wild grapes are distinguished from the cultivated varieties by their smaller, usually acid berries which are quite sour when eaten fresh. They are suitable, however, for the preparation of wine, juice, and vinegar. They can also be sun-dried and eaten as raisins. Grape leaves can be eaten as greens, and in various parts of the Balkans they are used for stuffing. In folk medicine, the cultivated wine tree and grapes are used in various ways. Juice from the grape stalks is used for speeding hair growth. Tea made from leaves relieves ear aches. Various parts of the grape tree are also used for protecting a child against evil, for inducing menstruation, and in preventing spontaneous abortions. In Balkan folk rituals, grapes represent chthonic plants. They are planted on graveyards and cause harm to anyone who sits under one after sunset. On the other hand, grapevines are holy plants among Christians and provoke fear in werewolves (Čajkanović 1985:64–66).

Archaeological Distribution

Vitis vinifera is one of the classical fruits of the Old World, together with olives, figs, and date palms. Although carbonized pips are brittle, often hollow, and disintegrate readily, the carbonized pips of grapes have been discovered in numerous prehistoric sites in Europe (Renfrew 1973:127; Riviera Núñes and Walker 1989:216; Zohary and Hopf 1993:143-150). Several dozens pips discovered in Neolithic sites (4500 b.c.) seem to represent wild grape vines. The earliest definite sign of *Vitis* cultivation comes from the Chalcolithic and from the Early Bronze Age sites in the Levant (Zohary and Hopf 1993:149). *Vitis* is a perennial climber that must be pruned yearly to regulate fruit production, which starts three years after planting (Zohary and Hopf 1993:143, 149). Discussions about grape pips found on prehistoric sites usually revolve around the question of whether the recovered pips are the wild grape subspecies (*Vitis vinifera* ssp. *sylvestris*) or the cultivated form (*Vitis vinifera* ssp. *vinifera*). Measurements of pips made by Stummer (1911) represent the fundamental work for distinguishing the wild and cultivated forms of grape pips, but since there are many intermediate forms between wild and domestic, we cannot make the typical distinction that thin pips represent cultivated forms and rounded pips represent wild (Janusevich 1976, cited in Kroll 1983:61–70).

In the Balkans, *Vitis vinifera* ssp. *sylvestris* was found already in the Late Paleolithic/ Mesolithic layers at Franchthi Cave in Greece (Hansen 1991). It is present in one of six Preceramic sites, in five of 20 Early Neolithic sites, in three of 12 Middle Neolithic sites, and in nine of 42 Late Neolithic sites referenced by Kroll (1991a:161–177). Wild grape pips were also found at Neolithic sites in Greece, the former Yugoslavia, and Bulgaria. *Vitis vinifera* (wild grape) seeds were identified from the Late Neolithic layer of the tell site Gomolava (van Zeist 1975:323, 2001/2002).

Opovo Findings

Only six uncarbonized but mineralized seeds of *Vitis vinifera*, weighing 0.13 g, were identified in the light fraction from Feature 46: Pit (Table 2.1). Thus, the percentage presence (ubiquity) is very low 0.7% from feature samples (Table 2.4). The outer layer of seeds was missing (Figure 2.16). Dimensions of four seeds ranged from 3.7 to 4.5 mm in length×2.8–3.5 mm in width (breadth). The index (width/length×100) ranged from 62 to 83, which falls strictly in Stummer's (1911) category characteristic for *Vitis vinifera* ssp. *sylvestris* (index 76–83). With such seed dimensions, is it rather clear that the Opovo seeds belong to the wild grapes and are not pips of cultivated grapes either from the past or the present, although positive identification cannot

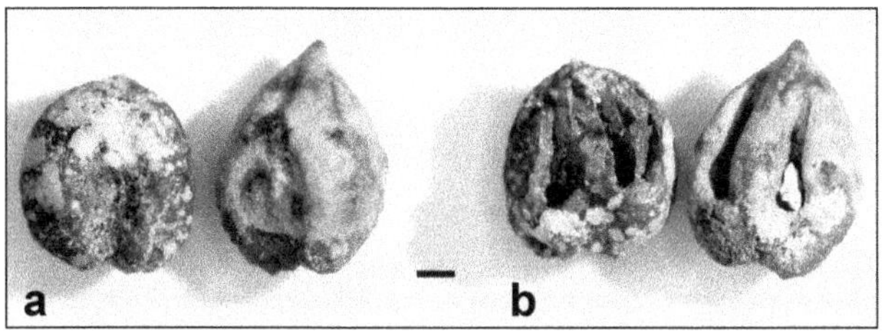

Figure 2.16. *Vitis vinifera* subsp. *sylvestris* mineralized seeds from Feature 46: **a)** dorsal; **b)** ventral side (scale 1 mm).

be made unless the complete, whole seeds are preserved (Helmut Kroll, personal communication 1997).

In the present day, there are no wine grapes in the immediate vicinity of the site because all the land is under crop cultivation. However, there are grapevines on the river banks of the Tamis and the Danube, a few kilometers away from the site. In these forests, wild grapes can be found on trees that do not have dense crowns and that permit light to filter through.

It is highly possible that before the modern system of water management was introduced, grape plants climbed the trees near the Opovo Neolithic settlement and that humans gathered and ate the fruit, or used it for other purposes. There is not enough evidence that they made wine or juice out of wild grapes, because only six pips were found in the Opovo samples. Usually when wine is made, hundreds or thousands of pips are found in samples. In addition, grape berries in the wild are eaten by various animals, including birds, and often only a few berries are left for human consumption (Kroll 1991b).

There is yet another possibility that grape plants collected together with other plants were part of animal fodder given to pigs. Considering the mode of arrival of mineralized seeds in the Opovo pit, they could have been contained in human or animal feces (although paleofecal material has not been observed as such in the Opovo site). Also, the seeds could have been spit onto the house floor and then swept into the pit.

Rubus L.
Blackberries (*Rubus*)
Rosaceae

Blackberry fruits are berries composed of numerous single-seeded drupelets set together on a conical, rounded receptacle. The seeds are ovate and compressed in shape. The dorsal side is rounded and the ventral side almost flat or slightly domed. The pericarp seed surface has deep distinctive pits. The *Rubus* genus has more than 300 species and many hybrid forms. Seeds of various species have the same general shape, but they differ slightly in size, surface pit-depth, and the shape of the hilar end. Therefore, it is difficult to determine the exact taxon of the seeds, and they often cannot be separated to the species level on the basis of seed morphology. In European archaeobotanical literature, the species most commonly discussed are *Rubus caesius*, *Rubus fruticosus*, and *Rubus idaeus*. *Rubus caesius* has the largest seeds ranging from 2.6–4.2 mm in length×1.8–2.1 mm in width×1.1–1.3 mm in thickness. *Rubus fruticosus* seeds range between 2.2–3.2 mm in length×1.4–2.2 mm in width×0.8–1.4 mm in thickness. The smallest seeds among the three mentioned species are of *Rubus idaeus*, ranging between 1.8–2.2 mm in length×1.0–1.3 mm in width×0.7–1.0 mm in thickness (Schoch et al. 1988:173–175).

The Modern Plant and Its Distribution

Rubus spp. plants vary from a low arching plant to a thorny bush that grows up to 2 m high. In the local flora of the former Yugoslavia, there are many species and variants of blackberries, such as *Rubus discolor*, and they are often imprecisely called by a general term *Rubus fruticosus* s. l. (Šilić 1988b:90). Plants flower in early summer and fruits ripen from the middle-to-late summer. They usually grow in hedges and forest margins, along the roadsides on both calcareous or silicate soil. Various species have different preferences, e.g., *Rubus caesius* thrives in alluvial forests, meadow thickets, and damp limey areas, but also as a weed in crops. On the other hand, *Rubus idaeus* prefers shady areas (Blamey and Blamey 1984:82–85).

The genus *Rubus* is widely distributed over temperate parts of Europe, Asia, and America. The plants are widely distributed from the low lands up to the mountains in Europe and West Asia. The domestication of *Rubus* began relatively late because it was easily gathered from the wild (Jennings 1976, cited in Zohary and Hopf 1993:199). In Europe, domestication did not start until the eighteenth century. Cultivated *Rubus idaeus* was introduced to North America at this time, and it was supplemented by selections from the native blackberry *Rubus occidentalis*, as well as by spontaneous hybrids between two species. The modern cultivated blackberries were developed by crossing several American and European *Rubus* species (Zohary and Hopf 1993:200).

Plant Use

Rubus fruits are among the most widespread and tasty wild berries in Europe and West Asia. The berries are eaten either fresh or dry. Preserves, juices, and wines can easily be made from the fruits. Fresh blackberries and their various products are recommended to anemic people. Teas are made from leaves and used to treat diarrhea, stomach aches, and ear aches. In folk medicine, tea made from the roots was used for diarrhea and chest pains. In Serbian folktales, the blackberry had both positive and negative connotations, and were sometimes referred to as "devil's grapes" (Radenković 1996:227). Among many European peoples, the thorny blackberry plant was used to trip negative characters in chases with good characters. Blackberry plants also had a very important role in rituals for the well-being of livestock (Čajkanović 1985:152–154).

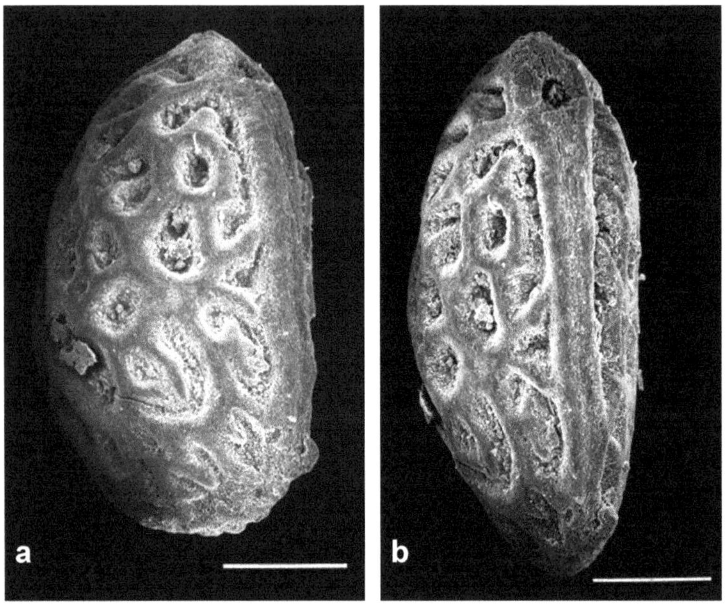

Figure 2.17. *Rubus fruticosus* from Feature 31: **a)** carbonized seed; **b)** detail showing hilum (SEM, scale 500 μm).

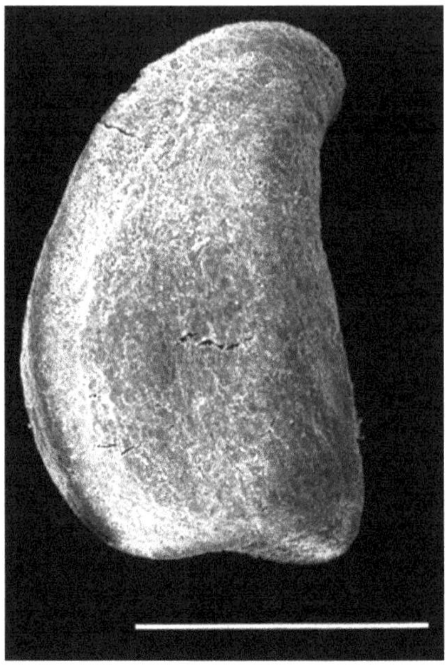

Figure 2.18. *Fragaria* cf. *vesca* carbonized seed from Feature 30 (SEM, scale 500 μm).

Archaeological Distribution

The small seeds, actually achenes, of *Rubus* spp. have been retrieved from archaeological sites from the Neolithic period onwards. They have been most commonly identified in European sites as *Rubus fruticosus* s.l., as *Rubus caesius*, or *Rubus idaeus*. *Rubus* sp. seeds were found in the Neolithic sites of Chevdar and Kazanluk in Bulgaria (Renfrew 1979:243–265). Seeds of *Rubus caesius* were identified in small quantities in the Neolithic site Selevac in the former Yugoslavia (McLaren and Hubbard 1990:247–254). A few seeds of *Rubus fruticosus* were identified at Gomolava, including the late Vinča layers (van Zeist 2002/2002).

Opovo Findings

Altogether, 17 *Rubus* seeds were identified in the Opovo light fraction samples. All of the seeds were carbonized except for one seed from a non-feature sample. There is no exact weight for *Rubus* seeds because three carbonized seeds weigh less than 0.01 g. Seeds were identified only in the light fraction in features 26, 31, 41, 27, 30, 34, and from a non-feature sample (Table 2.1), Thus, the percentage presence (ubiquity) is 7.3% from feature samples (Table 2.4).

Rubus seeds from the Opovo samples generally were not identified to the exact taxon because several *Rubus* species have very similar seed morphology. However, most of the seeds do not have pronounced beaks on the top, but rather show convex lateral sides and distinctive deep reticulation on surfaces resembling *Rubus fruticosus* s.l. (Figure 2.17). Measurements were taken for three whole seeds from Feature 30 and Feature 40. The dimensions of the seeds are 2.3–2.6 mm long×1.3 wide×0.8–1.6 mm thick, and the size range is of *Rubus fruticosus* seeds. One seed from Feature 41 does not have the seed coat preserved. In addition, one seed from Feature 34 resembles a *Rubus* seed, but since the identification was not positive it was included in the "Unknown seed" category.

The presence of seeds in the Opovo site, both in pits and in one instance on the house floor suggests that blackberries were intentionally gathered from the wild. However, it is difficult to explain their mode of arrival in a carbonized state in the Opovo features. Blackberries could have been eaten fresh or dry, so some of the uncarbonized seeds in pits may represent fecal contents that were carbonized later. Another possibility is that juice was extracted from the fresh fruits, and the seeds were thrown away afterwards. Several seeds were found on the floor of House 4. These seeds might have been carbonized when the whole house burned down if the blackberries had previously been dried or spilled on the floor.

***Fragaria vesca* L.**
Wild Strawberry (*Fragaria vesca*)
Rosaceae

The fruit of *Fragaria vesca* is a juicy, swollen receptacle with many minuscule achenes that are commonly called "seeds". The seeds are ovate in shape and have a slightly hook-shaped top. Infrequently, branched veins (clear in the dry state) run from the slightly prominent hilum. Dimensions range from ca. 0.8–1.5 mm in length×0.7–1.1 mm in width×0.6–0.8 mm in thickness (Schoch et al. 1988:161).

The Modern Plant and Its Distribution

Fragaria vesca is a perennial herb that propagates both vegetatively and by seeds. The fruits ripen from early summer until the middle of August, and they are among the first fruits available in the wild. The plants grow in open forests and forest clearings, often together with other pioneer species. The plants are widely distributed from the lowlands up to 1500 m above sea level. They prefer fresh, slightly acid humus soil, especially soils rich in nitrates, but they are also found in sandy and gravely soils. Two similar species of the same genus, *Fragaria moschata* and *Fragaria viridis*, also grow in forests and thickets in the former Yugoslavia (Šilić 1988a). *Fragaria vesca* is distributed in central and eastern Europe. *Fragaria moschata* occurs in the cooler and elevated parts of the Mediterranean region, while *Fragaria viridis* thrives in temperate Europe in meadows and along the edges of forests. The cultivation of strawberries started in Europe in the fourteenth and fifteenth century when the first attempts were made at domestication. Cultivation continued until the nineteenth century when pineapple strawberries, *Fragaria ananassa*, were bred from two American species, *Fragaria chiloensis* and *Fragaria virginiana* (Zohary and Hopf 1993:200).

Plant Use

The small, tasty fruits of wild *Fragaria* species are still collected from the wild and used for food. In the wild where there are dense stands of strawberry plants, one person can pick up to 2 kg of strawberries. Leaves and dried fruits are used for making aromatic tea, used to treat coughs. The fruits are eaten fresh and various preserves are made from them (Vračarić et al. 1977). The berries and leaves have a very high concentration of vitamin C, and can be used to make a special stimulating drink in spring for recovery from the winter's restricted vitamin diets. In the folklore of Serbia and other Slavic peoples, the strawberry has magical "love" powers. On the other hand, dreaming of strawberries is a sign of sorrow or sickness (Čajkanović 1985:119).

Archaeological Distribution

Strawberry seeds have been found in many prehistoric as well as historic sites. In archaeobotanical reports, the seeds are usually identified as those of *Fragaria vesca*. They have been identified in many Neolithic sites in Europe, and were even reported from the Mesolithic site of Muldbjerg, in Denmark (Renfrew 1973:149). Only a few *Fragaria vesca* seeds were found in the late Neolithic layer in the tell site Gomolava (van Zeist 1975, 2001/2002) and none from other Neolithic sites in the Balkans (Renfrew 1979:243–265; Kroll 1991a:161–177). Because the seeds of *Fragaria* are minuscule, one possibility for their absence from archaeobotanical samples is that the sieve-meshes were not fine enough to retrieve them.

Opovo Findings

Altogether, 70 seeds of *Fragaria* cf. *vesca* were found in the light fractions of Opovo samples and all from features. All the seeds were carbonized, except for one mineralized seed from Feature 41 (Table 2.1 and 2.2). The seeds were identified as *Fragaria* cf. *vesca*, although I was not able to see characteristic veins on the seeds' surfaces (Figure 2.18). I do not know whether the lack of observable veins was the consequence of bad preservation, or whether the seeds belong to another species. Their general shape and size, and the presence of the typical beak suggest *Fragaria vesca* seeds. Nevertheless, I do not exclude the possibility of the presence of other *Fragaria* species. *Fragaria* seeds were found from Features 26, 30, 31, and 41. The percentage presence (ubiquity) is 10.2% from feature samples, and not a single seed was found in non-feature samples (Table 2.4). The feature richest in *Fragaria vesca* seeds is Feature 30, a pit or well in which 40 carbonized seeds were found (density 53.33) In Feature 41, one uncarbonized and 22 carbonized seeds of *Fragaria vesca* were identified.

The presence of *Fragaria* seeds in the Opovo samples suggests that the berries were intentionally gathered from the wild, and in that case, the uncarbonized seeds might represent paleofecal content. It is difficult to conceive what kind of processing would yield carbonized strawberry seeds, unless they incidentally fell into the house fires. However, both *Fragaria vesca* and *Fragaria viridis* can be meadow and forest weeds that grow in hedges, and they could have been collected with other plants and brought to the site.

Water Chestnuts

***Trapa natans* L.**
Water Chestnut (*Trapa natans*)
Trapaceae

The fruit of *Trapa natans* is a large irregular stone with two-to-four horn-like projections. The endocarp consists of the hard part that has spine projections extending up to 10 mm in length. The surface of the fruit has longitudinal ridges. The rounded hilum is dimple like. The top of the fruit has a rounded opening (covered with dense hairs when fresh) and may be slightly arched from the main body of the fruit. The seed within the fruit is without endosperm, but it has two unequal cotyledons. The small cotyledon emerges with a young sprout from the top hole, while the large cotyledon remains in the fruit to nourish the sprout until the leaves are developed (Janković 1973:37). Dimensions of fruits of the genus *Trapa* range from 18–22×25–40 mm, depending on the species and ecological conditions under which they grow (Schoch et al. 1988:93).

The Modern Plant and Its Distribution

Trapa natans is an annual aquatic plant that grows in still, warm waters. The part of the stalk that grows in water is characterized by long internodes that become shorter and thicker as the stalk nears the water surface. The stalk is long and the length depends on the depth of the water. The leaves on the underwater part of the stalk are very narrow, whereas those floating on the surface become large. Populations of water chestnut plants may cover large areas, particularly in standing canals and shallow lakes. *Trapa* usually occurs side by side with populations of *Nymphaea alba* plants. The *Trapa natans* flowers are white. The plants are self-fertilizing, although some insects may take part in fertilization (Janković 1973:36).

The plant systematics of the genus *Trapa* are not final and are mainly based on morphological differences among fruits. *Trapa* is the only genus in the Trapaceae family, formerly called Hydrocaryaceae (Hegi 1965:882). According to Janković (1958:1–123), an expert on Yugoslav water chestnuts, the genus *Trapa* includes 15 to 20 species, although many species are imprecisely included in the broad term *Trapa natans* s.l. According to Janković, this term is an old name (also used by Hegi 1965:882–894) for the whole complex of European and west Asian *Trapa* species. In the flora of the former Yugoslavia, there are two sections of the genus *Trapa*, Longicarpa and Brevicarpa, based on morphological characteristics of the fruit. Janković (1958) made this taxonomic division by comparing the width to the length of the whole fruit as well as the shape of the upper horns in relation to the shoulder part. Section Longicarpa includes three species with long horns: *Trapa annosa*, *Trapa longicarpa*, and *Trapa europea*. Section Brevicarpa is represented by only one species with shorter horns *Trapa brevicarpa*. There are also forms of fruits, so called "*muzzanensis*" that besides have regular horns and additional knobs.

Trapa longicarpa and *Trapa brevicarpa* are commonly present in the calm waters of the Voyvodina region. Populations of these plants can grow together, but the dominant species is the one best suited for a particular habitat. Thus, *Trapa longicarpa* is dominant where there are higher concentrations of calcium carbonate in mud, while *Trapa brevicarpa* is not adapted to these conditions (Janković 1973:40–43).

Although fruits of *Trapa* ripen from later summer onwards, the harvest is best in the late summer in temperate areas, just before the fruits sink to the bottom. In other areas of the world with warmer climates, water chestnuts can be harvested for several months. Fruits of *Trapa* can be transported by water animals and water currents. Fruits are resistant to low temperatures and can germinate in spring; however, as long as the temperature reaches 10° C. The fruits are less resistant to drying than to freezing.

There are various species of genus *Trapa* in Europe, except in the northern regions from where water chestnuts have disappeared in the last centuries. Presently, the northern border of *Trapa* distribution consists of areas where there are at least 63 days with temperatures of at least 22° C. The presence of *Trapa natans* in still waters is an indicator of non-polluted waters, while the absence of *Trapa* from various regions could indicate climatic deterioration and/or water pollution. Water chestnuts are presently cultivated in southeast Asia. There are also *Trapa* species in North Africa. There is an interesting situation with water chestnuts in North America, where the oldest finds of *Trapa* species date to the upper Cretaceous and the lower Tertiary period, even though *Trapa* became extinct at the beginning of the Pleistocene (Hegi 1965:883).

Plant Use

The inside contents of water chestnuts (cotyledons) are consumed by humans and animals, including various aquatic animals and water fowl. The seeds are rich in nutrients, containing 52% starch, 15–20% proteins, 7% fat, and 3 % sugar (Janković 1973:38). Cleaned fruits can be eaten fresh, cooked, or roasted. Also, they can be dried and ground into flour.

While in some places of the world water chestnuts are used for food, in others they are considered aquatic pests. In North America, *Trapa natans* plants were introduced in the 1980s as ornamental plants and presently cover large areas of the Hudson and Mohawk rivers, hindering navigation, reducing recreational uses of the waters, and causing mosquito-control problems (Martin 1987:86).

In China, India, and Thailand water chestnuts are grown for their nutritive values. There are fenced expanses of water where water chestnuts are encouraged and are protected from other animals and reserved for humans. In south Asia, several species are cultivated and some have been selected for large fruits which measure up to 6.5 cm wide. In China and Thailand, cooked fruits are also canned and exported. Likewise, in Kashmir, 40,000 men collected fruits for five months every year in the nineteenth century. Similarly, in the former Soviet Union, on Lake Orehovoe, 160 tons of water chestnuts were produced in annually (Hegi 1965).

In middle and north west Europe, the use of water chestnuts was not so widespread, but there are several historical mentions of the use of the *Trapa* as food for humans. In most of southern Europe, south to Hungary, leaves were used as food for pigs. In addition to the use of *Trapa* fruits as human and animal food, and leaves as fodder, the plants are also used medicinally. Hippocrates describes the use of a beverage made from *Trapa* for inflammation, while various authors mention *Trapa* as being good for making tonics and a powder that was supposedly good for diarrhea. It is also mentioned that *Trapa* can be used as coffee substitute. There is a tradition in northern Italy, dating back to the sixteenth century of using non-spiky species of *Trapa* fruits for making rosaries. The fruits of these non-spiky forms are also used for making necklaces, and the tradition probably dates to Roman times (Hegi 1965:882–894).

Trapa was also used in Serbia, Moldavia, Hungary, and Slovakia. In the former Yugoslavia, in the ponds of the northern rivers Drava, Sava, the Danube, the use of water chestnuts has been documented ethnographically (Janković 1952). These areas in the Pannonian Plain have been successful in growing cereals, so water chestnuts are no longer used as human food. Water chestnuts were mainly used as animal food, especially for pigs, which are capable of releasing the seeds from the horny fruit shell (endocarp). Water chestnuts were gathered intentionally for animals, while in other cases, pigs were left unattended in forests and marshy areas to graze on them. Water chestnuts were known to be used for human food, and according to oral tradition, flour made from them was good for making cakes. The fruit for this purpose was peeled from the bottom diagonally and the starchy inside was released from its spiky skin. Afterwards, the starchy parts were ground into flour and used as food for humans in times of scarcity. Karg (1996a) cites the Hungarian folk tradition of collecting water chestnuts in a boot or a bag made from a sheep's tail.

In the Balkans, there was intensive use of water chestnuts around Lake Skadar in Montenegro where there was little agriculture. In this mountainous country, fishing in the lake and animal raising were the most important activities. In Lake Skadar, *Trapa natans longicarpa* spp. *scutarensis* thrive and have the biggest fruits (Janković 1952). The local name is *kasoronje*. Water chestnuts were used as animal and human food, often as a substitute for maize. The floating plants are also collected and fed fresh to animals. Ripe *Trapa* fruits have a sweet taste and can be eaten fresh or cooked in salty water. It is considered a very healthy food with positive effects on the digestion. The fruits ripen in batches. The harvest began already in August and continued from mid-September until the end of October. In some places, the stands were so dense that 50 fishing boats were filled with fruits. Usually, the water chestnuts were picked by Montenegrin Orthodox Christian women, although both young and old people could have participated in water chestnut harvests. The fruits were picked by hand and put into baskets in the boats. In Montenegro, everyone was allowed to harvest, including villagers from outside the immediate region around Lake Skadar (Stevan Lopičić, personal communication 1997). A line in a local folk song says that fruits can be harvested by anyone who wants them ("*Kasoronje bijelo voće nek vas bere ko god hoće*").

Water chestnuts from Lake Skadar had a small commercial value and were sold in the markets in Montenegro towns similarly to chestnuts (*Castanea sativa*). The use of water chestnuts was so important around Lake Skadar that plants have been wiped out in some places and only a few places were left where there were undisturbed populations and were suppressed by expanding populations of *Nymphaea alba* according to Janković (1952:307–309). According to recent information (Stevan Lopičić, personal communication 1997; personal observation 2001), the exploitation of water chestnuts in the Skadar region was important until the 1950–1960s, when the region was still densely populated, and the fruits were then used as food for pigs. Fruits were mostly gathered by women. Whole fruits were directly given to pigs. The floating leaves were also collected and fed to goats. Ripe fruits used for human consumption were usually boiled in water for three-to-four hours, while unripe floating fruits were eliminated from further processing. The horns from cooked fruits were peeled with knives or scissors (used in vineyards). The cooked inner parts were commonly eaten during gatherings of women and men as "snack food", and red wine of grape brandy was consumed with them.

Archaeological Distribution

Trapa species were much more widespread in Europe during the interglacials and in the postglacial warm periods than at present, indicating the deterioration of climate. Archaeobotanically, *Trapa* was found at Neolithic sites in Scandinavia. At the beginning of the Iron Age in Scandinavia, the climate started to cool and caused the elimination of *Trapa* species. Finds of *Trapa* were first made during the last century in the Neolithic Swiss Lake dwellings. *Trapa* fruits were also found at southern Scandinavian sites. Similarly, piles of broken water chestnut shells were found along with hazelnut shells in the lake settlement of Dullenried on the Federsee in Germany, in south Sweden, and in the site of Vanjavesi in Finland. In the book *Vergessene Nutzflanzen*, Brockman (1914) states *Trapa* that was already cultivated in the Neolithic times and was also known to Egyptians. Classical authors Dioskurides and Plinius mention its use in the Nile-land and in Thrace (Bulgaria), where bread was made from water chestnut flour. Disokurides also mentions that leaves of the whole plants were used as food for horses in Thrace. Mattioli wrote that water chestnuts were sold in the Venetian markets as well as in northern Italy, south France, and Dalmatia in the Middle Ages (Hegi 1965:882-894).

In more recent archaeological literature, *Trapa natans* fruit endocarp remains were recovered together with pitted stones from the Acheulian site Gesher Benot Ya'aqov in the Dead Sea Rift are were dated to 780,000 years ago (Goren-Inbar et al. 2000, 2002). This find represents the earliest evidence of *Trapa* processing. Water chestnut finds were discussed by Zvelebil (1994:35–74) for western and northern Europe. He cites the discovery of water chestnut remains from the Netherlands, Scandinavia, and especially from the east Baltic. At the site Sarnate in Latvia, layers of water chestnut shells were deposited around hearths inside dwellings, reaching 40 cm in thickness, and they were also found in ashes. There, the water chestnut remains were associated with wooden mallets that had fragments of water chestnut spiky endocarps imbedded in their working surface (Vankina 1970; Rimantienė 1979; cited in Zvelebil 1994). *Trapa natans* were also recovered in the Neolithic and Bronze Age lake dwellings in Germany (Karg 1996a, 1996b; Maier 1993). *Trapa* is not mentioned in the archaeobotanical surveys of Lisitzyna and Filipovich (1980) or Popova (1995) in Bulgaria, nor in the survey of Neolithic sites in southeast Europe compiled previously by Renfrew (1979). However, Kroll (1991a:161–177) mentions that

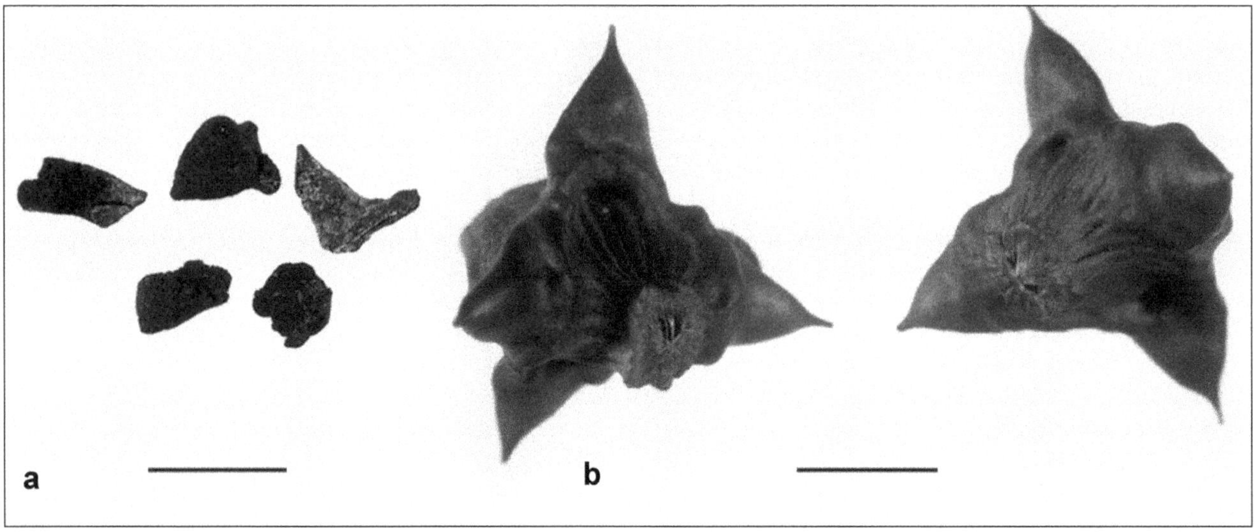

Figure 2.19. *Trapa natans*: **a)** carbonized fragments of the fruit endocarp from Features 26 and 46; **b)** two whole modern fruits from Voyvodina (scale 1 cm);

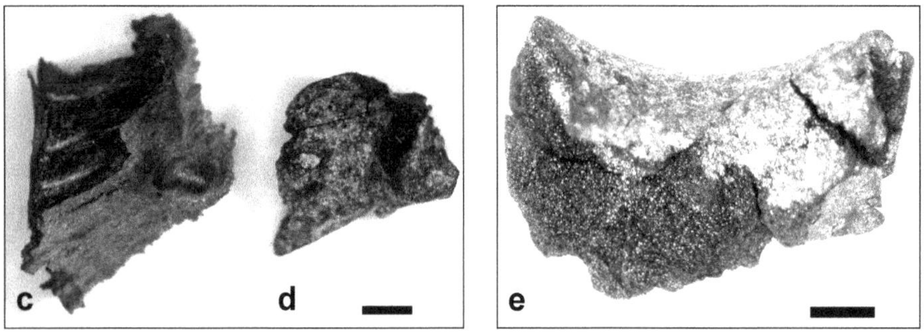

c) top fragment of a modern fruit;

d) carbonized top fragment from Feature 46;

e) carbonized horny projection fruit endocarp from Feature 46 (scale 1 mm).

the spectrum of gathered plants was broadened with strawberries and water chestnuts in the Early Bronze Age. And, indeed, water chestnuts were found in the Bronze Age layer at Feudvar, Voyvodina. My suspicion—that earlier *Trapa* fragments were overlooked—turned out to be right when remains were later identified at the Neolithic sites in the Pannonian Plain: Gomolava in Serbia (van Zeist 2001/2002), Uivar in Romania (Fischer and Rösch 2004), and Öcsöd-Kováshalom in Hungary (Judith Rasson, personal communication 2004).

However, *Trapa* is not mentioned in the archaeobotanical surveys of Lisitzyna and Filipovich (1980) or Popova (1995) in Bulgaria, nor in the survey of Neolithic sites in southeast Europe compiled previously by Renfrew (1979).

Opovo Findings

A total of 375 (1.94 g) carbonized endocarp fragments of *Trapa* fruit were found in the light fractions in all three building horizons in Features 42, 26, 31, 41, 52, 45, 46, and 61. In addition, 12 (0.26 g) endocarp *Trapa* fruit fragments were found in the non-feature light fraction samples (Table 2.1). Seven more fragments of endocarp were found in heavy fraction samples from Features 26, 31, and 41 (Table 2.3). The highest density of *Trapa* was found in Feature 46, in a pit from BH 3 (Table 2.2). Water chestnut fragments were found primarily in the feature samples; the percentage presence (ubiquity) is 28.5% from feature samples and 5.4% from non-feature samples (Table 2.4).

Most of the *Trapa natans* fruit fragments were rather small. I was not able to extrapolate the number of the whole fruits from the number and weight of fragments because not a single whole fruit was preserved. In one case the top part of the fruit remained, in two cases two whole lateral horns were preserved (Figure 2.19). The two horny fragments are rather wide and short, and on the basis of fruit drawings and reference collections, I tried to identify those fragments more precisely. The horns were perhaps parts of the fruit, *Trapa brevicarpa* (Janković 1973). Such identification seems plausible, because presently *Trapa brevicarpa* ssp. *callosa* is found along the rivers Tisa, Begej, and the Tamis, which is close to the Opovo site. Nowadays, *Trapa* occurs in a plant association with *Trapetun natans* Muller et Gors (1960), which includes very few other species of small floating plants (e.g., *Spirodela polyrrhiza*, *Salvina natans*, and *Lemna minor*).

The finds of *Trapa* in the Opovo samples are of particular interest, because initially there were the first finds of the *Trapa* from the Vinča sites (Borojević 1998).

Ethnographic and historical analogies cited above suggest that the fruits of *Trapa* were used for both human and animal consumption. Pigs were raised at the Opovo site (Russell 1993), and pigs were the animals typically fed on water chestnuts. During the late summer and fall, pigs could have eaten fresh water chestnuts. Therefore, the Opovo inhabitants need not have bothered with releasing the *Trapa* inner parts for animals during the ripening season. The Opovo inhabitants could have dried *Trapa* fruits and ground them into flour to use when there were no available fresh fruits. If water chestnuts were used during the summer, I suggest that they were consumed by humans in a cooked or roasted state similar to the ones eaten in the Skadar lake region in Montenegro. Because dense endocarp preserves well, the discarded charred *Trapa* horns remained in the Opovo refuse pits.

Weedy Herbaceous Plants

Altogether 339 seeds from the light fraction and 40 additional seeds from the heavy fraction samples (Tables 2.1–2.3) are included in this category of weedy herbaceous (wild/weed) plants. Although the majority of the seeds were carbonized, 29 small seeds were mineralized and retained their pale color. The weed category is the most arbitrary. Although *Bromus* sp. and *Lolium* sp. were certainly weeds in crops, other plants could have been intentionally gathered (such as *Chenopodium* for its green parts and starchy seeds, or Solanaceae and *Lithospermum officinale* for their medicinal properties).

There is no doubt that crop/weed communities changed over the millennia. In addition, harvesting methods, manuring, and grazing affected the occurrence of weed seeds of many species. For example, if the cereals were harvested near the ground, the seeds of the low weed plants would also occur in the archaeological record, and vice versa, if only the cereal spikes were harvested, then many of the lower weed plant seeds would not be present in archaeobotanical assemblages (Hillman 1984:1–41). Likewise, the weed species adapted to dissemination with cereal grains are the most abundant among the cereal grains; these weedy plants are usually present in the middle layer of the crop stand. On the other hand, the seeds of the plants that are higher or lower than the crop stand are usually adapted to spontaneous dissemination and appear in the cereal grain sporadically and in small numbers. Therefore, different weed communities develop on the same soil depending on the tilling techniques (Wasylikowa 1981:11–23). However, even if the same tilling and harvesting techniques are used, when the edaphic conditions are different, the weed flora change. In fact, special groups of weeds are known to be indicators of soil properties (Ellenberg 1979).[2]

Furthermore, reconstruction of past weed assemblages is difficult because tillage techniques have changed and the accompanying weedy flora of the crops have changed as well. Nevertheless, harvesting and grain-processing techniques remained essentially the same until the introduction of full mechanization in the twentieth century. Different tilling techniques affect the composition of the weed flora of crops. When an ard (scratch plow) is used for plowing, many biannual and perennial plants are present in crop fields. When a moldboard is used instead, most of the perennials and biannual weed plants are eliminated from the crop fields and annual weeds take over (Hillman 1991:27–40; Wasylikowa 1981:11–23). Taking this into consideration, along with the lack of archaeological evidence of ards in southeast Europe during the Neolithic period, it is likely that many of the biannuals and perennials were present as weeds in the Opovo crops.

Lithospermum officinale L.
White Gromwell (*Lithospermum officinale*)
Boraginaceae

Lithospermum officinale has distinctive smooth, white-yellowish, porcelain-like fruits. The "seeds" are, therefore, easily distinguished from other *Lithospermum* species seeds, which have specific reticulation or protuberances, e.g., *Lithospermum arvense*. The seeds of *Lithospermum officinale* are oval with pointed one end, and have a broad hilum on the other end. Along the seed width, several pale longitudinal margins are hardly observable. The dimensions of seeds range between 3.2–4.0 mm in length×2.0–2.25 mm in both width and thickness (Dobrohotov 1961:321).

The Modern Plant and Its Distribution

Lithospermum officinale is a perennial plant, 20–100 cm high. It grows in woods, thickets, and river banks. The plant prefers damp places, but is also present in crops and pastures. In the Balkans, it is widely distributed on the mainland. The fruits can stay on dried plants until winter. In recent times, this plant has been a common weed in crops in Eastern Europe.

Plant Use

The plant has medicinal properties, and hence the Latin name *officinale*. *Lithospermum officinale* leaves and roots are used both internally and externally for measles, chicken pox, rashes, and herpes (Bown 1995).

Archaeological Distribution

In archaeological literature, it has been mentioned that *Lithospermum officinale* was found in the burials at Brzesc Kujawaski (Gimbutas 1956:118, cited in Renfrew 1973), and it was claimed that 6,000 seeds were perforated and used as beads in a burial ascribed to the Gumelnita culture (Gaul 1948:133, cited in Renfrew 1973). The seeds (nutlets) are often found mineralized (uncharred) in archeological sites, a common preservational feature for nutlets of many species of the Boraginaceae family (Miller 1991; Pustovoytov et al. 2004).

Opovo Findings

Only five uncarbonized *Lithospermum officinale* seeds

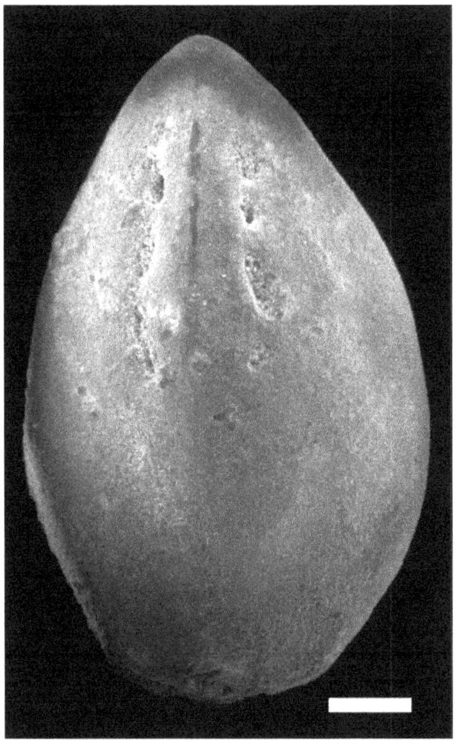

Figure 2.20. *Lithospermum officinale* mineralized seed from Feature 27 (SEM, scale 500 μm).

(nutlets) were found in the light fraction of a sample from Feature 27, floor of House 4 (Tables 2.1 and 2.4). The seeds (nutlets) were not carbonized but mineralized, and could have survived without being carbonized in the house rubble (Figure 2.20).

The isolated occurrence of these seeds is intriguing and could imply they were collected from the wild for medicinal or some other use. It is possible that the seeds were picked while cleaning of cereal grains, or that they were part of the ants' burrowing activity. Ecologically, the seed present imply the presence of wet moist places in the Opovo environs. In addition to this find, we recovered another uncarbonized seed of *Lithospermum* cf. *arvense* from a non-feature sample. I included this seed in the "Unknown Seed" category because it might have been a recent intrusion.

***Bromus* L.**
Brome grass (*Bromus* sp.)
Poaceae

The genus *Bromus* has more than 100 species. In the archaeobotanical literature of European sites, those most frequently mentioned are *Bromus arvensis*, *Bromus mollis*, *Bromus secalinus*, *Bromus sterilis*, and *Bromus tectorum*. The caryopses commonly called "seeds", are elongated and usually flat in cross-section. The tops of the seeds can be pointed or widely rounded. Some species have a pronounced margin on the ventral side, resembling a fine hem. *Bromus* seeds have particularly distinguishable cellular structures forming thin, narrow lines on the seed surface. The sizes of seeds of various species range from ca. 3–7 mm in length×0.5–1.5 mm in width×0.3–1.0 mm in thickness. It is sometimes impossible these seeds to the exact taxon, because the gross morphology of various *Bromus* seeds is often indistinguishable. Therefore, some authors, e.g., Kroll, are reluctant to identify *Bromus* seeds to the exact taxon despite good preservation of individual seeds. Instead, he classifies them in two categories: *Bromus* long-type and *Bromus* short-type seeds. In the long-type, *Bromus sterilis* and *Bromus tectorum* are included and *Bromus arvensis* and *Bromus secalinus* are included in the short category (Kroll 1983:89).

The Modern Plant and Its Distribution

Bromus species are grasses that are annual or perennial, or sometimes both annual and biannual, such as *Bromus mollis*. They vary in height (ca. 10–120 cm high) and in spike shape and arrangement. The long seed type, *Bromus sterilis* is an annual or biannual plant, 15–100 cm high. It is a common ruderal weed and also a weed in meadows, wastelands, cultivated lands, and in field margins, and usually thrives on well-drained soils. It is widespread in all of Europe and Southwest Asia (Hubbard 1968:63). *Bromus tectorum* also has a long seed. It is an annual plant, only 10–60 cm high, and the plants are much smaller than those of *Bromus sterilis*. *Bromus tectorum* is a meadow and ruderal weed, but also in wheat, barley, and millet fields (Dobrohotov 1961:129). It is native to the Mediterranean region, but it is widespread throughout Europe.

One of the bromes that has a short seed is *Bromus secalinus*. It is an annual or biannual plant and reaches a height of 20–120 cm. It is a ruderal weed, but also a common weed of cereal crops. *Bromus secalinus* is widespread throughout Europe (Hubbard 1968:89). Another type of short seed brome is *Bromus arvensis*. It is an annual plant 30–90 cm high. *Bromus arvensis* is a weed of meadows and it was occasionally grown in Europe on poor soils and used for hey. It can also be a weed in winter-sown crops. It is widespread throughout Europe and temperate Asia.

Archaeological Distribution

Seeds of various *Bromus* species have been found in archaeological sites. Because of the mentioned complexity and overlapping morphological characteristics of various species, often the complete seeds were not often identified to the exact taxon. *Bromus* sp. seeds and *Bromus arvensis* seeds were reported from the Neolithic and Eneolithic sites in Bulgaria (Lisitzyna and Filipovitch 1980:55). *Bromus arvensis* was also recovered from the late Vinča layers at Gomolava (van Zeist 2001/2002). *Bromus mollis*, which has a short seed with two ridges on the dorsal side, was identified in the Neolithic settlement of Kakanj in Bosnia. *Bromus secalinus* was identified in Butmir, a Neolithic settlement in Bosnia, and also in a Neolithic settlement in Hungary (Renfrew 1973:167–168). More recently, also in Poland, a plaster made of wood tar with the ftuit of *Lithospermum officinale* was discovered in a woman's grave dated to the Early Bronze Age (Baczyñska and Lityñska-Zajic 2005). A few *Bromus* seeds were recovered from the Neolithic settlement Selevac, in the former Yugoslavia (McLaren and Hubbard 1990:247–254).

Opovo Findings

Altogether, 19 carbonized *Bromus* seeds have been found in the Opovo samples from all three building horizons. *Bromus* seeds were identified only in the light fractions of Features 26, 41, 42, 52, and 59 (Table 2.1). The percentage presence (ubiquity) is 8.8 % from features samples, and not a single seed was found in non-feature samples (Table 2.4). The weight of one seed was considerably less than 0.01 g, while seven seeds weighed 0.02 g. Fifteen seeds came from Feature 41. In the Opovo samples, both long and short-type *Bromus* seeds were identified. However, more of the seeds belong to the short-type, probably to *Bromus* cf. *secalinus*. The dimensions of the five whole seeds range between 3.8–5.7 mm in length×0.8–1.2 mm in width×0.3–0.8 mm in thickness. One seed from Feature 41 resembles *Bromus arvensis* (Figure 2.21).

In addition, several *Bromus* seeds belong to the long-type, most likely to *Bromus sterilis*. The dimensions of two almost complete seeds range from 7–8 mm in length×0.8–1 mm in width×0.3–0.8 mm in thickness. It is interesting to note that both *Bromus* long and *Bromus* short-type seeds were recovered together from Feature 41. If the identification of *Bromus* short-type seeds is precise, then *Bromus secalinus* could have been a weed in a winter wheat crop. *Bromus sterilis*, on the other hand, could have been a weedy intrusion from the field margins.

Lolium L.
Rye Grass (*Lolium* sp.)
Poaceae (Gramineae)

The caryopsis or "seed" of the *Lolium* species has a broad basal part and narrower truncated top. Their truncated tops cause some *Lolium* seeds to resemble *Avena* seeds. On the ventral side, besides the common central furrow (hilum), *Lolium* seeds have additional shallow lateral lines on each side, parallel to the hilum. On the dorsal side, the embryo can sometimes be preserved in archaeological specimens, though it is often missing. The radical depression where the embryo is placed is small and rounded in shape. The seeds have asymmetrical cross-sections with oval-shaped sides and rather flattened ventral sides. *Lolium temulentum*, *Lolium remotum*, and *Lolium perenne* are the most commonly mentioned species in archaeobotanical samples. The dimensions for *Lolium temulentum* are ca. 5–6 mm long×2.0–2.5 mm wide×1.25–2.0 mm thick. The seeds of the other two species are somewhat smaller. The dimensions for *Lolium remotum* seeds range from 3.5–5.0 mm in length×1.5–1.75 mm in width×0.75–1.00 mm in thickness. The dimensions for *Lolium perenne* range from 3.5–4.5 mm in length×1.0–1.25 mm in width×0.5 mm in thickness.

Figure 2.21. *Bromus* cf. *arvense* carbonized seed from Feature 41: **a)** ventral side; **b)** dorsal side (scale 1 mm).

These dimensions include both the seeds with glumes, so the grains themselves are somewhat smaller (Dobrohotov 1961:131–132, 135).

The Modern Plant and Its Distribution

Lolium temulentum and *Lolium remotum* are annual grasses, while *Lolium perenne* is a perennial grass. These species also differ in their height and habitat. *Lolium* species are grasses and common weeds in crops. *Lolium temulentum* is a weed of arable lands, especially in wheat fields, and it is native to the Mediterranean region. *Lolium temulentum* seeds are often infected with mycelia threads (hyphae) of a fungus *Stromatinia temulenta.* This fungus produces narcotic alkaloid *temulin*, which is poisonous to humans and causes, apathy, nausea, and even death. *Lolium remotum* is a weed in spring crops, especially in flax, and it is common in central Europe. In contrast, *Lolium perenne* is a meadow grass that thrives in the heavy soils of lowlands but also grows in wastelands. *Lolium perenne* hybridizes with other *Lolium* species and *Festuca* species, and therefore its spikes are is highly variable. The plant is widespread in Europe, temperate Asia, and north Africa (Hubbard 1968:151).

Archaeological Distribution

Lolium temulentum, *Lolium remotum*, and *Lolium perenne* were often identified in archaeological sites in Europe. A few seeds of *Lolium perenne* were found in the Neolithic site Selevac, which can be explained as the success of ruderal plant expansion. However, *Lolium* seeds are not mentioned in any reviews of archaeobotanical finds in any other Neolithic sites in the Balkans (Kroll 1991a; Renfrew

1979; Lisitzyna and Filipovitch 1980). In contrast, there were numerous seeds of *Lolium temulentum* identified in the site Kastanas in Greece, from the Late Bronze Age layer (Kroll 1983:82–86). At the same time, at Kastanas, the number of barley seeds decreased. The appearance of a large number of *Lolium temulentum* seeds is connected with more intensive wheat cultivation. The frequency of *Lolium temulentum* seeds diminishes in the later periods at Kastanas. Kroll (1983:83-87) explains this trend as a shift from einkorn-emmer to bread wheat fields. A similar high concentration of *Lolium temulentum* was found in a storage context at an Iron Age site in Israel (Kislev 1980). The possibility of finding *Lolium temulentum* seeds infected with the poisonous fungus has been neglected in the archaeobotanical literature, or explained as people developing an immunity for the poison (Hopf 1978, cited in Kroll 1983:85). In folk tradition, the bread made from wheat containing large amounts of *Lolium temulentum* seeds is called "drunk bread". The name shows recognition of the delirium caused by infected *Lolium temulentum* seeds (Dobrohotov 1961).

Opovo Findings

Only two carbonized seeds of *Lolium* spp. were found in the Opovo light fraction samples, from Features 41 and 52 (Table 2.1). Both of the seeds were incomplete, and it was not possible to identify them to the exact taxon. The seed of one appeared small, although incomplete, and it is possible that it was a seed of *Lolium remotum* which is a typical weed of flax. If the identification is correct, then both the occurrence of *Linum usitatissimum* seeds and *Lolium remotum* seeds in the same features may indicate the cultivation of flax in the Opovo site. On the other hand, if the *Lolium* seeds were *Lolium temulentum*, then the occurrence of both *Lolium temulentum* and *Bromus* cf. *secalinus* seeds can point to fall-sown cereal crops. If that is the case, then both *Linum* seeds occur in the features where a few seeds of *Linum usitatissimum* were identified. That might imply that the *Lolium* seeds were weeds of flax that might have been grown in Opovo rather than ruderal intrusions.

Solanaceae (Nightshade family)

The Solanaceae family includes several genera, among which *Solanum*, *Hyoscyamus* and *Physalis* are commonly cited in archaeobotanical literature. All three genera have berry fruits that contain several seeds, All of which are small, flat, and similar. In overall shape, they may be rounded (e.g., *Hyoscyamus niger*), kidney-shaped (e.g., *Physalis alkekengi* and *Solanum dulcamara*), and pointed (e.g., *Solanum nigrum*). All seeds have distinctive reticulation on the surface, which varies from species to species, but they are difficult to distinguish even with a scanning electron microscope. In all seeds, the hilum is apparent on the seed margin, and can vary in shape, length, and position. The embryo of the seeds is elongated and placed within the endosperm. The seeds vary from ca. 1–2.6 mm in diameter and usually measure less than 1 mm in thickness (Schoch et al. 1988:180–182).

The Modern Plant and Its Distribution

There are only a few Solanaceae species in the local flora of Voyvodina (Čanak et al. 1978), in contrast to more than 1,000 Solanaceae species in the New World. Most species are annual, while some are biannual such as *Solanum dulcamara*. They are widely distributed throughout Europe except in the northern parts. They are common ruderal plants and weeds in gardens, crops, and hedges. *Physalis alkekengi* is a weed in orchards and vineyards, while *Solanum nigrum* is a weed of spring-sown crops. Many of the species, e.g., *Solanum nigrum*, like humus nitrogen-rich soils, while others, e.g., *Solanum dulcamara*, prefer wet places. *Solanum dulcamara* usually thrives in oak forests or among other perennial plants. *Hyoscyamus niger* is a typical ruderal plant that often grows in wastelands.

Plant Use

Most of the species have poisonous berries, especially *Solanum nigrum*. On the other hand, *Physalis alkekengi* has sweet fruits that can readily be eaten. There is ethnographic evidence (e.g., Okiek in Kenya) that some Solanaceae plants, considered poisonous by botanical taxonomists, are eaten as greens, boiled in soups (Fiona Marshall, personal communication 1997). Some of Solanaceae have medicinal properties, e.g., *Solanum dulcamara* whose fruits have been used as a remedy for dizziness. *Hyoscyamus niger* seeds have 31% oil which contains alkaloid hyosciamin, a powerful narcotic. In Finland, burnt seeds are used in folk medicine for severe toothache (Lempaiäinen 1992:238). In Turkey, they are used to treat itching in the eyes caused by worms (Sezik et al. 1997). Legends credit these plants with special powers to ward off evil spirits (Blamey and Blamey 1984:186).

Archaeological Distribution

Mostly uncarbonized *Hyoscyamus niger* seeds were found in archaeological sites in Finland, dating from the Viking Age onwards. Northern Europe is outside the natural distribution of this species; thus, *Hyoscyamus* plants must have been introduced to the area. Lempaiäinen (1992:227-239) suggests the plants were grown for medicinal purposes because the *Hyoscyamus niger* seeds were often found close to Medieval hospitals and were distributed by monks. The discovery of a small cache of henbane seeds (*Hyoscyamus niger*) and pollen were identified from the Neolithic ceremonial/ritual site Balfarg in Scotland. The discovery of these henbane seeds together with a Grooved Ware sherd, the astronomical alignments, and ceremonial enclosures, evoked for some authors (see Sherratt 1996) the analogy of prehistoric "hippies" in Scotland using mind-bending drug *Hyoscyamus niger*. However, the use of *Hyoscyamus niger* as a hallucinogen during Neolithic ritual activities in Scotland was questioned by Long et al. (2000), as they re-evaluated other sherds and found no additional evidence of henbane presence.

The seeds of other Solanaceae species were also identified from several archaeological sites in the Balkans where flotation was employed, e.g., Kastanas (Kroll 1983). Usually they were interpreted as crop weeds. At the multilayer site Gomolava, several Solanaceae species were identified, including *Solanum nigrum*, *Solanum dulcamara*, and *Physalis alkekengi*. Considerable number of *Physalis alkekengi* seeds were retrieved from the late Vinča layers but not in the Iron Age levels and included in gathered fruits (van Zeist 2001/2002).

Figure 2.22. Solanaceae carbonized seeds from Feature 48 (scale 1 mm).

Opovo Findings

A total of 224 Solanaceae seeds were recovered from the light fractions of the Opovo samples (Table 2.1, Figures 2.22–2.25). An additional 38 seeds were recovered from the heavy fraction samples (Table 2.3). Most of the seeds were carbonized, but 22 seeds were mineralized (Figure 2.25). The majority of uncarbonized seeds are from the heavy fraction samples. The seeds are minuscule, and 18 carbonized small-type seeds weigh only 0.02 g. The seeds have been identified from Features 26, 27, 30, 31, 41, 42, 46, 48, 52, 59, and 61, and in non-feature samples (Tables 2.1–2.3). The seeds of Solanaceae are the most ubiquitous in the wild/weed category, present in 44.5% of feature samples, and in 9.2% of non-features samples (Table 2.4).

It seems likely that there are at least two Solanaceae species present in the Opovo samples. One of the present species could belong to the genus *Solanum*. The seeds are kidney shaped and have distinctive though fairly shallow reticulation. The oval hilum is placed in the central part of the kidney-shaped lateral side. The seeds of this type occur both in carbonized and mineralized forms (Figures 2.22–2.25). Dimensions of these seeds are on the average 1.32 mm (1.0–1.5 mm) long×1.22 mm (0.9–1.5) wide×0.3–0.4 mm thick. On the other hand, there are seeds that may belong to different *Solanum* species or to another species of the Solanaceae family. These seeds are smaller, rounder, and have slightly deeper reticulation. The oval hilum is placed towards the upper part of the lateral margin. These seeds are generally thicker and look more inflated (Figure 2.23) than the other *Solanum* species seeds described above. The rounder and smaller seeds seem to appear only in a carbonized form, and resemble *Hyoscyamus niger* seeds (Helmut Kroll, personal communication 1997). Because the exact identification of the Solanaceae seeds was impossible, it was difficult to discuss their modes of arrival to the Opovo site and preservation of both carbonized and mineralized forms. There are various possibilities. The seeds might have arrived as weeds in cereal crops. In that case, the plants probably grew in the field margins and were harvested with the main crop, then the Solanaceae berries were later sieved out in the settlement and thrown into the pits. On the other hand, if the seeds, belong to palatable Solanaceae species, the berries could have been collected

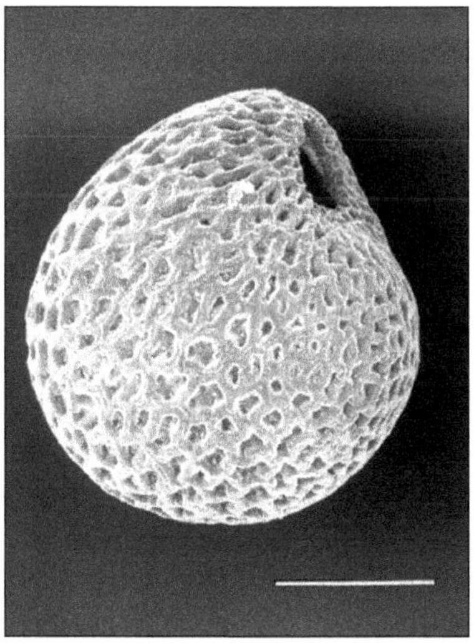

Figure 2.23. Solanaceae carbonized seed from Feature 48 (SEM, scale 500 μm).

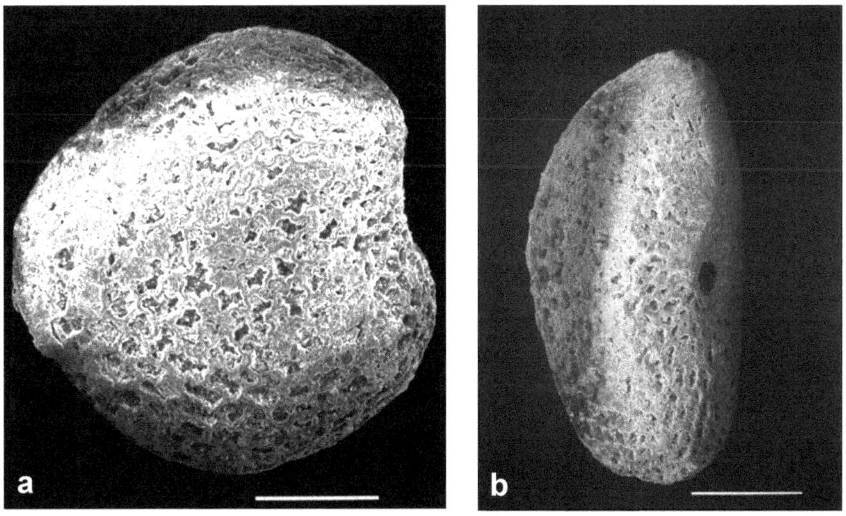

Figure 2.24. Solanaceae from Feature 31: **a)** carbonized seed; **b)** detail showing hilum (SEM, scale 500 μm).

Figure 2.25 Solanaceae from Feature 41: a) Mineralized seed; b) Detail showing hilum (SEM, scale 500 μm).

from the wild and eaten afterwards in the settlement. Many Solanaceae plants are ruderals, such as *Hyoscyamus niger*, which is a highly poisonous plant. These plants could have infested dump heaps in the settlement and weeded out and then the whole weeded plants with berries could have been thrown into fires and dumped into the pits. They also could have been gathered and used medicinally.

***Polygonum* L.**
Knotweed (*Polygonum* sp.)
Polygonaceae

There are several *Polygonum* species, but I describe here only those that commonly occur in archaeobotanical samples. Several species have three-sided or trigonous achenes ("seeds") such as *Polygonum convolvulus*, *Polygonum aviculare*, and *Polygonum hydropiper*. There are also species with two sided achenes, e.g., *Polygonum lapathifolium* and in most cases *Polygonum persicaria*. Various *Polygonum* achenes often resemble *Rumex* achenes, the latter being generally smaller in size with more prominent margins on lateral edges. In both genera, the bottom part of a perianth can be preserved in archaeological specimens. *Polygonum convolvulus* has three-sided fruits pointed at both ends. The lateral surfaces can be indented, especially if a fruit is not fully ripe. In archaeological samples, fruits are often left without brittle pericarps and thin testae, leaving bare endosperm with an embryo groove. The dimensions of the *Polygonum convolvulus* achenes range from 2.7–4.3 mm in length×1.8–2.7 mm in width×1.8–2.7 mm in thickness (Schoch et al. 1988:137). *Polygonum aviculare* is another species that has three-sided fruits. Its achenes are smaller and elongated, and its edges are more rounded than those of *P. convolvulus*. Both *Polygonum convolvulus* and *Polygonum hydropiper* have pearly nodules arranged in longitudinal rows on the outer surface of the achenes. The achene dimensions of *P. aviculare* range between 2.2–3.0 mm length×1.2–1.7 mm width×1.2–1.7 mm length (Schoch et al. 1988:138). *Polygonum lapathifolium* achenes are two-sided or very rarely three-sided. They are oval and rounded in shape with blunt edges. They are flat on both sides and have pointed tops. The pericarp of *P. lapathifolium* does not have pearly nodules, but is smooth with fine ripples. The dimensions range between 1.7–3.0 mm in length×1.5–3.0 mm in width×0.5–1.0 mm in thickness (Schoch et al. 1988:138).

The Modern Plant and Its Distribution

Polygonum convolvulus is an annual and a prostrate plant, but also can be a binding weed reaching up to 100 cm in height. It is a common, hardy weed in spring-sown cereals, meadows, and gardens. As one of the most common cereal weeds, it often aggravates the crop growth and harvests. *Polygonum convolvulus* is also a ruderal weed that does not like very cold climates and prefers humus, moist, and slightly acid soils.

Polygonum aviculare is also an annual prostrate plant. It reaches a height of 30–60 cm. It thrives in closely packed soil such as roadways and pathways. The light-loving species prefers temperate and warmer climates. *Polygonum aviculare* is an aggressive colonizer, and can thrive on very poor soils, including roadways and pathways. As a weed, it usually infests spring-sown cereals because it is not resistant to frost. It prefers carbonate or sandy humus soils and it is very difficult to eliminate from the cereal fields because the seeds can sprout after many years. *Polygonum aviculare* is distributed in almost all of Europe except in the northern, cold climates.

Polygonum lapathifolium is an annual plant, 20–200 cm high that prefers warm summers and moist soils. It is also a weed in spring-sown cereals and especially prefers gardens. The fruits of both species ripen from summer to late fall (Schoch et al. 1988:137–140; Šarić 1989:130–134).

Plant Use

Both seeds and leaves of *Polygonum convolvulus* can be eaten by humans and animals. From young leaves, soups

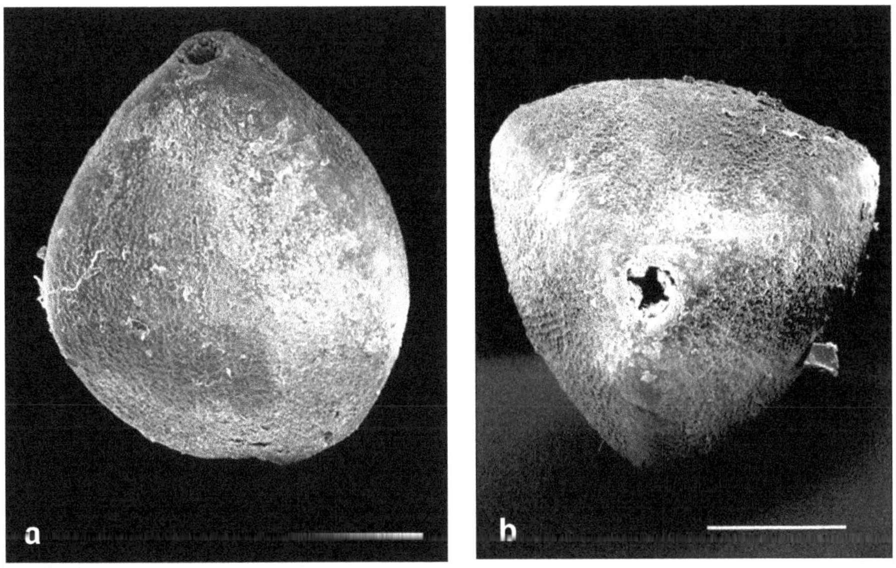

Figure 2.26. *Polygonum convolvulus* from Feature 46: **a)** carbonized seed; **b)** detail showing hilum (SEM, scale 500 μm).

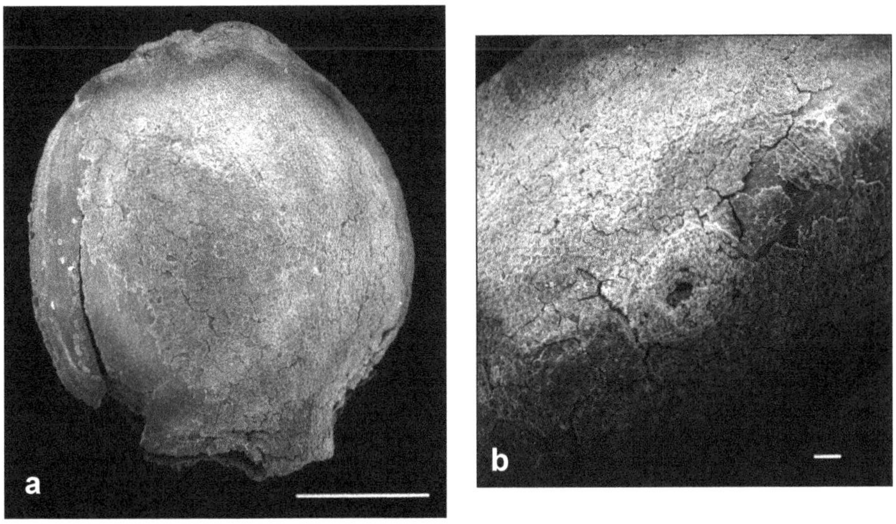

Figure 2.27. *Polygonum* cf. *lapathifolium* from Feature 41: **a)** carbonized seed (SEM, scale 500 μm); **b)** detail showing detail of a broken tip (SEM, scale 50 μm).

and vegetable stews can be made. The starchy seeds contain carbohydrates, proteins, and fats, and are used to make porridges. The seeds can be ground into flour. Similarly, *Polygonum aviculare* is utilized for making soups and vegetable stews. In dense stands, one person can pick up to 1 kg of young leaves in one hour (Vračarić et al. 1977). It is noted in several places that *P. aviculare* has many medicinal properties, and in Greece it is used as a diuretic (Schoch et al. 1988:137; McLaren and Hubbard 1990:247–254). Livestock can graze on both *Polygonum convolvulus* and *Polygonum aviculare*. In contrast, cattle will not eat *Polygonum lapathifolium* because it contains glycosides, which have a bad taste and can be harmful. Songbirds commonly feed on upland *Polygonum* seeds, while waterfowl eat marshland *Polygonum* species (Martin 1987:40–42).

Archaeological Distribution

The appearance of certain *Polygonum* species in archaeobotanical samples is commonly interpreted as weed intrusion in cereal crops, or plants that were used as animal fodder, or as medicinal plants. Sometimes when the seeds of *Polygonum convolvulus* are found in larger quantities, e.g., from the Early Iron Age site Ermello in Holland, they are interpreted as deliberately collected (van Zeist 1970:97). Similarly, the finds of *Polygonum erectum* in North America are interpreted as seeds that were used for their starchy contents, because caches of these seeds have been found in archaeological sites (Cowan 1985; Fritz 1987). *Polygonum* seeds have been identified since the Neolithic period in many European sites. In the Balkans, *Polygonum aviculare* was found at the Neolithic settlements, Sitagroi in Greece and Obre in Bosnia (Renfrew 1979:243–265). They were found also in Butmir in Bosnia (Schröter 1895:37–46), and Selevac in Serbia (McLaren and Hubbard 1990:247–254). Additionally, *Polygonum convolvulus* was recovered from the tell site Gomolava (van Zeist 1975, 2001/2002). *Polygonum aviculare* and *Polygonum convolvulus* and other undetermined *Polygonum* seeds were identified from Neolithic sites in Bulgaria (Lisitzyna and Filipovitch 1980:5–95; Popova 1995:193–209).

Opovo Findings

Altogether, 66 *Polygonum* seeds were found in the light and heavy fractions of the Opovo samples (Tables 2.1–

2.3). All seeds were carbonized except one uncarbonized *Polygonum* sp. seed from Feature 41. The highest density (21 seeds) was from Feature 46 (Figure 2.26). The weight of 16 seeds is 0.02 g (Tables 2.1 and 2.2). Because many flotation samples were taken from these two features, the percentage presence (ubiquity) of *Polygonum* is relatively high, 21% from feature samples (Table 2.4).

Most of the seeds have no preserved perianth, and in some instances only *Polygonum convolvulus* endosperms with an embryo groove were recovered. Two out of the total of 24 *Polygonum* seeds identified from Feature 41 are flat and probably belong to *Polygonum lapathifolium* (Figure 2.27). One *Polygonum* seed from the same feature is very elongated and could possibly be *Polygonum aviculare*. In another sample from the same feature, two *Polygonum* seeds are very small and could belong to *Rumex* species. Identification of one other seed resembling *Polygonum* from the same feature is even less positive, and hence it was placed in the "Unknown Seed" category. Because, no caches or concentrations of *Polygonum* seeds were found at Opovo, and in all the cases *Polygonum* seeds were found in pits in which cereals were present, it seems plausible to assume that they represent weedy intrusions in cereal crops.

Galium aparine L.
Bedstraw (*Galium aparine*)
Rubiaceae

The fruit of *Galium* species splits into two dry, one-seeded portions of various sizes and shapes. *Galium aparine* seeds are covered with hooked bristles that are not typically preserved in archaeobotanical samples. The seeds of *Galium aparine* are almost spherical in contrast to, e.g., *Galium spurium*, which has kidney-shaped seeds. The ventral side of *Galium aparine* is flattened with a distinctive rounded aperture leading to the interior of the seed. The dimensions of the fresh seeds range between 1–3 mm in diameter, and carbonized seeds are somewhat smaller (Renfrew 1973:173).

The Modern Plant and Its Distribution

Galium aparine is an annual and sometimes a biannual plant, 40–150 cm high. Stems have prickly hooks. The plants can thrive in various kinds of habitats such as humus, moist soils, but also in thickets and near roadsides from lowlands up to the mountains. *Galium aparine* is a very aggressive weed that can germinate almost year around. It is most commonly a weed in crops, especially in autumn-sown wheat, and it aggravates the harvest. *Galium aparine* is found across Europe, Asia, and North America.

Plant Use

Galium aparine is a very aggressive weed in cereal crops in Europe. In contrast, its appearance in some North American archaeological samples is interpreted as a deliberate use for cushioning, food, or drink (Fritz 1987). It is known that *Galium aparine* has medicinal properties and is used as a diuretic and a laxative in reducing weight (Fernald and Kinsey 1958:342–343).

Archaeological Distribution

Galium aparine has been found in numerous European sites. It was identified in several Neolithic settlements in Bulgaria (Lisitzyna and Filipovitch 1980:5–95; Popova 1995:193–203). *Galium aparine* was recovered from the Neolithic settlement of Obre in Bosnia (Renfrew 1979:243–265). At the tell site Gomolava, three types of *Galium* seeds were distinguished: *Galium aparine*, *G. spurium*, and *G. tricornutum*. From the late Vinča layers at this site, only *G. aparine* was identified (van Zeist 2001/2002). In European archaeobotanical literature, *Galium aparine* is cited among weeds and among plants used for fodder (Popova 1995).

Opovo Findings

Only four carbonized seeds and one uncarbonized seed of *Galium aparine* were found (Figure 2.28). The seeds were recovered in the light fraction in Features 41, 46, and 61 (Tables 2.1 and 2.4).

With such few *Galium aparine* seeds in all Opovo samples, it is difficult to determine how the seeds arrived in the pits. It is very probable that they arrived as weed intrusions among wheat seeds that were found in the same pits. They would have been sieved out from the main crop, thrown into a fire, carbonized, and later swept into the refuse pits. One uncarbonized *Galium* seed was recovered from Feature 61.

Chenopodium L.
Goosefoot (*Chenopodium* sp.)
Chenopodiaceae

Chenopodium has small rounded seeds, which together with the pericarps are properly called fruits. The seed has a distinctive "beak" that is a part of an embryo surrounding the perisperm. Often the surfaces of the seeds have specific reticulations that facilitate determination. However, the seeds are often similar in size and shape to many seeds of the Amaranthaceae family. Therefore in the North American archaeobotanical literature, the ambiguous seeds are often placed in a "Cheno-Amaranth" category. Several *Chenopodium* species, e.g., *Chenopodium album*, *Chenopodium polyspermum*, and others, have been found in European sites. Seed size varies from very small, less than 1 mm in diameter (e.g., *Chenopodium rubrum* and *Chenopodium glaucum*) to over 2 mm in diameter (e.g., *Chenopodium hybridum*). However, the most commonly cited *Chenopodium* in the European archaeobotanical literature is *Chenopodium album*. It has rounded seeds with a projecting short radical tips and clear, sharp borders. A polymorphism of *Chenopodium* plants is evident, not only in the green parts of the plants but also in the seeds (Dobrohotov 1961:192). The upper surface of *Chenopodium* seeds is rounded, with centrally located small pistil scars from

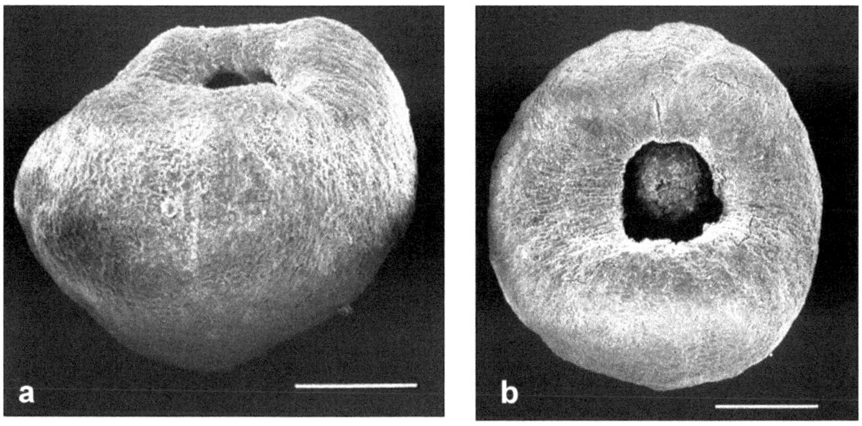

Figure 2.28. *Galium aparine* from Feature 46:
a) carbonized seed; **b)** detail showing hilum (SEM, scale 500 μm).

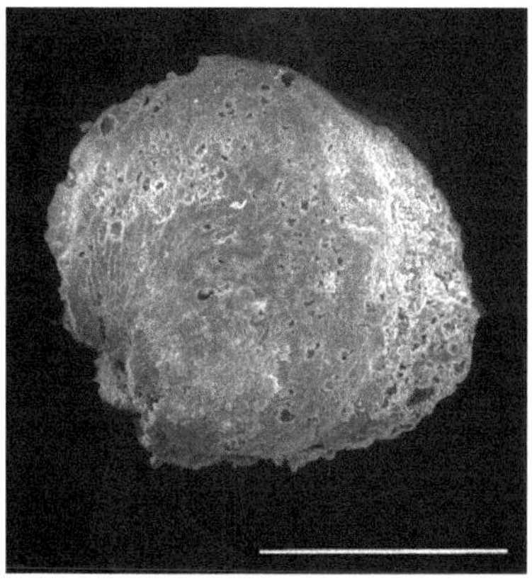

Figure 2.29. *Chenopodium* sp. carbonized seed from Feature 26 (SEM, scale 500 μm).

which fine grooves radiate. A short furrow runs from the outer indentation toward the middle. Carbonized and uncarbonized fruits are very similar, both black and lustrous. Dimensions range from 1.1–1.5 mm in length×1.0–1.1 mm in width×0.3–0.9 mm in thickness (Schoch et al. 1988:53).

The Modern Plant and Its Distribution

Chenopodium species are annual plants, 20–180 cm high. The fruit ripens from summer onwards. Most of the chenopods are ruderals and are considered to be undesirable weeds in gardens or crop fields. *Chenopodium album* can be a common weed of spring-sown cereals. It is a cosmopolitan plant that is widely distributed across Europe. *Chenopodium album* grows on humus- and and nitrogen-rich soils as well as in sand or clay soils, and it uses many of the nutritious elements from the soil.

Plant Use

In modern times, in areas where there are dense stands of *Chenopodium* plants, one person can pick between 3 and 5 kg of greens in one hour (Vračarić et al. 1977). The seeds are very nutritious and high in protein due to the large embryo. Each plant produces roughly 3,000 seeds and large quantities of the seeds can easily be collected. The seeds of *Chenopodium album* can be ground into flour and young leaves and stems of many *Chenopodium* species can be used to make soups and vegetable dishes. The young leaves are readily eaten by domestic animals. In the European botanical literature, the palatability and food potential of green parts of *Chenopodium album*, *Chenopodium murale*, and *Chenopodium polyspermum*, is mentioned; however, the use of seeds for food is not always fully explored and is commonly associated with famine foods.

Archaeological Distribution

Chenopodium seeds were found in many European sites, and they all mature in late summer or fall, even *Chenopodium album*. Usually they were believed to be weeds in sown crops or greens even when caches of *Chenopodium* seeds were found. At Feudvar, a whole cache of *Chenopodium polyspermum* seeds was recovered from a house attributed to the Bronze Age (Kroll 1990a:46–48). Although in this and similar cases, analogies from historic and ethnographic use were drawn, the analogy from prehistoric use and the domestication of *Chenopodium berlandieri* spp. *jonesianum* in North America (Fritz 1986) has not been explored for the European *Chenopodium* species. In the Balkans, *Chenopodium* seeds were identified from Neolithic sites in Bulgaria (Lisitzyna and Filipovitch 1980:88-98; Popova 1995:193-203). In the former Yugoslavia, *Chenopodium album* seeds were identified at the Neolithic settlement of Anza (Renfrew 1976:300–312) and Selevac (McLaren and Hubbard 1990:247–254). From the tell site Gomolava, three species of *Chenopodium* are identified *Chenopodium album*, *C. hybridum*, and *C. polyspermum*, and the seeds of all three species appear in the samples from the late Vinča layers (van Zeist 2001/2002).

Opovo Findings

A total of 20 carbonized seeds of *Chenopodium* spp. were recovered from the pits in the Opovo samples. *Chenopodium* seeds were found in non-feature samples and in the following Features: 26, 31, 41, 52, 59, and 61 (Tables 2.1–2.4). Because most of the seeds were not preserved with intact and distinctive testa, one seed from Feature 26, two seeds from Feature 31, one seed from Feature 41, one seed from Feature 52, and one seed from a non-feature sample are included in the "Cheno-Amaranth" category. It is interesting to note that two *Chenopodium* sp. seeds from Feature 59 are from the same sample, but the seeds have different shapes and could represent two species or two morphs of the same species. One seed has a truncated margin (Figure 2.29), a diagnostic trait of a North American domesticated chenopod seed, *Chenopodium berlandieri* ssp. *jonesianum* (Fritz 1986), while the other does not have such a feature.

All chenopod seeds from the Opovo samples were recovered from pits from BH 2/3 and BH 3. It is difficult to determine with certainty, whether the seeds ended up in

Figure 2.30. Unknown (unidentified) carbonized seed from Feature 27 (SEM, scale 500 μm).

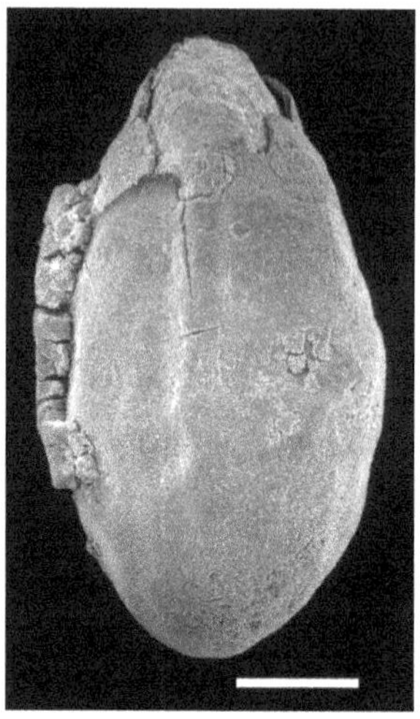

Figure 2.31. Unknown (unidentified) carbonized seed from Feature 52 (SEM, scale 500 μm).

pits as sieved byproducts of cereals or whether the chenopod plants were deliberately collected. Because no larger concentration of chenopod seeds was found in the Opovo site, and because the percentage presence (ubiquity) is low (8%), one might exclude the possibility that the chenopod seeds were gathered for their starchy content, at least in the Opovo site.

Unknown Seeds

In the "Unknown Seed" category, I included the whole or almost-whole seeds that I was not able to identify to the specific taxon. Altogether, 61 seeds were not identified from the Opovo light fraction samples (Table 2.1). In addition, five more seeds were unidentified from the heavy fraction samples (Table 2.3). Some of the seeds were preserved well enough so that they could possibly be identified, but I was not able to classify them more precisely (Figures 2.30 and 2.31). Furthermore, I was not positive whether several seeds were of modern origin or not because they were not carbonized, and I included those possibly modern seeds among "Unknown Seeds" as well. Among those uncarbonized seeds of probably modern origin are seeds that possibly belong to the Boraginaceae, Asteraceae, and Polygonaceae families, and to *Lithospermum* cf. *arvense* and *Fumaria* species.

Unidentifiable Fragments

When seeds or some other parts of fruits were fragmented with no diagnostic characteristic of even being seeds, I included these fragments in the category "Unidentifiable fragments". Altogether 160 fragments (1.19 g) in the Opovo light fraction were considered not identifiable (Table 2.1). Additionally, 49 fragments (0.26 g) were unidentifiable from the heavy fraction samples (Table 2.3). Sometimes there were so many unidentified fragments per sample that I was not able to count them all, and I had to extrapolate the number of fragments from the weight of other counted fragments found in the same sample.

Table 2.1. Count and weight of the plant macro remains from the Opovo light fraction samples.

Feature number:description	F42:Bedding trench (House 1)		F31:Pit with rubble top		F26:Pit with rubble top		F52:Pit	
Building Horizon (BH 1–3)	1		2/1		2		2	
Number of flotation samples/ Volume of floated sediment per feature (liters)	7	80	15	195	19	230	9	90
Wood charcoal abundance (min 1–max 3)	1		2		1		2	
Count (Ct)/Weight (Wt) in grams	**Ct**	**Wt**	**Ct**	**Wt**	**Ct**	**Wt**	**Ct**	**Wt**
Taxa								
Triticum dicoccum	15	0.12	13	0.11	25	0.19	13	0.12
Triticum dicoccum spikelet bases			6		13			
Triticum monococcum	3	0.03	12	0.09	15	0.1	11	0.1
Triticum monococcum spikelet bases			353				490	
Triticum sp. fragments		0.08		0.24		0.27		0.1
Hordeum vulgare								
Cerealia fragments		0.46		1.43		0.08		0.92
Poaceae	1	0.01			2	0.02		
Total cereal grains (Ct)/ Cereal grains and fragments (Wt)	**19**	**0.7**	**25**	**1.87**	**42**	**0.66**	**24**	**1.24**
Total *Triticum* spp. spikelet bases (chaff)			**359**		**13**		**490**	
Lens culinaris					2	0.02	2	
Fabaceae	1				1	0.01	2	0.01
Total legume seeds	**1**				**3**	**0.03**	**4**	**0.01**
Total oil-fiber seeds (*Linum usitatissimum*)					**1**		**1**	
Cornus mas stone fragments	3	0.04	21	0.3	7	0.33	7	0.1
Cornus mas whole stones (extrapolated)	1		4		4		1	
Sambucus cf. *nigra*					1			
Sambucus ebulus								
Vitis sylvestris								
Rubus sp.			3		1			
Fragaria cf. *vesca*			5		2			
Total fruit seeds/stones	**1**		**12**		**8**		**1**	
Total *Trapa natans* fragments	**5**	**0.08**	**20**	**0.14**	**16**	**0.19**	**4**	**0.06**
Lithospermum officinale								
Bromus sp.	1				1		1	
Lolium sp.							1	
Solanaceae	4		39		12		9	
Polygonum sp.	1		4		4		3	
Galium aparine								
Chenopodium sp.			3		4		1	
Total wild/weed seeds	**6**		**46**		**21**		**15**	
Unknown seeds	4		3		3		6	
Unidentified various plant fragments	4	0.13	22	0.12	9	0.15	9	0.06
Total all seeds/grains/stones (Ct)/Total seeds/grains/stones and fragments (Wt)	**31**	**0.97**	**86**	**2.43**	**78**	**2.16**	**51**	**1.48**

Table 2.1. Count and weight of the plant macro remains from the Opovo light fraction samples. (Continued)

Feature number:description	F41:Pit (or two pits)		F48:Pit		F27:Floor (House 4)		F30:Pit/Well	
Building Horizon (BH 1–3)	2/3		2/3		3		3	
Number of flotation samples/ Volume of floated sediment per feature (liters)	41	610	4	41	16	160	7	75
Wood charcoal abundance (min 1–max 3)	2		2		1		2	
Count (Ct)/Weight (Wt) in grams	**Ct**	**Wt**	**Ct**	**Wt**	**Ct**	**Wt**	**Ct**	**Wt**
Taxa								
Triticum dicoccum	529	5.10	3	0.03	10	0.1	2	0.02
Triticum dicoccum spikelet bases	2				1			
Triticum monococcum	44	0.40	2	0.02	5	0.05		
Triticum monococcum spikelet bases	4				1			
Triticum sp. fragments		2.98				0.17		
Hordeum vulgare	26	0.21						
Cerealia fragments		4.56		0.09		0.03		0.04
Poaceae	2	0.02			2	0.02		
Total cereal grains (Ct)/ Cereal grains and fragments (Wt)	**601**	**13.26**	**5**	**0.14**	**17**	**0.02**	**2**	**0.06**
Total *Triticum* spp. spikelet bases (chaff)	**6**				**2**			
Lens culinaris	5	0.04					1	0.01
Fabaceae	16	0.11	1	0.01				
Total legume seeds	**21**	**0.15**	**1**	**0.01**			**1**	**0.01**
Total oil-fiber seeds (*Linum usitatissimum*)	**3**							
Cornus mas stone fragments	40	1.28	1	0.01	5	0.03	2	0.02
Cornus mas whole stones (extrapolated)	16							
Sambucus cf. *nigra*	3							
Sambucus ebulus	2						2	
Vitis sylvestris								
Rubus sp.	7				1		2	
Fragaria cf. *vesca*	23						40	
Total fruit seeds/stones	**51**				**1**		**44**	
Total *Trapa natans* fragments	**43**	**0.27**						
Lithospermum officinale					5			
Bromus sp.	15							
Lolium sp.	1							
Solanaceae	100		18				6	
Polygonum sp.	24						3	
Galium aparine	2							
Chenopodium sp.	6							
Total wild/weed seeds	**148**		**18**		**5**		**9**	
Unknown seeds	17		1		6		1	
Unidentified various plant fragments	40	0.33	5	0.01	7	0.05	2	0.01
Total all seeds/grains/stones (Ct)/Total seeds/grains/stones and fragments (Wt)	**841**	**15.49**	**25**	**0.17**	**29**	**0.06**	**57**	**0.12**

Table 2.1. Count and weight of the plant macro remains from the Opovo light fraction samples. (Continued)

Feature number:description	F34:Oven Floor		F45:Ash pit		F46:Pit		F51:Posthole/Pit (House 4)	
Building Horizon (BH 1–3)	3		3		3		3	
Number of flotation samples/ Volume of floated sediment per feature (liters)	1	10	2	20	8	105	1	10
Wood charcoal abundance (min 1–max 3)			3		2			
Count (Ct)/Weight (Wt) in grams	**Ct**	**Wt**	**Ct**	**Wt**	**Ct**	**Wt**	**Ct**	**Wt**
Taxa								
Triticum dicoccum			1	0.01	10	0.07		
Triticum dicoccum spikelet bases								
Triticum monococcum					27	0.17		
Triticum monococcum spikelet bases			2					
Triticum sp. fragments				0.01		0.47		
Hordeum vulgare								
Cerealia fragments						0.08		0.03
Poaceae					1	0.01		
Total cereal grains (Ct)/ Cereal grains and fragments (Wt)			**1**	**0.02**	**38**	**0.8**		**0.03**
Total *Triticum* spp. spikelet bases (chaff)			**2**					
Lens culinaris					13	0.09		
Fabaceae					1	0.01		
Total legume seeds					**14**	**0.1**		
Total oil-fiber seeds (*Linum usitatissimum*)								
Cornus mas stone fragments					3	0.07		
Cornus mas whole stones (extrapolated)					1			
Sambucus cf. *nigra*								
Sambucus ebulus								
Vitis sylvestris					6	0.13		
Rubus sp.								
Fragaria cf. *vesca*								
Total fruit seeds/stones					**7**			
Total *Trapa natans* fragments			**2**	**0.01**	**272**	**1.11**		
Lithospermum officinale								
Bromus sp.								
Lolium sp.								
Solanaceae					16			
Polygonum sp.					21		1	
Galium aparine					2			
Chenopodium sp.								
Total wild/weed seeds					**39**		**1**	
Unknown seeds	1				5			
Unidentified various plant fragments			14	0.02	8	0.03		
Total all seeds/grains/stones (Ct)/Total seeds/grains/stones and fragments (Wt)	**1**	**0.01**	**1**	**0.05**	**103**	**2.15**	**1**	**0.03**

Table 2.1. Count and weight of the plant macro remains from the Opovo light fraction samples. (Continued)

Feature number:description	F59:Pit		F61:Pit/posthole (under House 4)		Total Non-feature (Routine) samples		Total Feature and Non-feature samples	
Building Horizon (BH 1–3)	3		3					
Number of flotation samples/ Volume of floated sediment per feature (liters)	3	25	2	30	130	1235	265	2916
Wood charcoal abundance (min 1–max 3)	3		3					
Count (Ct)/Weight (Wt) in grams	**Ct**	**Wt**	**Ct**	**Wt**	**Ct**	**Wt**	**Ct**	**Wt**
Taxa								
Triticum dicoccum	8	0.07	1		6	0.05	636	5.99
Triticum dicoccum spikelet bases			1				23	
Triticum monococcum	5	0.04			3	0.01	127	1.01
Triticum monococcum spikelet bases	76		20		1		947	
Triticum sp. fragments		0.03		0.04		0.26		4.66
Hordeum vulgare							26	0.21
Cerealia fragments		0.14		0.23		0.43		8.52
Poaceae							8	0.06
Total cereal grains (Ct)/ Cereal grains and fragments (Wt)	**13**	**0.28**	**1**	**0.27**	**9**	**0.75**	**797**	**20.11**
Total *Triticum* spp. spikelet bases (chaff)	**76**		**21**		**1**		**970**	
Lens culinaris	1				2	0.02	26	0.18
Fabaceae	1						23	0.15
Total legume seeds	**2**				**2**	**0.02**	**49**	**0.33**
Total oil-fiber seeds (*Linum usitatissimum*)							**5**	**2.87**
Cornus mas stone fragments			2	0.01	10	0.55	101	2.67
Cornus mas whole stones (extrapolated)					7		34	
Sambucus cf. *nigra*					1		5	
Sambucus ebulus					1		5	
Vitis sylvestris							6	0.13
Rubus sp.					3		17	
Fragaria cf. *vesca*							70	
Total fruit seeds/stones					**12**		**137**	
Total *Trapa natans* fragments			**1**	**0.01**	**12**	**0.07**	**375**	**1.94**
Lithospermum officinale							5	
Bromus sp.	1						19	
Lolium sp.							2	
Solanaceae	3		2		15		224	
Polygonum sp.					3		64	
Galium aparine			1				5	
Chenopodium sp.	3		1		2		20	
Total wild/weed seeds	**7**		**4**		**20**		**339**	
Unknown seeds			1		12	0.05	61	1.17
Unidentified various plant fragments	3	0.01	4	0.02	33	0.25	160	
Total all seeds/grains/stones (Ct)/Total seeds/grains/stones and fragments (Wt)	**22**	**0.29**	**6**	**0.31**	**55**	**1.69**	**1338**	**27.42**

Table 2.2. Density by count and weight of the plant macro remains from the Opovo light fraction samples.[1]

Feature number: description	F42:Bedding trench (House 1)	F31:Pit with rubble top	F21:Floor (House 6)	F26:Pit with rubble top	F52:Pit	F41:Pit (or two pits)
Building Horizon (BH 1–3)	1	2/1	2	2	2	2/3
Volume of floated sediment per feature (liters)	80	195	20	230	90	610
Density by Count (Ct/100 liters) and Weight (g/100 liters)						
Triticum dicoccum Ct	18.75	6.67		10.87	14.44	86.72
Triticum dicoccum spikelet bases Ct		3.08		5.65		0.33
Triticum monococcum Ct	3.75	6.15		6.52	12.22	7.21
Triticum monococcum spikelet bases Ct		181.03			544.44	0.66
Triticum sp. fragments Wt	0.10	0.12	0.05	0.12	0.11	0.49
Hordeum vulgare Ct						4.26
Cerealia fragments Wt	0.58	0.73		0.03	1.02	0.75
Poaceae Ct	1.25			0.87		0.33
All cereal grains Ct	**23.75**	**12.82**		**18.26**	**26.67**	**98.52**
All cereal grains and fragments Wt	**0.88**	**0.96**	**0.05**	**0.29**	**1.38**	**2.17**
All *Triticum* spp. spikelet bases Ct (chaff)		**184.10**		**5.65**	**544.44**	**0.98**
Lens culinaris Ct				0.87	0.20	5.77
Fabaceae Ct	1.25			0.43	2.22	2.62
All legume seeds Ct	**1.25**			**1.30**	**4.44**	**3.44**
All oil fiber seeds (*Linum usitatissimum* Ct)				**0.43**	**1.11**	**0.49**
Cornus mas fragments Ct	3.75	10.77		3.04	7.78	6.56
Cornus mas stones Ct	0.63	1.92		1.79	1.39	2.62
Sambucus nigra Ct				0.43		0.49
Sambucus ebulus Ct						0.33
Vitis sylvestris Ct						
Rubus sp. Ct		1.54		0.43		1.15
Fragaria cf. *vesca* Ct		2.56		0.87		3.77
All fruit seeds/stones Ct	**0.63**	**6.03**		**3.53**	**1.39**	**8.36**
All *Trapa natans* fragments Ct	**6.25**	**10.26**		**6.96**	**4.44**	**7.05**
Lithospermum officinale Ct						
Bromus sp. Ct	1.25			0.43	1.11	2.46
Lolium sp. Ct					1.11	0.16
Solanaceae Ct	5.00	20.00		5.22	10.00	16.39
Polygonum sp. Ct	1.25	2.05		1.74	3.33	3.93
Galium aparine Ct						0.33
Chenopodium sp. Ct		1.54		1.74	1.11	0.98
All wild/weed seeds Ct	**7.50**	**23.59**		**9.13**	**16.67**	**24.26**
Unknown seeds Ct	5.00	1.54	5.00	1.30	6.67	2.79
All seeds/grains/stones Ct	**38.13**	**43.98**	**5.00**	**33.53**	**55.84**	**137.38**

[1] Density by weight was only computed for cereals, because for other taxa it was insignificant.

Table 2.2. Density by count and weight of the plant macro remains from the Opovo light fraction samples. (Continued)

Feature number: description	**F48:Pit**	**F27:Floor (House 4)**	**F30:Pit/Well**	**F34:Oven Floor**	**F45:Ash pit**	**F46:Pit**
Building Horizon (BH 1–3)	2/3	3	3	3	3	3
Volume of floated sediment per feature (liters)	41	160	75	10	20	105
Density by Count (Ct/100 liters) and Weight (g/100 liters)						
Triticum dicoccum Ct	7.32	6.25	2.67		5.00	9.52
Triticum dicoccum spikelet bases Ct		0.63				
Triticum monococcum Ct	4.88	3.13				25.71
Triticum monococcum spikelet bases Ct		0.63			10.00	
Triticum sp. fragments Wt		0.11			0.05	0.45
Hordeum vulgare Ct						
Cerealia fragments Wt	0.22	0.02	0.05			0.08
Poaceae Ct		1.25				0.95
All cereal grains Ct	**12.20**	**10.63**	**2.67**		**5.00**	**36.19**
All cereal grains and fragments Wt	**0.34**	**0.01**	**0.08**		**0.10**	**0.76**
All *Triticum* spp. spikelet bases Ct (chaff)		**1.25**			**10.00**	
Lens culinaris Ct			0.20			0.87
Fabaceae Ct	2.44					0.95
All legume seeds Ct	**2.44**		**1.33**			**13.33**
All oil-fiber seeds (*Linum usitatissimum* Ct)						
Cornus mas fragments Ct	2.44	3.13	2.67			2.86
Cornus mas stones Ct	0.30	0.23	0.33			0.83
Sambucus nigra Ct						
Sambucus ebulus Ct			2.67			
Vitis sylvestris Ct						5.71
Rubus sp. Ct		0.63	2.67			
Fragaria cf. *vesca* Ct			53.33			
All fruit seeds/stones Ct	**0.30**	**0.63**	**59.00**			**6.55**
All *Trapa natans* fragments Ct					**10.00**	**259.05**
Lithospermum officinale Ct		3.13				
Bromus sp. Ct						
Lolium sp. Ct						
Solanaceae Ct	43.90		8.00			15.24
Polygonum sp. Ct			4.00			20.00
Galium aparine Ct						1.90
Chenopodium sp. Ct						
All wild/weed seeds Ct	**43.90**	**3.13**	**12.00**			**37.14**
Unknown seeds Ct	2.44	3.75	1.33	10.00		4.76
All seeds/grains/stones Ct	**61.28**	**18.14**	**76.33**	**10.00**	**5.00**	**97.97**

Table 2.2. Density by count and weight of the plant macro remains from the Opovo light fraction samples. (Continued)

Feature number: description	F51:Posthole/ Pit (House 4)	F59:Pit	F61:Pit/ posthole (under House 4)	Total non-feature (Routine) samples	Total feature and non-feature samples
Building Horizon (BH 1–3)	3	3	3		
Volume of floated sediment per feature (liters)	10	25	30	1235	2936
Density by Count (Ct/100 liters) and Weight (g/100 liters)					
Triticum dicoccum Ct		32.00	3.33	0.49	21.66
Triticum dicoccum spikelet bases Ct			3.33		0.78
Triticum monococcum Ct		20.00		0.24	4.33
Triticum monococcum spikelet bases Ct		304.00	66.67	0.08	32.25
Triticum sp. fragments Wt		0.12	0.13	0.02	0.16
Hordeum vulgare Ct					0.89
Cerealia fragments Wt	0.30	0.56	0.77	0.03	0.29
Poaceae Ct					0.27
All cereal grains Ct		**52.00**	**3.33**	**0.73**	**27.15**
All cereal grains and fragments Wt	**0.30**	**1.12**	**0.90**	**0.06**	**0.69**
All *Triticum* spp. spikelet bases Ct (chaff)		**304.00**	**70.00**	**0.08**	**33.04**
Lens culinaris Ct		0.10		0.20	7.70
Fabaceae Ct		4.00			0.78
All legume seeds Ct		**8.00**		**0.16**	**1.67**
All oil-fiber seeds (*Linum usitatissimum* Ct)					**0.17**
Cornus mas fragments Ct			6.67	0.81	3.44
Cornus mas stones Ct			0.42	0.56	1.17
Sambucus nigra Ct				0.08	0.17
Sambucus ebulus Ct				0.08	0.17
Vitis sylvestris Ct					0.20
Rubus sp. Ct				0.24	0.58
Fragaria cf. *vesca* Ct					2.38
All fruit seeds/stones Ct			**0.42**	**0.96**	**4.67**
All *Trapa natans* fragments Ct			**3.33**	**0.97**	**12.77**
Lithospermum officinale Ct					0.17
Bromus sp. Ct		4.00			0.65
Lolium sp. Ct					0.07
Solanaceae Ct		12.00	6.67	1.21	7.63
Polygonum sp. Ct	10.00			0.24	2.18
Galium aparine Ct			3.33		0.17
Chenopodium sp. Ct		12.00	3.33	0.16	0.68
All wild/weed seeds Ct	**10.00**	**28.00**	**13.33**	**1.62**	**11.55**
Unknown seeds Ct			3.33	0.97	2.28
All seeds/grains/stones Ct	10.00	88.00	20.41	4.44	47.32

Table 2.3. Count and weight of the plant macro remains from the Opovo heavy fraction samples.

Feature number: description	**F42**		**F31**		**F26**		**F52**	
Building Horizon (BH 1–3)	1		2/1		2		2	
Volume of floated sediment per feature (liters)/ Volume (l) of sediment analyzed	80	80	195	115	230	210	90	10
Weight (g) of heavy fraction samples / Wt (g) of subsamples	1200	456	5500	537	2900	799	1100	75
Wood charcoal Weight (g)	0.58			0.1	3.59			
Count (Ct)/Weight (Wt) in grams	**Ct**	**Wt**	**Ct**	**Wt**	**Ct**	**Wt**	**Ct**	**Wt**
Taxa								
Triticum dicoccum								
Triticum dicoccum spikelet bases	2							
Triticum monococcum								
Triticum sp. spikelet bases	1				13			
Cerealia fragments		0.03				0.02		
Total cereal grains (Ct)/ grains and fragments (Wt)		**0.03**				**0.02**		
Total *Triticum* spp. spikelet bases (chaff)	**3**				**13**			
Fabaceae								
Cornus mas stone fragments	1	0.01						
Cornus mas whole stones (extrapolated)								
Total fruit/stones seeds	**1**							
Total *Trapa natans* fragments			**1**	**0.01**	**3**	**0.02**		
Solanaceae	2				12			
Polygonum sp.								
Total wild/weed seeds	**2**							
Unknown seeds	1	0.01						
Unidentified various plant fragments	4	0.01	3	0.01	15	0.07		
Total seeds/grains/stones (Ct)/Total seeds/grains/stones and fragments (Wt)	**4**	**0.06**		**0.02**	**12**	**0.11**		

Table 2.3. Count and weight of the plant macro remains from the Opovo heavy fraction samples.

Feature number: description	**F41**		**F48**		**F27**		**F30**	
Building Horizon (BH 1–3)	2/3		2/3		3		3	
Volume of floated sediment per feature (liters)/ Volume (l) of sediment analyzed	610	525	41	35	160	60	75	35
Weight (g) of heavy fraction samples / Wt (g) of subsamples	16500	2481	3700	205	275	275	400	153
Wood charcoal Weight (g)		5.95				0.08		5.81
Count (Ct)/Weight (Wt) in grams	**Ct**	**Wt**	**Ct**	**Wt**	**Ct**	**Wt**	**Ct**	**Wt**
Taxa								
Triticum dicoccum	96	1.46						
Triticum dicoccum spikelet bases	1							
Triticum monococcum	9	0.09						
Triticum sp. spikelet bases	15				2			
Cerealia fragments		0.22				0.03		0.01
Total cereal grains (Ct)/ grains and fragments (Wt)	**105**	**1.77**				**0.03**		**0.01**
Total *Triticum* spp. spikelet bases (chaff)	**16**				**2**			
Fabaceae	2							
Cornus mas stone fragments	11	0.08	1	0.01	1	0.01	4	0.37
Cornus mas whole stones (extrapolated)	1						5	
Total fruit/stones seeds	**1**				**1**		**5**	
Total *Trapa natans* fragments	**3**	**0.02**						
Solanaceae	20				1		2	
Polygonum sp.	2							
Total wild/weed seeds	**22**				**1**		**2**	
Unknown seeds	4	0.02						
Unidentified various plant fragments	25	0.15	1	0.01				
Total seeds/grains/stones (Ct)/Total seeds/grains/ stones and fragments (Wt)	**134**	**2.04**		**0.02**	**2**	**0.04**	**7**	**0.38**

Table 2.3. Count and weight of the plant macro remains from the Opovo heavy fraction samples.

Feature number: description	**F46**		**F51**		**Total non-feature (Routine) samples**		**Total feature and non-feature samples**	
Building Horizon (BH 1–3)	3		3					
Volume of floated sediment per feature (liters)/ Volume (l) of sediment analyzed	105	95	10	10	1235	15	2831	1190
Weight (g) of heavy fraction samples / Wt (g) of subsamples	4600	419	800	80	100	100	37075	5580
Wood charcoal Weight (g)		2.18						14.12
Count (Ct)/Weight (Wt) in grams	**Ct**	**Wt**	**Ct**	**Wt**	**Ct**	**Wt**	**Ct**	**Wt**
Taxa								
Triticum dicoccum							96	1.46
Triticum dicoccum spikelet bases			6				9	
Triticum monococcum							9	0.09
Triticum sp. spikelet bases							31	
Cerealia fragments								0.31
Total cereal grains (Ct)/ grains and fragments (Wt)							**105**	**1.86**
Total *Triticum* spp. spikelet bases (chaff)			**6**				**40**	
Fabaceae							2	
Cornus mas stone fragments							18	
Cornus mas whole stones (extrapolated)							6	
Total fruit/stones seeds							8	
Total *Trapa natans* fragments	**48**	**0.21**					**55**	**0.26**
Solanaceae	1						38	
Polygonum sp.							2	
Total wild/weed seeds	**1**						**40**	
Unknown seeds							5	0.03
Unidentified various plant fragments	1	0.01					49	0.26
Total seeds/grains/stones (Ct)/Total seeds/ grains/stones and fragments (Wt)	**1**	**0.22**					**160**	**2.89**

Table 2.4. Ubiquity (percentage presence) of the plant species in the Opovo flotation samples.

Taxa	Non-feature samples N=130	Feature samples N=137	Total samples N=267
Triticum dicoccum	3.8	48.2	26.6
T. dicoccum spikelet bases		7.3	3.7
T. monococcum	1.5	27.0	14.6
T. monococcum spikelet bases	0.8	15.3	8.2
Hordeum vulgare		6.6	3.4
Poaceae		5.1	2.6
Lens culinaris	1.5	10.9	6.4
Fabaceae		10.9	5.6
Linum usitatissimum		2.9	1.5
Cornus mas stone fragments	6.2	37.2	22.1
Sambucus cf. *nigra*	0.8	1.5	1.1
Sambucus ebulus	0.8	2.9	1.9
Vitis sylvestris		0.7	0.4
Rubus sp.	2.3	7.3	4.9
Fragaria cf. *vesca*		10.2	5.2
Trapa natans fragments	5.4	28.5	17.2
Lithospermum officinale		0.7	0.4
Bromus sp.		8.8	4.5
Lolium sp.		1.5	0.8
Solanaceae	9.2	44.5	27.3
Polygonum sp.	0.8	21.9	11.6
Galium aparine		2.9	1.5
Chenopodium sp.	1.5	8.0	4.9

Opovo Plant Remains in Regional Perspective

I compared the identified plant remains from the Opovo site with the spectrum of plant remains from other Vinča and Starčevo sites (Table 2.5). The results of identified plant remains from a total of 26 Neolithic sites in the region (the eastern part of the former Yugoslavia) are grouped chronologically to the Middle Neolithic (the sixth millennium B.C.) and the Late Neolithic (the fifth millennium B.C.) periods and are presented in Tables 2.5 and 2.6. The plant information was compiled from archaeobotanical reports, as cited in Table 2.5. The limited information allowed only the presence/absence survey of identified plant remains from 12 sites attributed to the Starčevo culture (Middle Neolithic) and from 15 sites attributed to the Vinča culture (Late Neolithic) from the territory of the former Yugoslavia and the Vinča site Uivar in Romania (Figure 2.32).

Furthermore, the identified plant remains from the Vinča sites, including the Opovo site, are compared in general with the spectrum of identified plant remains from sites of the contemporaneous late Neolithic/Eneolithic (Chalcolithic) cultures in southeast Europe (Figures 1.4 and 2.32), including the Pannonian Plain (the Carpathian Basin) and the Balkan region (Cârciumaru 1996; Fischer and Rösch 2004; Hartyányi and Nováki 1975; Kreuz et al. 2005; Kroll 1991a; Lisitzyna and Filipovitch 1980; Marinova 2001; Popova 1995; Renfrew 1979; Valamoti 2004; van Zeist 2001/2002; Ferenc Gyulai personal communication, 2004; and Brigitta Berzsényi personal communication, 2004).

The most ubiquitous plants identified from the Middle (Starčevo) and Late Neolithic (Vinča) sites were einkorn (*Triticum monococcum*) and emmer (*Triticum dicoccum*), followed by less abundant barley grains (Table 2.5). Grains of hexaploid wheat were also identified in small quan-

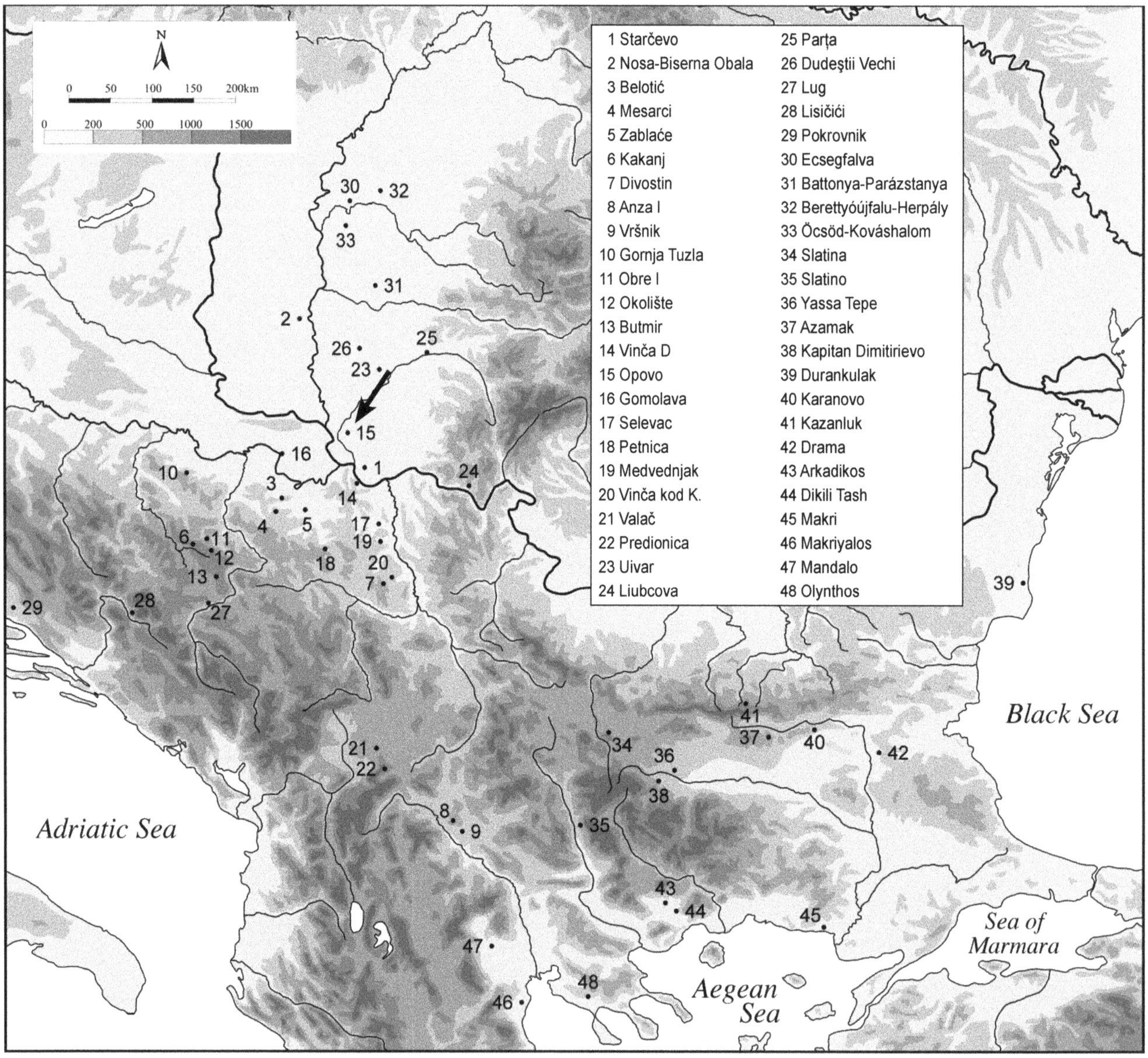

Figure 2.32. Map of Southeast Europe showing the Middle and Late Neolithic/Chalcolithic sites from which plant macro remains were analyzed. The Opovo site indicated with an arrow.

tities. Most of the hexaploid wheat grains are short and belong to the so-called club wheat (*Triticum compactum* or *Triticum aestivo-compactum*), a type of grain now more commonly classified as *Triticum aestivum/durum/turgidum* in the absence of rachis fragments for more precise identification (cf. Maier 1996). Recently, the "new type of tetraploid glume wheat" documented by Jones et al. (2000) was recognized from the Vinča layers at the tell site Uivar in Romania, and identified as *Triticum* cf. *timpheevi* by Fischer and Rösch (2004). Barley (*Hordeum vulgare*) grains are of the hulled form, although in several instances naked barley grains were also identified. Among legumes, lentils (*Lens culinaris*) were the most ubiquitous, followed by peas (*Pisum sativum*), and in a few instances, vetches (*Vicia* species).

Most of the plant remains were charred and only few were preserved in mineralized form. Because of a general lack of comparability of data, including the different number of sites and various retrieval methods for collecting plants used, it is difficult to infer the importance of particular crops and their possible change during the Neolithic period, from the Starčevo to the Vinča culture.

Cereals

Triticum monococcum (Einkorn) and *Triticum dicoccum* (Emmer)

At Opovo, emmer and einkorn were the principal crops. From multi-layer sites from which there are data available, it seems that emmer was the principal cultivated crop at Anza and Obre (Table 2.5). Emmer was the chief cultivated crop throughout the Anza occupation, which included earlier Starčevo and later Vinča culture phases. However, in one pit from Anza II (the earlier phase) unthreshed emmer and einkorn were almost equally represented and mixed with a small quantity of unthreshed six-row barley and fruit remains (Renfrew 1976). Similarly, from Obre I (Starčevo culture), Obre II (Butmir culture), emmer was ubiquitously present in samples from the older period and from the younger period, and was the principal crop during Obre I and Kakanj (Renfrew 1974). Two deposits out of four, however, contained exclusively einkorn grains that were dated to the later occupation, Obre II (Renfrew 1974). Emmer and einkorn were identified at the site Okolište near Butmir (Bittman and Kučan 2004), Lug and Lisičići in Bosnia (Hopf 1958) and Pokrovnik in Dalmatia (Karg and Müller 1990). Throughout the occupation of the Gomolava site (Neolithic-Iron Age), einkorn was of greater importance than emmer (van Zeist 2001/2002). Both emmer and einkorn were present in ca. 90% of the samples (out of 41 analyzed samples) dating to the late Vinča period at Gomolava. Einkorn was the main cereal grain stored in a silo from Staro Selo/Selevac (Hopf 1974) attributed to Vinča B period and directly dated to 4790–4925 B.C. (see Table 2.6). Einkorn is present in 94% of the samples, and emmer in 85% of 47 samples attributed to various Vinča phases at Selevac (analyzed by McLaren and Hubbard 1990). In the Vinča sites in Romania, the main cereals were also einkorn and emmer. At the Vinča site of Liubcova, the major cereal was emmer accompanied by einkorn in a significant percentage. From the Vinča phases at the site Parța, einkorn and emmer were also identified (Cârciumaru 1996). At the tell site Uivar in the Banat region, einkorn and emmer are the most ubiquitous in the Vinča phases (Fischer and Rösch 2004). Large quantities of einkorn and emmer chaff were retrieved by sieving and flotation along with smaller amounts of grain. The single cereal grain storage discovered at Uivar consisted of a mixture of charred *Triticum monococcum* and *Triticum* cf. *timopheevi*, i.e., the new type of glumed tetraploid wheat (Fischer and Rösch 2004). The large proportion of einkorn grains in relation to emmer grains identified in many samples from several sites in the West-central Balkans could support the suggestion by Hopf (1974) and later by Van Zeist (2001/2002) that emmer wheat was a minor admixture to einkorn throughout the region.

Emmer and einkorn were the principal crops in late Neolithic Hungary (Ferenc Gyulai, personal communication 2004). From the site Battonya-Parázstanya in east Hungary, emmer was the major cereal and einkorn and barley were less common (Hartyányi 1988/89), though in some sites they are represented equally to emmer (Hartyányi and Máthé 1980). Emmer and einkorn were ubiquitously present at the Neolithic sites of Bulgaria and were often found as mixtures in storages in the late Neolithic; only during the Chalcolithic period are there really pure einkorn finds, as at Kazanluk (Dennell 1978; Hopf 1973), Karanovo V (Arnaudov 1937; Hopf 1973), Kapitan Dimitirievo, Durankulak, and other sites (Marinova 2001; Popova 1995). The co-occurrence of emmer and einkorn is explained as cultivation of in maslin fashion (as a risk-reducing strategy) because einkorn is hardier and more resistant to cold and drought.

In the second half of the Early Neolithic sites of southern Bulgaria, emmer dominated the plant assemblage, whereas during late Chalcolithic, einkorn was the dominant cereal in northern Bulgaria and Bulgarian Thrace (Marinova 2001; Popova 1995). In northern Greece, the new evidence from five late Neolithic sites, Mandalo, Makri, Arkadikos, Makriyalos, and Dikili Tash indicates the predominance of einkorn in four of the five sites (Valamoti 2004). However, in the same region where einkorn dominates, emmer is the principal crop in some plant assemblages, e.g., Mandalo. It is unlikely that in the same climate, different environmental factors could explain the preference for the hardier einkorn. According to Valamoti (2004), perhaps cultural factors underlie these differences. One of the possible explanations for the predominance of einkorn, particularly of einkorn chaff in the Late Neolithic sites in north Greece, could be its use as fodder and subsequent preservation in dung used as fuel (Valamoti 2004). The use of dung as fuel is not yet attested in the Vinča sites where flotation was employed and where sizeable quantities of wood charcoal and of einkorn grains and/or einkorn chaff were recovered (e.g., Opovo and Gomolava). The use of dung as fuel is often explained as a consequence of scarce wood to burn (Miller 1996). Perhaps, there was no need to use dung as fuel at the mentioned Vinča sites given adequate firewood supplies during the Neolithic period.

Rich remains of einkorn are often found in the Balkan and Pannonian region during the Neolithic and the Bronze Age (Hopf 1974; Renfrew 1979). The longevity of the tradition of einkorn use could be differently explained. The distribution of the wild einkorn (*Triticum boeticum*) includes the southern Balkans (Zohary and Hopf 2000) and may be a factor in the argument for the local domestication of einkorn. *Cerealia*-type pollen was identified from the Mesolithic sites in the Iron Gates region of the Danube, including seven specimens of human feces from Vlasac (Cârciumaru 1973, 1978). According to Cârciumaru, the increasing percentage of *Cerealia* pollen from the lower to upper strata demonstrates the existence of certain primitive cereal crops and cereal cultivation (Cârciumaru 1996). At the Vinča site Liubcova, also on the Danube, the identified *Triticum dicoccum* resulted from the cultivation of wild species of *Triticum dicoccoides*, but it remains unclear whether the grains were actually from the wild species or from the abundant emmer fields (Cârciumaru 1996).

Wild einkorn (*Triticum boeticum*), hulled six-row barley (*Hordeum vulgare* ssp. *hexastichum*), lentil (*Lens* sp.), and vetch (*Vicia ervilia*) were identified in the Mesolithic layers at the Theopetra Cave in Thessaly in Greece (Kyparissi-Apostolika 2000). The same plants were found as cultivated species in the Neolithic material, which "provide definite proof" for the indigenous development of farming in Thessaly, according to the principal investigator Kyparissi-Apostolika (2000:137). Further research of plant remains is necessary from this site to provide definite proof. According to Kotsakis (2001), the present-day distribution of wild einkorn in the nearby Grevena region (Zamanis et al. 1988, cited in Kotsakis 2001:66) would provide further evidence for domestication *in loco*. However, the DNA studies of modern cultivated and wild einkorn by Heun et al. (1997) support the assumption of a monophyletic origin of cultivated einkorn in the Karakadağ region in Turkey, and would negate the local domestication of einkorn in the Balkans. The indigenous development of agriculture was also rejected by Hansen (1991) in her later studies of the plant macro remains from Franchthi Cave in Greece where the Mesolithic–Neolithic layers exist. Her statements were also supported by Perlès (2001).

From the later periods, Kroll (1983) interpreted the predominance of einkorn over other cereals at the Bronze Age sites in northern Greece as surplus exchanged for Mycenaean pottery in the case of Kastanas. Halstead (1994) interpreted the predominance as a provision against famine and a staple finance for exchange with settlements in the south in the case of Assiros.

Spelt wheat (*Triticum spelta*) another glume wheat, was not identified on any of the Starčevo or the Vinča culture sites. In the Neolithic and Chalcolithic sites in Bulgaria, spelt wheat was rare, its occurrence attested solely on the basis of a few grains; spikelet bases of spelt wheat, which are more reliably identified, were not recovered (Marinova 2001). According to Cârciumaru (1996:194), during the fourth millennium B.C. in eastern Romania, in the Bărăgan region, *Triticum spelta* is dominant and *Triticum aestivum* along with einkorn and emmer are less prevalent.

The newly recognized type of tetraploid glumed wheat resembling *Triticum* cf. *timopheevi* was identified only from three Neolithic sites in Romania, Uivar, Parţa (Tell II), (both dated to the Vinča C period), and from Dudeştii Vechi, dated to Starčevo–Criş and Tiszapolgár periods (Fischer and Rösch 2004). As archaeobotanists become more familiar with its characteristic morphological features, perhaps the newly identified wheat will be documented at other Neolithic sites in the Pannonian Plain, as it has been retroactively identified from the Bronze Age site Feudvar in Voyvodina (Helmut Kroll, personal communication 2004).

Triticum aestivum (Bread Wheat)

At Opovo, a few of the shorter emmer grains could perhaps be attributed to the type of bread wheat *Triticum aestivo-compactum*. Free-threshing bread wheat (*Triticum aestivum*), including the compact form, is present in the Starčevo and the Vinča culture sites (Table 2.5); however, in small quantities and in the absence of rachis fragments, the identification remains ambiguous. At Gomolava, *Triticum aestivum* is present in ca. 50% of 41 samples from the Vinča layers, but again, in small quantities (van Zeist 2001/2002). In Romania, at Uivar, *Triticum aestivum/durum* was identified in small quantities from the Vinča layers. On the basis of rachis fragments it appears that the majority of seeds belongs to the *aestivum*-type (Fischer and Rösch 2004), as well as from the site Liubcova where it constitutes 19% of grains (Cârciumaru 1996).

Similarly, bread wheat is only sporadically encountered in the Neolithic sites in Hungary (Ferenc Gyulai, personal communication 2004) and Greece (Kroll 1991a). Its limited representation in Greece can be explained as the Neolithic farmers reluctance to grow naked grain wheat, which is more susceptible to bird, fungal, and insect attacks, and less resistant to unfavorable growing conditions (Valamoti 2004). However, in Bulgaria the presence of naked wheat increases during the late Neolithic, and especially during the late Chalcolithic period. The intensification is associated with more advanced agriculture (Marinova 2001).

Hordeum vulgare **(Barley)**
A small quantity of hulled barley grains was found at Opovo. At Gomolava, a rather small number of hulled barley grains was present in less than 50% of the samples from the Vinča layers, but the quantities increased in post-Neolithic occupation of the site (van Zeist 2001/2002). A single deposit of barley grains was recovered from Obre II attributed to the Butmir culture (Renfrew 1974). It seems that barley was of minor importance at the Starčevo and Vinča sites, including at the Romanian sites where naked barley is rare but more common than hulled (Fischer and Rösch 2004).

Likewise, barley was considerably less represented than glume wheats in the late Neolithic sites in northern Greece (Valamoti 2004). Barley seems to be less frequent in the late Neolithic sites in Hungary. Barley (two-row naked) is a dominant cereal at least at the late Neolithic site of Berettyóújfalu-Szilhalom, a tell in northeast Hungary (Hartányi and Máthé 1980). In the Neolithic and Chalcolithic sites in Bulgaria, barley played a more important role than at the sites of Serbia and Greece, though it remained less ubiquitous than glume wheats. Hulled barley seems to have played a more important role in the Early Neolithic sites of Bulgaria. Naked barely predominates in the late Neolithic in east Thrace and north Bulgaria (Marinova 2001).

Panicum miliaceum **(Millet)**
Millet grains were not identified in the Opovo samples. However, of special interest are finds of millet (*Panicum miliaceum*) from the late Vinča layers at Gomolava, where millet grains were recovered both from the flotation samples analyzed by Zeist (2001/2002) and independently in the samples analyzed by Bottema and Ottaway (1982). Five *Panicum*-type grains were recovered from the flotation samples at the ongoing excavation of the Vinča-Belo Brdo site (Borojević and Filipović, analysis in process). Four *Panicum miliaceum* grains were identified from a pit dated to Vinča C at the site Uivar in Romania (Fischer and Rösch 2004) and a few grains cf. *Panicum miliaceum* from Okolište in Bosnia (Bittman and Kučan 2004).

The earlier find of millet grains (along with acorns and beech-nuts) from a clay-lined pit at Nosa-Biserna Obala (Starčevo-Körös culture) in north Voyvodina was first reported by Garašanin (1961:305), without any further details of the find (not included in Table 2.5). The single seed of common millet (*Panicum miliaceum*) from Ecsegfalva in Hungary was also recovered from the Starčevo-Körös culture site and, together with the uncertain find of millet from Nosa-Biserna Obala, presents the earliest finds of this crop in the Carpathian basin, predating the finds of millet in the Linearbandkeramik cultures in central Europe (Amy Bogaard, personal communication 2004).

Millet was also found at the late Neolithic settlement Berettyóújfalu-Herpály (Herpály culture) in the northeast Hungary (Ferenc Gyulai, personal communication 2004). In Bulgaria, the earliest positive find of millet is from the early Chalcolithic site Durankulak in east Bulgaria (Hamangia culture), contemporaneous with the late Vinča culture (Marinova 2001). Millet grains were also identified from a Chalcolithic site (Tripolje culture) in Moldavia (Janushevich 1978). From the Neolithic sites in northern Greece, millet grains were reported only from the early excavations of Olynthos (Halstead 1994).

Legumes

Lentils (*Lens culinaris*) are present at Opovo, whereas no (positively identified) pea grains were found. Seeds of vetches (*Vicia* sp.) were recovered at Opovo, but are attributed to the wild/weed species. Among legumes, lentils are more common than peas in the Vinča sites. The only vetch crop identified so far was *Vicia sativa* in Anza II, from the Starčevo culture layer (Renfrew 1976). Bitter vetch (*Vicia ervilia*), broad bean (*Vicia faba*) and grass peas (*Lathyrus sativus*) were not found on the Vinča sites, except for a find of two *Lathyrus sativus* seeds from Divostin (Grüger and Beug 1988). The absence of these crops from the Vinča sites cannot be attributed solely to the vagaries of sampling, because at the sites where flotation was employed, e.g., Opovo and Gomolava, there were no cultivated vetches or grass peas recovered. Nor can it be attributed to the inadequacy of the plant identification because, at Gomolava, bitter vetch (*Vicia ervilia*), broad bean (*Vicia faba*), and grass peas (*Lathyrus sativus*) were identified from the samples, post-dating the Vinča layers (van Zeist 2001/2002). At the Vinča sites on the Romanian side, Uivar (Fischer and Rösch 2004), and Liubcova (Cârciumaru 1996), legumes were identified but also in small quantities. Lentils (*Lens culinaris*) were the most ubiquitous among legumes at both sites, followed by peas (*Pisum sativum*) at Uivar. Very rarely, *Vicia* sp. seeds were found. A single seed of bitter vetch *Vicia ervilia* was identified at Uivar, and a single seed of broad bean *Vicia faba* was identified in the layers post-dating the Vinča Period (Fischer and Rösch 2004). Two pots from the site Liubcova contained a mixture of cereal caryopses and leguminous seed. In one pot, the two types of seeds were almost equally represented, and the content was interpreted as intended for human consumption. In another pot, cereal grains constituted only 7% and the mixture was interpreted as a result of crop rotation, because it also contained seeds of weeds, *Galium spurium* and *Rumex acetosa* (Cârciumaru 1996).

Lathyrus sativus, *Vicia ervilia*, *Pisum sativum*, and *Lens culinaris* have been found at several Late Neolithic sites in Hungary, sometimes in significant quantities, e.g., peas at Berettyóújfalu-Szilhalom and Berettyóújfalu-Herpály (Ferenc Gyulai, personal communication 2004). Legumes are also frequently present in the Neolithic and Chalcolithic sites of Bulgaria (Marinova 2001; Popova 1995). *Vicia ervilia*, which has not been identified from a single site of the Starčevo or the Vinča culture, is an important crop found also in storage at several sites dating to the Neolithic and Chalcolithic periods in Bulgaria (e.g., Azamak, Karanovo, Slatino, Yassa Tepe, and Kapitan Dimitreivo). Similarly, seeds of Lathyrus species are commonly found at the sites in Bulgaria; they seem to be more important during the early than the late Neolithic, when *Vicia ervilia* becomes more significant (Marinova 2001; Popova 1995). In the Late Neolithic sites in Greece, the presence of pulses is prominent, including peas, grass pea, bitter vetch, and lentils; on some sites, e.g., Arkadikos, pulses outnumber cereals (Valamoti 2004).

Oil and Fiber Plants

A few seeds of flax (*Linum usitatissimum*) were recovered from the Opovo site, from the Late Vinča layers at Gomolava (van Zeist 2001/2002), and from two Vinča pits at the site of Uivar in Romania (Fischer and Rösch 2004). These are the only three sites from which flax seeds were identified, probably due to flotation/sieving of the large number of samples employed at the sites which facilitated their recovery. Additionally, at Opovo, a fragment of a textile and a cord made from flax fibers were discovered (Tringham et al. 1992), adding to the evidence that this crop was used for fiber production in the late Vinča culture.

Flax seeds were identified from at least one site in Hungary (Hartyányi 1988/89). *Linum* sp. seeds were also recovered from the Early Neolithic sites (e.g., Slatina) in Bulgaria (Dontscheva 1990), and from several late Neolithic sites and two Chalcolithic sites (Marinova 2001). Concentrations of flax seeds were found in the assemblages at the Late Neolithic sites in northern Greece, and may indicate in the absence of other oil producing plants, the use of this crop for oil in Late Neolithic Greece (Valamoti 2004).

Fruits and Nuts

Remains of several fruits were found at Opovo (including pits of Cornelian cherry, pips of wild grapes, seeds of elder, blackberry, and wild strawberry), in addition to fragments of water chestnuts. Stones of Cornelian cherry drupes are the most ubiquitous at the Opovo site. In a few instances of more detailed analyses, at the sites of Gomolava (van Zeist 1975, 2001/2002), Selevac (McLaren and Hubbard 1990), Anza (Renfrew 1976), and Uivar (Fischer and Rösch 2004) other seeds, stones, and nut shell fragments were identified, including, e.g., *Cornus mas* (Cornelian cherry), *Corylus avellana* (hazelnut), *Fragaria vesca* (strawberry), *Malus* sp. (apple), *Pyrus* sp. (pear), *Rubus* sp. (blackberry), *Sambucus* sp. (elder), *Vitis vinifera* (grape), *Quercus* sp. (acorn), and *Tilia* sp. (linden fruit). *Sambucus ebulus* and *S. nigra* seeds were commonly found mineralized. Among the fruit remains, the stones of Cornelian cherry were the most common at the Vinča sites. Generally, the presence of *Cornus mas* might be somewhat overestimated, because the characteristic fruit stones structure allows identification of even small fragments. *Cornus mas* stones were also found in a Late Neolithic site in Hungary (Hartyányi 1988/89). More recent studies confirmed the importance of *Cornus mas* in the earlier phase of the Neolithic in Bulgaria (Marinova 2001). A storage of Cornelian cherry was found from the Chalcolithic site of Drama in south Bulgaria (Küster 1989). *Cornus mas* seems less ubiquitous during the Late Neolithic and Chalcolithic in Greece, a phenomenon already noticed by Kroll (1991a:161–177).

At Opovo, only a few mineralized wild grape pips (*Vitis vinifera* spp. *sylvestris*) were discovered. A single-seed wild grape seed was discovered from the Vinča layers at Gomolava (van Zeist 2001/2002). At Uivar, a single mineralized seed of wild grape was discovered from the Eneolithic layer, post-dating Vinča C layers (Fischer and Rösch 2004). Grape is sporadically present in the Late Neolithic sites in Hungary (Ferenc Gyulai, personal communication 2004). In Bulgaria, grape pips are found more ubiquitously, but always as isolated inclusions among other seeds (Marinova 2001). In the Late Neolithic sites of northern Greece, grape pips were found at several sites. Additionally, in the Late Neolithic layers at the site Dikili Tash, remains of wine pressing were found (grape pips in skins), possibly indicating production of must/wine (Valamoti 2004).

Seeds attributed to *Physalis alkekengi* (Chinese lantern) are palatable and interpreted as fruits and were probably collected (McLaren and Hubbard 1990; van Zeist 2001/2002). Few *Physalis alkekengi* seeds were also recovered from Vinča houses and pits at Uivar (Fischer and Rösch 2004). It is interesting that *Physalis alkekengi* seeds were identified from several Neolithic sites in Bulgaria (Marinova 2001), but not yet from a single Chalcolithic site that was contemporaneous with the Vinča culture sites, perhaps because flotation was employed on a very few Chalcolithic sites.

Many water chestnut (*Trapa natans*) fruit fragments were identified from the Opovo site. Few fragments of water chestnut fruits were identified from the late Vinča layers at Gomolava (Van Zeist 2001/2002) and from Uivar (Fischer and Rösch 2004). These are the only three Vinča sites where water chestnuts were identified. Water chestnut fragments were also found at the Middle Neolithic site of the Tisza culture, Öcsöd-Kováshalom (Judith Rasson, personal communication 2004). Water chestnuts have not yet been recovered and/or identified from the Neolithic/Chalcolithic sites in Greece and Bulgaria.

Weedy Herbaceous (Wild/Weed) Plants

Seeds of weeds including *Polygonum* sp. (knotweed), *Bromus* sp. (brome grass), *Lolium* sp. (rye grass), *Chenopodium* sp. (goosefoot), and *Galium aparine* (bedstraw) were identified at the Opovo site, also present at Gomolava (van Zeist 2001/2002) and Uivar (Fischer and Rösch 2004). Plant remains other than crops were correspondingly interpreted as either fruits collected from the wild or as crop weeds, fodder, or ruderals.

Seeds of the Solanaceae family were recovered from Opovo, Gomolava, and Selevac. Some of the Solanaceae seeds from Opovo could be attributed to the species of *Physalis*, *Solanum*, or *Hyoscyamus*. *Physalis alkenengi* berries were probably collected as fruits, while I would suggest *Hyoscyamus* and *Lithospermum officinale* might have been used medicinally and in rituals.

***

Emmer and einkorn are the ubiquitous cereals in the area including the Vinča sites. The statement by Hubbard (1980), that einkorn attained peak popularity in the Balkans during the Late Neolithic period, while emmer become dominant in many areas of Europe, is supported by new evidence from archaeobotanical investigations in Serbia, Romania, Bulgaria, and Greece. Although free-threshing wheats were known, glume wheats were favored. Bread wheat is present in very small quantities throughout the region but is absent from storage contexts. Barley is overall less prevalent, except on a few sites in Bulgaria. Millet is found at sites in the Pannonian Plain (Carpathian basin) in the Early Neolithic, and it continues to be sporadically present in the northern Vinča sites and Tisza sites, but it was even more rare in the cultures of Bulgaria and Greece.

The differences in legume exploitation seem more obvious. Lentils are the most ubiquitous legumes in the Vinča culture, followed by peas, but are present in rather small quantities. Cultivated vetches and grass peas, e.g., *Vicia ervilia* and *Lathyrus sativus*, are for some reason completely absent from the Starčevo and Vinča sites so far, except for the single seed of *Vicia ervilia* from a Vinča house at Uivar. However, they are present in considerable amounts in all surrounding cultures.

Flax is present sporadically in the Vinča and Tisza culture sites, but no concentrations of flax seeds were found that would point with certainty to oil extraction, a likely practice in the Late Neolithic sites in Greece. The finds at Opovo, an impression of textile and a cord made of flax, in addition to flax seeds, would point to the growing of flax for fibers in these northern regions during the Late Neolithic.

In terms of wild plant procurement, the spectrum of identified wild plants is limited due to the lack of waterlogged sites and better preservation. Clearly the spectrum of available wild plants would vary in the different natural environments of the southern and northern regions of southeastern Europe. Water chestnuts were gathered in the northern regions, and the evidence of their procurement is found in the north Vinča and Tisza culture sites. Cornelian cherry stones are frequent in the Vinča culture sites, and in the Late Neolithic of Bulgaria and Hungary, but not in Greece. Grapes were gathered from the wild in the northern part of southeastern Europe, but were probably cultivated in northern Greece, where large quantities of grape pips have been dated to the Late Neolithic (Valamoti 2004). Gathering of wild fruits continued during the Neolithic, but seemed to gain in importance again in the Vinča C phase, as noted by Fischer and Rösch (2004) and as documented by the numerous water chestnuts and Cornelian cherry finds at Opovo.

At Opovo, no small seeds of fodder plants were found that would indicate the use of dung as fuel, although all the crops can be used for fodder. It could be argued that similarities in crop cultivation used as fodder may reflect similarities in animal husbandry. Cattle, sheep, and goats were the main animals raised in the Late Neolithic, but the number of bones of certain species varies among sites, making it difficult to discern which animals can be associated with certain crops used as fodder in the wider region.

Accounting for all results of plant remains from various sites in the Pannonian Plain and the Balkan region, it is apparent that a diverse but similar spectrum of crops was present in different cultures dating to the Late Neolithic/Chalcolithic period. The identified plant remains from the Opovo site (Table 2.5) fall into the typical spectrum of plant species identified from the Vinča sites and other Late Neolithic/Chalcolithic (Eneolithic) sites of the region (Figure 2.32). Based on a limited number of sites from which plant remains were systematically collected, it is not apparent that there was some kind of specialization of certain Vinča sites for growing particular crops.

Variations of the identified plant remains exist among different sites of the same culture, some of which reflect the different plant material deposited and different recovery methods. However, some generalizations about the region and some peculiarities of the specific regions seem plausible, albeit difficult to explain: Why was einkorn the dominant wheat at several sites in the whole region regardless of the local environment or cultural affiliation? Why was bread wheat ubiquitously present but not yet an important crop? Why was barley more important in the east than in the rest of the region? Why was millet only sporadically present? Why were there no cultivated vetches and grass peas in the Vinča culture when they were important crops in all surrounding cultures? Finally, why did fruit gathering seem to become more important during the Late Neolithic phases in the Vinča culture? All these questions cannot be answered yet and necessitate further investigations.

The Vinča farmers had all of the main staple crops already available in domesticated form. The Vinča settlers, including the Opovo inhabitants, grew crops but also continued or even intensified gathering of plants in the later phases of the Vinča culture.

Table 2.5. Presence of identified plant macro remains from the Starčevo (the sixth millennium B.C.) and Vinča culture sites (the fifth millennium B.C.)[1].

Site	Starčevo	Belotić	Mesarci	Zablaće	Divostin
Cultural affiliation[2]	Starčevo	Starčevo II	Starčevo II–III	Starčevo II	Starčevo
Reference	Renfrew unpublished, in Renfrew 1979	Borojević 1990	Borojević 1990	Borojević 1990	Grüger and Beug 1988
Type of samples[3]	impressions and few carbonized seeds	few carbonized seeds from a pit	few carbonized seeds from a pit	few carbonized seeds from a pit and house daub	only two carbonized seeds
Taxa[4]					
Cereals					
Triticum monococcum (1-grain)	X				
T. monococcum (2-grain)					
T. monococcum spikelet bases/glumes					
Triticum dicoccum	X				X
T. dicoccum spikelet bases/glumes					
Triticum compactum			X		
Triticum cf. *timopheevi*					
Triticum aestivum/durum		X	X		
Triticum sp.					
Triticum (hulled spp.) spikelet bases/glumes					
Triticum aestivum/durum rachises					
Hordeum vulgare (hulled)	X	X	X	X	
Hordeum vulgare var. *nudum*		X			
H. vulgare hulled (6-row)					
H. vulgare rachis fragments			X		
Panicum miliaceum					
Unidentified Gramineae					
Legumes					
cf. *Lathyrus sativus*					
Lens culinaris					
Pisum sativum	X				
Vicia sativa					
Vicia sp.					
Oil and Fiber Plants					
Linum usitatissimum					
Fruits					
Cornus mas	X				
Corylus avellana					
Fragaria vesca					
Malus sylvestris					
Physalis alkekengi					
Prunus insititia					
Prunus spinosa					
Pyrus communis					

[1] There are only a few direct dates on grains (see Table 2.6 for ^{14}C dates).
[2] Cultural affiliation assigned according to the referenced publications.
[3] The majority of samples consist of carbonized seeds/grains or impressions on pottery sherds or daub.
[4] Plant names in this table follow nomenclature used by respective authors as referenced.

Table 2.5. Presence of identified plant macro remains from the Starčevo (the sixth millennium B.C.) and Vinča culture sites (the fifth millennium B.C.). (Continued)

Site	Gornja Tuzla	Obre I	Kakanj	Anza I	Anza II
Cultural affiliation	Starčevo	Starčevo/Kakanj	Kakanj /Starčevo	Starčevo-Körös (Anzabegovo-Vršnik)	Starčevo-Körös
Reference	Hopf 1966/67	Renfrew 1974	Renfrew 1974	Renfrew in Gimbutas 1976; Beug in Gimbutas 1976	Renfrew in Gimbutas 1976; Beug in Gimbutas 1976
Type of samples	impressions in a pot sherd	less than 10 seeds	flotation	flotation and wet sieving	flotation and wet sieving
Taxa					
Cereals					
Triticum monococcum (1-grain)		X	X	X	X
T. monococcum (2-grain)			X		
T. monococcum spikelet bases/glumes					
Triticum dicoccum		X	X	X	X
T. dicoccum spikelet bases/glumes					X
Triticum compactum		X	X	X	X
Triticum cf. *timopheevi*					
Triticum aestivum/durum					
Triticum sp.					
Triticum (hulled spp.) spikelet bases/glumes	X				X
Triticum aestivum/durum rachises					
Hordeum vulgare (hulled)			X		
Hordeum vulgare var. *nudum*					
H. vulgare hulled (6-row)				X	X
H. vulgare rachis fragments					X
Panicum miliaceum					
Unidentified Gramineae					
Legumes					
cf. *Lathyrus sativus*					
Lens culinaris		X		X	X
Pisum sativum		X	X	X	X
Vicia sativa					X
Vicia sp.					X
Oil and Fiber Plants					
Linum usitatissimum					
Fruits					
Cornus mas		X	X		X
Corylus avellana					
Fragaria vesca					
Malus sylvestris					
Physalis alkekengi					
Prunus insititia					
Prunus spinosa					
Pyrus communis					

Table 2.5. Presence of identified plant macro remains from the Starčevo (the sixth millennium B.C.) and Vinča culture sites (the fifth millennium B.C.). (Continued)

Site	Anza III	Vršnik	Anza IV	Gomolava	Gomolava
Cultural affiliation	late Starčevo-Körös	Starčevo-Vršnik III	Vinča	Vinča C and C/D	Vinča C and C/D
Reference	Renfrew in Gimbutas 1976; Beug in Gimbutas 1976	Hopf 1961	Renfrew in Gimbutas 1976; Beug in Gimbutas 1976	Van Zeist 2001–2002	Bottema and Ottaway 1982
Type of samples	flotation and wet sieving	unknown, carbonized seeds sent to Hopf by Garašanin	flotation and wet sieving	flotation	flotation from 2x2 m trench
Taxa					
Cereals					
Triticum monococcum (1-grain)	X	X	X	X	X
T. monococcum (2-grain)					
T. monococcum spikelet bases/glumes				X	
Triticum dicoccum	X	X	X	X	X
T. dicoccum spikelet bases/glumes			X	X	
Triticum compactum		X			
Triticum cf. *timopheevi*					
Triticum aestivum/durum				X	X
Triticum sp.					
Triticum (hulled spp.) spikelet bases/glumes	X				X
Triticum aestivum/durum rachises					
Hordeum vulgare (hulled)		X			X
Hordeum vulgare var. *nudum*					
H. vulgare hulled (6-row)	X		X	X	
H. vulgare rachis fragments				X	
Panicum miliaceum				X	X
Unidentified Gramineae				X	
Legumes					
cf. *Lathyrus sativus*					
Lens culinaris	X		X	X	X
Pisum sativum			X	X	
Vicia sativa					
Vicia sp.				X	
Oil and Fiber Plants					
Linum usitatissimum				X	
Fruits					
Cornus mas				X	
Corylus avellana	X				
Fragaria vesca				X	
Malus sylvestris				X	
Physalis alkekengi				X	
Prunus insititia					
Prunus spinosa					
Pyrus communis					

Table 2.5. Presence of identified plant macro remains from the Starčevo (the sixth millennium B.C.) and Vinča culture sites (the fifth millennium B.C.). (Continued)

Site	Opovo	Selevac	Divostin	Staro Selo/Selevac	Valač
Cultural affiliation	**Vinča C/D**	**Vinča B/C**	**Vinča D**	**Vinča B**	**late Vinča**
Reference	Borojević 1998	McLaren and Hubbard 1990	Grüger and Beug 1988	Hopf 1974	Hopf 1974
Type of samples	flotation across the site	flotation	seeds extracted from pollen cores	seeds from two grain silos submitted to Hopf by Galović	seeds from the Museum collection in Priština submitted to Hopf
Taxa					
Cereals					
Triticum monococcum (1-grain)	X	X		X	
T. monococcum (2-grain)				X	
T. monococcum spikelet bases/glumes	X	X		X	
Triticum dicoccum	X	X			
T. dicoccum spikelet bases/glumes	X	X			
Triticum compactum			X		
Triticum cf. *timopheevi*					
Triticum aestivum/durum					
Triticum sp.	X				
Triticum (hulled spp.) spikelet bases/glumes	X				
Triticum aestivum/durum rachises					
Hordeum vulgare (hulled)	X	X			
Hordeum vulgare var. *nudum*					
H. vulgare hulled (6-row)					
H. vulgare rachis fragments					
Panicum miliaceum					
Unidentified Gramineae					
Legumes					
cf. *Lathyrus sativus*			X		
Lens culinaris	X	X		X	
Pisum sativum		X			X
Vicia sativa					
Vicia sp.	X				
Oil and Fiber Plants					
Linum usitatissimum	X				
Fruits					
Cornus mas	X	X			
Corylus avellana		X			
Fragaria vesca	X				
Malus sylvestris					
Physalis alkekengi		X			
Prunus insititia					
Prunus spinosa					
Pyrus communis		X			

Table 2.5. Presence of identified plant macro remains from the Starčevo (the sixth millennium B.C.) and Vinča culture sites (the fifth millennium B.C.). (Continued)

Site	Predionica	Vinča kod Kragujevaca	Petnica	Gornja Tuzla	Vinča D
Cultural affiliation	Vinča	Vinča	Vinča B	Vinča C	Vinča
Reference	Hopf 1974	Hopf 1974	Borojević 1990	Hopf 1966/67, 1974	Renfrew unpublished, in Renfrew 1979
Type of samples	impressions on pottery sherds	impressions on pottery sherds and daub	carbonized seeds from a pit	impressions from house rubble	carbonized seeds from the Vinča collection at the University of Belgrade
Taxa					
Cereals					
Triticum monococcum (1-grain)			X	X	X
T. monococcum (2-grain)					
T. monococcum spikelet bases/glumes				X	
Triticum dicoccum	X		X		X
T. dicoccum spikelet bases/glumes	X	X		X	
Triticum compactum					
Triticum cf. *timopheevi*					
Triticum aestivum/durum					
Triticum sp.			X		
Triticum (hulled spp.) spikelet bases/glumes	X				
Triticum aestivum/durum rachises					
Hordeum vulgare (hulled)			X		
Hordeum vulgare var. *nudum*			X		
H. vulgare hulled (6-row)					
H. vulgare rachis fragments					
Panicum miliaceum					
Unidentified Gramineae					
Legumes					
cf. *Lathyrus sativus*					
Lens culinaris					
Pisum sativum					
Vicia sativa					
Vicia sp.					
Oil and Fiber Plants					
Linum usitatissimum					
Fruits					
Cornus mas					X
Corylus avellana					
Fragaria vesca					
Malus sylvestris					
Physalis alkekengi					
Prunus insititia					
Prunus spinosa					
Pyrus communis					

Table 2.5. Presence of identified plant macro remains from the Starčevo (the sixth millennium B.C.) and Vinča culture sites (the fifth millennium B.C.). (Continued)

Site	Medvednjak	Uivar[5]	Early and Middle Neolithic (the sixth millenium B.C.)		Late Neolithic (the fifth millenium B.C.)	
Cultural affiliation	Vinča	Vinča C	Total 12 Starčevo sites		Total Vinča 15 sites	
Reference	Renfrew unpublished, in Renfrew 1979	Fisher and Rösch 2004	Number of sites present	Percentage (%) of sites present	Number of sites present	Percentage (%) of sites present
Type of samples	unknown, submitted to Renfrew by Galović	flotation and sieving across the site				
Taxa						
Cereals						
Triticum monococcum (1-grain)	X	X	7	58%	11	73%
T. monococcum (2-grain)			1	8%	1	7%
T. monococcum spikelet bases/glumes		X			6	40%
Triticum dicoccum	X	X	8	67%	10	67%
T. dicoccum spikelet bases/glumes		X	1	8%	8	53%
Triticum compactum			6	50%	1	7%
Triticum cf. *timopheevi*		X			1	7%
Triticum aestivum/durum	X	X	2	17%	4	27%
Triticum sp.		X			3	20%
Triticum (hulled spp.) spikelet bases/glumes		X	3	25%	4	27%
Triticum aestivum/durum rachises		X			1	7%
Hordeum vulgare (hulled)		X	6	50%	5	33%
Hordeum vulgare var. *nudum*		X	1	8%	2	13%
H. vulgare hulled (6-row)			3	25%	2	13%
H. vulgare rachis fragments		X	2	17%	2	13%
Panicum miliaceum		X			3	20%
Unidentified Gramineae		X			2	13%
Legumes						
cf. *Lathyrus sativus*					1	7%
Lens culinaris		X	4	33%	7	47%
Pisum sativum		X	5	42%	5	33%
Vicia sativa			1	8%		
Vicia sp.		X	1	8%	3	20%
Oil and Fiber Plants						
Linum usitatissimum		X			3	20%
Fruits						
Cornus mas		X	4	33%	5	33%
Corylus avellana		X	1	8%	2	13%
Fragaria vesca		X			3	20%
Malus sylvestris					1	7%
Physalis alkekengi		X			3	20%
Prunus insititia		X			1	7%
Prunus spinosa		X			1	7%
Pyrus communis					1	7%

[5] The list of wild/wild seeds is more extensive from Ulivar, but the single finds are not included in this table.

Table 2.5. Presence of identified plant macro remains from the Starčevo (the sixth millenium B.C.) and Vinča culture sites (the fifth millenium B.C.). (Continued)

Site	Starčevo	Obre I	Kakanj	Anza II
Cultural affiliation	Starčevo	Starčevo/Kakanj	Kakanj /Starčevo	Starčevo-Körös
Reference	Renfrew unpublished, in Renfrew 1979	Renfrew 1974	Renfrew 1974	Renfrew in Gimbutas 1976; Beug in Gimbutas 1976
Type of samples	impressions and few carbonized seeds	less than 10 seeds	flotation	flotation and wet sieving
Taxa				
Fruits (continued)				
Pyrus malus	X			X
Rubus caesius				
Rubus idaeus				
Rubus fruticosus				
Sambucus ebelus				
Sambucus nigra				
Vitis sylvestris				X
Trapa natans (fruit)				
Weedy Herabaceous Plants (Wild/Weed)				
Agrostemma githago				
Ajuga chamaepitys				
Atropa belladonna				
Avena fatua				
Avena sp.				X
Bromus arvensis				
Bromus mollis			X	
Bromus secalinus				
Bromus sp.				
Carex sp.				
Chenopodium album				X
Chenopodium hybridum				
Chenopodium polyspermum				
Chenopodium sp.				
Euphorbia helioscopia				
Galium aparine				
Galium spurium				
Lathyrus sativus				
Lithospermum officinale				
Lolium perenne				
Lolium sp.				
Plantago lanceolata				
Polygonum arenastrum				
Polygonum aviculare				X
Polygonum convolvulus		X		X

Table 2.5. Presence of identified plant macro remains from the Starčevo (the sixth millenium B.C.) and Vinča culture sites (the fifth millenium B.C.). (Continued)

Site	Anza III	Anza IV	Gomolava	Gomolava	Opovo
Cultural affiliation	late Starčevo-Körös	Vinča	Vinča C and C/D	Vinča C and C/D	Vinča C/D
Reference	Renfrew in Gimbutas 1976; Beug in Gimbutas 1976	Renfrew in Gimbutas 1976; Beug in Gimbutas 1976	Van Zeist 2001-2002	Bottema and Ottaway 1982	Borojević 1998
Type of samples	flotation and wet sieving	flotation and wet sieving	flotation	flotation from 2x2 m trench	flotation across the site
Taxa					
Fruits (continued)					
Pyrus malus					
Rubus caesius					
Rubus idaeus					
Rubus fruticosus			X		
Sambucus ebelus				X	
Sambucus nigra			X		X
Vitis sylvestris			X		X
Trapa natans (fruit)			X		X
Weedy Herabaceous Plants (Wild/Weed)					
Agrostemma githago			X		
Ajuga chamaepitys			X		
Atropa belladonna			X		
Avena fatua			X		
Avena sp.					
Bromus arvensis			X		
Bromus mollis					
Bromus secalinus					
Bromus sp.					X
Carex sp.			X		
Chenopodium album	X	X	X		
Chenopodium hybridum			X		
Chenopodium polyspermum			X		
Chenopodium sp.			X		X
Euphorbia helioscopia			X		
Galium aparine					X
Galium spurium			X		
Lathyrus sativus					
Lithospermum officinale					X
Lolium perenne					
Lolium sp.					X
Plantago lanceolata			X		
Polygonum arenastrum					
Polygonum aviculare					
Polygonum convolvulus			X		

Table 2.5. Presence of identified plant macro remains from the Starčevo (the sixth millenium B.C.) and Vinča culture sites (the fifth millenium B.C.). (Continued)

Site	**Opovo**	**Selevac**	**Divostin**	**Vinča D**	**Uivar**
Cultural affiliation	**Vinča C/D**	**Vinča B/C**	**Vinča D**	**Vinča**	**Vinča C**
Reference	Borojević 1998	McLaren and Hubbard 1990	Grüger and Beug 1988	Renfrew unpublished, in Renfrew 1979	Fisher and Rösch 2004
Type of samples	flotation across the site	flotation	seeds extracted from pollen cores	carbonized seeds from the Vinča collection at the University of Belgrade	flotation and sieving across the site
Taxa					
Fruits (continued)					
Pyrus malus					
Rubus caesius		X			X
Rubus idaeus					X
Rubus fruticosus					
Sambucus ebelus					X
Sambucus nigra	X	X			X
Vitis sylvestris	X				X
Trapa natans (fruit)	X				X
Weedy Herabaceous Plants (Wild/Weed)					
Agrostemma githago					X
Ajuga chamaepitys					
Atropa belladonna					
Avena fatua					
Avena sp.		X		X	
Bromus arvensis					X
Bromus mollis					
Bromus secalinus					
Bromus sp.	X	X			
Carex sp.					
Chenopodium album		X			X
Chenopodium hybridum					X
Chenopodium polyspermum					X
Chenopodium sp.	X	X			
Euphorbia helioscopia					
Galium aparine	X				X
Galium spurium					X
Lathyrus sativus			X		
Lithospermum officinale	X				
Lolium perenne		X			
Lolium sp.	X				
Plantago lanceolata					
Polygonum arenastrum		X			
Polygonum aviculare		X			X
Polygonum convolvulus					X

Table 2.5. Presence of identified plant macro remains from the Starčevo (the sixth millenium B.C.) and Vinča culture sites (the fifth millenium B.C.). (Continued)

Site	Early and Middle Neolithic (the sixth millenium B.C.)		Late Neolithic (the fifth millenium B.C.)	
Cultural affiliation	Total 12 Starčevo sites		Total Vinča 15 sites	
Reference	Number of sites present	Percentage (%) of sites present	Number of sites present	Percentage (%) of sites present
Type of samples				
Taxa				
Fruits (continued)				
Pyrus malus	2	17%		
Rubus caesius			2	13%
Rubus idaeus			1	7%
Rubus fruticosus			1	7%
Sambucus ebelus			2	13%
Sambucus nigra			4	27%
Vitis sylvestris	1	8%	3	20%
Trapa natans (fruit)			3	20%
Weedy Herabaceous Plants (Wild/Weed)				
Agrostemma githago			2	13%
Ajuga chamaepitys			1	7%
Atropa belladonna			1	7%
Avena fatua			1	7%
Avena sp.	1	8%	2	13%
Bromus arvensis			2	13%
Bromus mollis	1	8%		
Bromus secalinus				
Bromus sp.			2	13%
Carex sp.			1	7%
Chenopodium album	2	17%	4	27%
Chenopodium hybridum			2	13%
Chenopodium polyspermum			2	13%
Chenopodium sp.			3	20%
Euphorbia helioscopia			1	7%
Galium aparine			2	13%
Galium spurium			2	13%
Lathyrus sativus			1	7%
Lithospermum officinale			1	7%
Lolium perenne			1	7%
Lolium sp.			1	7%
Plantago lanceolata			1	7%
Polygonum arenastrum			1	7%
Polygonum aviculare	1	8%	2	13%
Polygonum convolvulus	2	17%	2	13%

Table 2.5. Presence of identified plant macro remains from the Starčevo and Vinča culture sites. (Continued)

Site	Divostin	Anza I	Anza II	Anza I	Anza II
Cultural affiliation	Starčevo	Starčevo-Körös (Anzabegovo-Vršnik)	Starčevo-Körös	Starčevo-Körös (Anzabegovo-Vršnik)	Starčevo-Körös
Reference	Grüger and Beug 1988	Renfrew in Gimbutas 1976; Beug in Gimbutas 1976	Renfrew in Gimbutas 1976; Beug in Gimbutas 1976	Renfrew in Gimbutas 1976; Beug in Gimbutas 1976	Renfrew in Gimbutas 1976; Beug in Gimbutas 1976
Type of samples	only two carbonized seeds	flotation and wet sieving	flotation and wet sieving	flotation and wet sieving	flotation and wet sieving
Taxa					
Weedy Herabaceous Plants (Wild/Weed) (continued)					
Polygonum minus					
Polygonum lapathifolium					
Polygonum sp.					
Rumex acetosella					
Rumex sp.					
Sambucus ebulus					
Setaria viridis/verticillata					
Silene sp.					
Solanaceae					
Solanum dulcamara					
Solanum nigrum					
Teucrium chamaedrys/scordium					
Triflolium campestre					
Weedy *Secale*					
Utricularia vulgaris			X		X
Wood charcoal[6]					
Abies pecitinata					
Acer					
Castanea		X		X	
Evonymus		X		X	
Fraxinus					
Juniperus		X	X	X	X
Ligustrum		X		X	
Pinus		X	X	X	X
Platanus		X		X	
Populus					
Quercus	X	X	X	X	X
Quercus (fruit)					
Rosaceae	X				
Salix		X	X	X	X
Tilia					
Tilia (fruit)					
Ulmus		X	X	X	X

[6] Trees are represented by wood charcoal of trees, except in two cases of tree fruits.

Table 2.5. Presence of identified plant macro remains from the Starčevo and Vinča culture sites. (Continued)

Site	Anza III	Anza IV	Gomolava	Gomolava	Predionica
Cultural affiliation	late Starčevo-Körös	Vinča	Vinča C and C/D	Vinča C and C/D	Vinča
Reference	Renfrew in Gimbutas 1976; Beug in Gimbutas 1976	Renfrew in Gimbutas 1976; Beug in Gimbutas 1976	Van Zeist 2001-2002	Bottema and Ottaway 1982	Hopf 1974
Type of samples	flotation and wet sieving	flotation and wet sieving	flotation	flotation from 2x2 m trench	impressions on pottery sherds
Taxa					
Weedy Herabaceous Plants (Wild/Weed)					
Polygonum minus					
Polygonum lapathifolium					
Polygonum sp.					
Rumex acetosella					
Rumex sp.		X	X		
Sambucus ebulus			X	X	
Setaria viridis/verticillata					
Silene sp.			X		
Solanaceae					
Solanum dulcamara					
Solanum nigrum			X		
Teucrium chamaedrys/scordium			X		
Triflolium campestre					
Weedy *Secale*			X		
Utricularia vulgaris					
Wood charcaol					
Abies pecitinata					
Acer			X		
Castanea					
Evonymus					
Fraxinus	X	X			
Juniperus	X	X			
Ligustrum					
Pinus					
Platanus					
Populus			X		
Quercus	X		X		
Quercus (fruit)					
Rosaceae					
Salix					
Tilia			X		
Tilia (fruit)					X
Ulmus			X		

Table 2.5. Presence of identified plant macro remains from the Starčevo and Vinča culture sites.

Site	Uivar	Early and Middle Neolithic (the sixth millenium B.C.)		Late Neolithic (the fifth millenium B.C.)	
Cultural affiliation	Vinča C	Total 12 Starčevo sites		Total Vinča 15 sites	
Reference	Fisher and Rösch 2004	Number of sites present	Percentage (%) of sites present	Number of sites present	Percentage (%) of sites present
Type of samples	flotation and sieving across the site				
Taxa					
Weedy Herabaceous Plants (Wild/Weed) (continued)					
Polygonum minus	X			1	7%
Polygonum lapathifolium	X			1	7%
Polygonum sp.				1	7%
Rumex acetosella	X			1	7%
Rumex sp.				3	20%
Sambucus ebulus				4	27%
Setaria viridis/verticillata	X			2	13%
Silene				1	7%
Solanaceae				1	7%
Solanum dulcamara				1	7%
Solanum nigrum	X			2	13%
Teucrium chamaedrys/scordium	X			2	13%
Triflolium campestre	X			1	7%
Weedy *Secale*				1	7%
Utricularia vulgaris		1	8%		
Wood charcoal					
Abies pecitinata					
Acer				1	7%
Castanea		1	8%		
Evonymus		1	8%		
Fraxinus		1	8%	1	7%
Juniperus		3	25%	1	7%
Ligustrum		1	8%		
Pinus		2	17%		
Platanus		1	8%		
Populus				1	7%
Quercus		4	33%	2	13%
Quercus (fruit)				1	7%
Rosaceae		1	8%	1	7%
Salix		2	17%		
Tilia				2	13%
Tilia (fruit)				1	7%
Ulmus		2	17%	1	7%

Table 2.6. Radiocarbon dates on carbonized cereal grains from the Vinča culture sites.

Lab Number	C-14 Dates B.P.	Calibrated Age Ranges B.C.	Sample Type	Site	Provenience	Reference
GrN 1537	5845 ± 160	4939 (4750) 4520	Carbonized wheat grains	Vinča	Find 326 C, depth 2.98–3.48 m, Late Vinča-Vinča D2 collected by Vasić given to Waterbolk by Gavela	Gimbutas (1991:450)[1]
GrN 1542	5710 ± 90	4717 (4553) 4465	Carbonized wheat grains	Banjica	Unknown	Gimbutas (1991:450)
GrN 1537	5845 ± 160	4934 (4750) 4520	Possibly carbonized wheat grain	Valač	Unknown	Gimbutas (1991:450)[2]
HAR 3222	6230 ± 100	5000	Carbonized wheat grain – emmer and einkorn	Staro Selo/ Selevac	Trench 15/ Quad. 1/2 boundary, Level 10, Feature 60	Tringham and Krstić (1990:50–53)[3]
LJ 2521	6080 ± 70	4790–4870	Carbonized grain from Silo A	Staro Selo/ Selevac	Trench 7, Level 2, Silo A	Tringham and Krstić (1990:50–53)[4]
Z 233A	6113 ± 80	4900	Carbonized grain from Silo A	Staro Selo/ Selevac	Trench 7, Level 2, Silo A	Tringham and Krstić (1990:50–53)
Z 233B	6152 ± 90	4925	Carbonized grain from Silo A	Staro Selo/ Selevac	Trench 7, Level 2, Silo A	Tringham and Krstić (1990:50–53)
Z 233	6366 ± 100	5110–5130	Broken carbonized grains	Staro Selo/ Selevac	Trench 7, Level 2, Silo A	Tringham and Krstić (1990:50–53)

[1] Cited in Todorović and Cermanović (1961) as Lab Number GRO 1.537 dates 5.605 ± 160 B.P. from the letter by Waterbolk 1957.
[2] Cited in Todorović and Cermanović (1961) as Lab Number GRO 1.536 dates 5.430 ± 120.
[3] Grains identified by Hubbard and McLaren in Tringham and Krstić (1990:50–53).
[4] Some grains identified by Hopf (1974).

CHAPTER 3. CONTEXTUAL ANALYSIS OF PLANT REMAINS

The spatial distribution of the plant remains within a site is a key component for assessing context in which plant remains were found. The identified plant taxa from the Opovo archaeobotanical samples are grouped according to the features in which they were found and discussed within their archaeological context. The results represent the analysis from 15 features consisting mostly of pits, but also of floors of two houses (House 3 and 4), an oven floor, a bedding trench, and a posthole (Figure 1.9). Most plant remains were recovered from the deeper layers, plant remains from the upper layers having been destroyed. The most prolific samples were from pits. Several of the larger pits probably originated as borrow pits for daub needed for house construction. Probably later, the pits were filled with garbage and in many cases followed by in situ burning, then capping with house rubble. Very few plant remains were recovered from house floors despite flotation of large number of samples from houses. Initially, samples were analyzed according to their feature provenience, regardless of building horizons (BH 1, BH 2, BH 3). The features were later grouped in the tables according to the feature assignments as suggested by Ruth Tringham (personal communication 1995, 2004), although for some pits it remained unclear whether they belonged to the second or third building horizon.[1]

In the "Archaeological Description" section, I provide basic information about feature provenience (Figure 1.9). I selectively emphasize certain layers and loci and what was found in them, providing additional information about other material from the loci that yielded the most prolific plant samples.

In the following section "Archaeobotanical Description", I give information about the amount of sediment floated per feature and percentages of the samples analyzed, in addition to a few general remarks about plant remains found in features (Tables 2.1–2.4; Figures 3.11–3.13).

In the section "Taphonomy and Interpretation", I attempt to reconstruct the mode of arrival of plant remains to answer how these plants came to rest in each of the features. Further, I proceed with the interpretation of the results. I also compute the distribution percentages for each large plant category per feature. In addition, I reconstruct past agricultural activities within each feature's detailed context, including the possible location of the field, hedges, or marshlands. In this section, repetition of plant ecology cannot be avoided because often a similar plant spectrum was found in various features. Also, I attempted to determine the function of the pit and its use-life, including the seasonality pattern. It was very difficult to assign seasonality of the features on the basis of plant remains, especially in continental Europe where first fruits cannot be picked before late May, most crops are harvested in June-July, and wild plant gathered by September-October. During winter, only storable plant food can be consumed. Therefore, if we do not take into account deposition of stored food, all the features exhibit rather similar seasonality use-patterns: from late spring until early fall.

In the last section "Animal Bone", using Russell's data (1993), I briefly enumerate the main animal taxa that were found in each feature, and the seasonality pattern that she reconstructed on the basis of bone analysis. For bone frequencies, Russell mostly relied on diagnostic zone (DZ) quantifications defined by J. Watson (cited in Russell 1993:107) as the best approximation for the species balance. The number of diagnostic zones is sometimes considerably different from the number of identified specimens (NISP) and the minimum number of individuals (MNI). Thus, in some instances, it seemed to the excavators that a feature contained quite a large quantity of animal bones, while in Russell's (1993) description of the same feature a small number of diagnostic zones per taxa were identified. These discrepancies are sometimes reflected between the description of the bone abundance in the archaeological section and the bone section of certain features. I include this information about animal remains for each feature to determine whether there is any correlation between bone and plant deposition and seasonality patterns.

Features

Feature 21: Floor (House 3)

Archaeological Description

Feature 21 is a floor and superstructure (primary deposition of burned rubble) of House 3 dating to BH 2 (Figure 1.9). It is located in the middle part of the excavation block in the quadrants 64-65/113-114. In this house, a fair amount of broken pottery was found that may have been deposited after the house was destroyed.

Archaeobotanical Description

From Feature 21, 20 liters were floated from the floor and under the house superstructure from excavation of layer 6 (Table 2.1). All light fraction samples were analyzed. One sample from locus 1605 of layer 6 contained only a small amount of *Triticum* sp. seed fragments and no other archaeobotanical remains. A sample from locus 1591, also from layer 6, contained only one unknown seed that resembled a *Spergula*, but the seed could have been of modern origin and was hence included in the Unknown category. No wood was recovered either.

Animal Bone
In House 3, the most abundant bones were of red deer, followed by cattle, pig, roe deer, sheep/goat and dog bones.

Taphonomy and Interpretation
In sum, there were very few plant remains left to aid understanding of any taphonomic processes that took place in House 3. It appears either few crops were processed within the house, or that postdepositional did not preserve the plant remains.

Feature 26: Pit with house rubble topping

Archaeological Description
Feature 26 is a pit abutting House 3, and both house and the pit probably date to BH 2 (Figure 3.1). Feature 14, which capped the center of the pit, may have been house rubble topping this pit. The pit is about 4 m long, and touches the eastern profile of the excavation block. It is located in quadrants 64-65/111-112 (Figure 1.9). There is no evidence of in situ burning, but it contains a dozen layers of ash that were mixed with garbage. Therefore, it seems that the cleanings of ovens and kilns were regularly dumped into this pit. The pit was rich in archaeological material that included flint flakes, axe fragments, bone tools, and ceramics. House rubble was mostly concentrated in layer 7. A human ulna was found in layer 2, but no archaeobotanical material was recovered from the upper layers.

Archaeobotanical Description
From Feature 26, 230 liters of sediment were floated, and it is among the pits from which the largest samples of sediment were floated. All the light fraction samples were analyzed, and heavy samples from 210 liters (91%) out of the total 230 liters were also examined (Tables 2.1–2.3). All the samples came from layers 7–10, with the richest ones were from layers 8 and 9, from loci 1654, 1845, and 1696. All the plant remains were carbonized.

Wood
Wood was also present in most of the light fractions and wood abundance was rated 1 on a scale of 1–3. However, there was also a large quantity (3.59 g) in the heavy fraction samples.

Crops
In this feature, the most abundant seeds were cereal grains. There were 25 *Triticum dicoccum* and 15 *Triticum monococcum* seeds identified from the light fraction samples. In addition, 13 *Triticum dicoccum* spikelet forks were recovered from the light fraction samples, and 13 *Triticum* spp. spikelet forks from the heavy fraction samples. Despite the number of retrieved cereals, the density of *Triticum dicoccum* (10.87 per 100 liters) and *Triticum monococcum* (6.52 per 100 liters) is not high considering the large quantity of sediment that was floated from this feature (Tables 2.1 and 2.2). Among the Poaceae, one-half of an elongated seed was preserved.

Figure 3.1. Feature 26: pit with house rubble topping (Tringham ©Creative Commons License).

Dimensions were taken for five whole seeds of *Triticum dicoccum*. The average dimensions were length 5.78 (5.50–6.00) mm, breadth 2.40 (2.00–2.80) mm, and thickness 1.88 (1.88–1.50) mm. Four *Triticum monococcum* seeds were also measured. The average dimensions were length 5.40 (5.00–6.00) mm, breadth 1.53 (1.20–1.80) mm, and thickness 2.48 (2.00–2.80) mm. The dimensions of the few measured seeds were within the expected ranges of seeds for these species, and were also comparable to the seeds from other Opovo features.

Two *Lens culinaris* seeds and one Fabaceae seed were identified. The latter one was probably *Pisum sativum*. The *Lens culinaris* seeds were recovered from the same locus (1654) from level 8. Although this locus yielded most of the cereal grains, in the samples in which the lentil seeds were found, there were very few cereal seeds and abundant of wood. One seed of *Linum usitatissimum* was recovered, but from a different locus (1696). In the locus where the flax seed was found, no cereal grains were recovered; however, there was a large quantity of wood.

Fruits
Only eight fruit seeds were identified. The most abundant were *Cornus mas* stone fragments, but when extrapolated to the whole number of stones, the equivalent of only four drupes were recovered. One *Sambucus* sp., one *Rubus* sp., and two *Fragaria* cf. *vesca* seeds were also identified.

Water Chestnuts
Sixteen *Trapa natans* fragments were found in the light fraction samples and three fragments in the heavy fraction samples (Figure 2.19).

Weedy Herbaceous Plants
In total, 21 seeds of weedy herbaceous plants were recovered from this feature. Among the weeds, the most

numerous were Solanaceae seeds. They were similar to the Solanaceae seeds from other features, although no mineralized seeds were recovered. From the light fraction, 12 Solanaceae seeds were identified, and from the heavy fraction, 12 additional seeds were found. Only one *Bromus* sp. seed and four *Polygonum* sp. seeds were found. All four *Polygonum* seeds were the three-sided type and three belonged to the *Polygonum convolvulus* species. The fourth *Polygonum* seed could not be identified to species, but probably was not *Polygonum convolvulus*.

Unknown

The Unknown category included three specimens from the light fraction samples. Among the unknown seeds, a fragment of a fruit type drupe ca. 6×6 mm large was found, but I was not able to identify it more precisely. One seed may belong to the Astereceae family, and one grass seed resembling the *Panicum*-type was also included in this category. In one of the samples from locus 1696, small, rounded, carbonized items reassembling spores were found. One of the light fraction samples from locus 1654 contained the largest number of various unidentifiable fragments, totaling 0.06 g.

Taphonomy and Interpretation

In sum, plant remains from Feature 26 consisted mostly of wheat seeds (40%) and wheat glumes (13%), seeds of weedy herbaceous plants (20%), and small quantities of water chestnut fragments, fruit seeds, and legume seeds (Figure 3.11). This combination of plant taxa is difficult to interpret. In total, there was a low density of both *Tritium monococcum* and *Triticum dicoccum*, but, proportionate to the number of cereal grains, however, the density of wheat spikelet forks was high (Table 2.2). It is difficult to say, whether these wheat seeds where charred while being hulled during parching process, or whether chaff fragments were thrown into the pit separately from cereal grains. The samples that contained a majority of *Triticum dicoccum* seeds did not contain the majority of *Triticum monococcum* seeds. However, most *Triticum monococcum* seeds were recovered from level 7, and most *Triticum dicoccum* seeds as well as most spikelet forks came from level 8. Whether the present *Triticum monococcum* and *Triticum dicoccum* seeds were harvested in the same field, or were cultivated separately, is difficult to determine.

The presence of lentil seeds and a pea seed is indicative that Opovo inhabitants consumed legumes, but because the legumes were not found in the same sample with cereal grains, it is possible that the legumes were also intentionally cultivated in separate fields.

One *Bromus* seed as a weedy intrusion is not strong enough evidence for suggesting possible field infestation. In contrast, quite a few Solanaceae seeds were recovered from the heavy and light fraction samples compared with the number of other identified seeds. These seeds came mostly from two samples, and in one case, only one *Triticum dicoccum* seed was recovered together with the Solanaceae seeds. The lack of wheat seeds in the samples where Solanaceae seeds were recovered might suggest that Solanaceae seeds were deposited separately from the cereal grains in this pit. It is unclear, whether and how Solanaceae plants were used. On the other hand, four *Polygonum* seeds were found in the samples where cereal grains were noted, thus implying that they were weeds harvested with the emmer crop.

The presence of carbonized Cornelian cherry stones and water chestnut fragments in the pit where there was no evidence of in situ burning might indicate that processing of these two types of fruits were exposed to heat prior to their disposal. Thus, these indigestible fruit parts were possibly thrown into house fires and later dumped with ashes into the pit.

In sum, it seems that a little of everything was thrown into this pit. As already mentioned, probably ashes from cleanings of ovens and random sweepings (discrete events) were regularly dumped into this pit. The density of recovered specimens is not high for any of the identified taxa (Table 2.2). On the basis of the taxa present, I was not able to determine whether the cereals were autumn or spring-sown, nor what the ecology of the cereal fields might have been. Also, in terms of seasonality, the presence of strawberry seeds would indicate late spring or early summer deposition, while the presence of Cornelian cherry and water chestnuts would indicate an autumn deposition. However, in one sample, seeds of *Rubus* sp. and *Fragaria vesca* and a fragment of a *Cornus mas* stone were found together. The fact that these three taxa were present in the same sample from the same layer does not allow us to speculate about seasonality use of one layer of the pit, but rather suggests that this pit was used from late spring to autumn. It could have been used during the winter as well, since cereals and dried fruits could have been consumed and deposited year around.

Animal Bone

Among the bones identified in this pit, the most numerous were of red deer. Quantities of fish bones were also found in layer 8. The following most represented taxa were pigs and cattle. Bones of roe deer, sheep/goat, and aurochs were also identified (Russell 1993:Figure 8.18). Russell was not able to determine any seasonal pattern of use of this pit; she believes there were many separate episodes of deposition in accordance with the numerous ash layers.

Feature 27: Floor of House 4

Archaeological Description

Feature 27 is designated as floor and superstructure (primary deposition of burned rubble) of House 4, which belongs to BH 3. This is a house ca. 6×6 m in size, located in the center of the excavation block and oriented northeast-southwest (Figures 1.9. and 1.10). Since two clay floors were discovered, the excavators suggested that the house might have been two stories tall or that the second floor represented some sort of mezzanine. Part of the lower floor was painted with several layers of pale yellow paint, and the walls were decorated with red triangle-and-spiral designs in relief. There were no internal divisions or an indoor oven, otherwise characteristic for Vinča houses. In typical Vinča houses, a large quantity of pottery is found, and that was also the case in Opovo House 4. Since excavators found numerous pots but no oven or internal divisions, it was suggested that the house could represent a facility for storage of food and other commodities (Tringham et al. 1992).

Archaeobotanical Description

During the analysis of plant remains, the material from this feature was treated as two separate sets of samples coming from Feature 19 and Feature 27, respectively. In the final analysis, following the excavator's suggestion (Ruth Tringham, personal communication 1995), the material from these two features was merged into one feature: Feature 27. In total, 160 liters of sediment were floated (Tables 2.1 and 2.2). The matrix that was floated consisted predominantly of house rubble from the floor, and several samples had to be pre-sieved before floating in order to extract big chunks of rubble. In addition, sediment from inside and around the pots was also floated without pre-sieving. All the light fraction samples were analyzed, and heavy fraction samples from 60 liters (37%) of the total 160 liters were also examined (Table 2.3). Several samples contained some kind of vitrified clay. From the western part of the feature (quadrants 60-61/112-115), the samples were taken from the upper layers 5-7. For the central part (quadrants 62-63/113-114), the samples were taken from the deeper layers 6-7. The house in general contained very few plant remains and wood. All the seeds were carbonized, except for five mineralized *Lithospermum officinale* seeds. Some of the samples contained no archaeobotanical remains at all, and some contained only modern weed seed intrusions.

Crops

In total, only 10 *Triticum dicoccum* and five *Triticum monococcum* seeds were found. *Triticum monococcum* seeds, like many other einkorn seeds from the Opovo samples, did not have pronounced dorsal ridges; in that respect, they resembled *Triticum dicoccum* seeds. One *Triticum monococcum* and one *Triticum dicoccum* glume base were recovered from the light fraction samples, and an additional two *Triticum* sp. spikelet forks from the heavy fraction samples were found. Two Poaceae seeds were also retrieved, but I was not able to determine whether they were seeds of some other cereal or merely wild grass seed intrusions.

Fruits

Only five *Cornus mas* stone fragments and one *Rubus* seed were found from the light fraction. In addition, one *Cornus mas* fragment was identified from the heavy fraction samples.

Weedy Herbaceous Plants

In this category, I included five mineralized *Lithospermum officinale* seeds from locus 1492 (60/114) from layer 5. This locus is described as inside and around the whole pot. The seeds were not carbonized, but they were clearly not fresh and hence were determined to be mineralized (Figure 2.20). However, the seeds may not have been weedy intrusions, but rather collected for some other purpose, possibly medicinal or ritual.

Unknown

In this category, six specimens were classified. One small seed was ca. 1.0 mm in length and has a distinctive texture on the surface, but it was unknown to me (Figure 2.30). The other two seeds resembled Boraginaceae seeds, but because they looked fresh, they were classified in this category.

Taphonomy and Interpretation

In sum, despite the large amount of sediment floated from this feature, very few plant remains were recovered. However, proportionally, cereal grains (65%) were the most abundant category (Figure 3.11). These few cereal remains do not imply any kind of storage or cereal processing that might have taken place within the house. Thus, even if House 4 had been a storage place, cereal grains either were all used up, or not stored there, or else were removed in light containers (bags) before the house was destroyed. Even if the seeds were removed prior to the destruction, one would expect more seeds spilled onto the house floor, unless the Opovo inhabitants were exceptionally careful about the cereal grains, or unless postdepositional processes allowed little to remain on the house floor.

The discovery of mineralized *Lithospermum officinale* seeds is especially interesting. If these seeds were contemporaneous with House 4 contents, then we can suggest that the seeds were deliberately collected. As the species name *officinale* indicates, the plant has medicinal properties and it used for eczema and other kinds of skin rashes (Bown 1995). Also, the plant has very attractive ivory-colored seeds that could have been collected deliberately for ornaments, or for some kind of ritual use (one of the local names is *biserka*, meaning pearl). *Lithospermum officinale*, however, could have been a ruderal perennial plant, and its seeds may have accumulated in the deep layers due to animal activity. According to Ellenberg's indicator values (1979:79), *L. officinale* thrives in non-acid, almost basic soils and thus could have grown in the vicinity of the site. Nowadays, the plants can nowadays be found growing in open, oak forests, and such forests may have existed in the Opovo environs.

Animal Bone

Relatively few bones were found in this house. Among the recovered bones, red deer was the most abundant, followed by a small quantity of pig and dog bones (Russell 1993: Figure 8.5). Russell suggested that bones were secondarily deposited together with house rubble and probably mixed with midden. The identified plant remains might have come from the same secondarily deposited provenience.

Feature 30: Pit/Well

Archaeological Description

Feature 30 is a 3 meter-deep and 2 meter-wide cylindrical pit in the northeast part of the excavation block and belongs to BH 3 (Figure 3.2). It was filled from top to bottom with house rubble. The upper, eastern part of the pit had a heavy concentration of bones with some garbage that was sealed with unburnt clay below House 2 (quadrants 63/115), dating to BH 1 (Figure 1.9). The pit is interpreted by the excavators as a well because it was deep and extended below the present water table. A part of a wooden log found near the bottom of the pit may have been the wooden ledge of the well. The pit was filled with unweathered loose fragments of burnt clay rubble, presumably originating from the destruction of House 4 (previously labeled House 5 in Tringham et al. 1992:365). In the deeper layers, fragments of oven floor and a cupola were found. In the upper layers, grindstone fragments, a copper fragment, obsidian flakes, and a clay amulet were found. From layers 6 and 7, flint sickles, cores, grindstones, polished axes, and bone tools were recovered. Layer 16 consisted of a lens of gray clay that probably represents primary deposition in a well. Russell (1993:78-79, 83) believes that this pit might have been a well dug from BH 1 rather than BH 3. Russell also thinks that the well was located in the interior of House 2, which served as a protective structure around the well. Archaeologically, House 2 consisted of a concentration of house rubble and did not have interior wall divisions or ovens. This interpretation seems logical because normally houses were not built over wells, according to Russell (1993). However, the well could have been filled up decades or centuries before the BH 1 structures were built, and perhaps the Opovo inhabitants forgot where the well was located.

Archaeobotanical Description

The floated samples from this feature come from two distinct areas that were both designated as Feature 30 during excavations. The samples from the eastern part of the pit were from layers 6 and 7 in quadrants 64/115-116. The samples from the western part that were interpreted as the well were quadrants 63-64/115, from deeper layers 10, 11, and 16. In total, 75 liters of sediment were floated, and all the light fraction samples were analyzed (Table 2.1). In addition, the heavy fraction samples from 10 liters (13%) of the total 75 liters were also examined (Table 2.3). The eastern part of the pit had very little plant material, consisting of a few cereals and unidentified fragments. The most prolific samples were from loci 1756 and 1782, from layers 10 and 11, respectively, from which mostly fruit seeds were recovered. On the other hand, from locus 1910 in layer 16 only wood was recovered. Most of the seeds were carbonized, except one *Sambucus ebulus* seed and one Solanaceae seed, which were mineralized.

Wood

This feature in general had very little charcoal preserved, except from level 11 in locus 1782, and from locus 1920 from level 16, which were located in the western part of the pit. Due to the large quantity of the wood retrieved from these samples, the average abundance for the entire feature was 2. Most of charcoal (5.69 g) was from a heavy fraction sample in locus 1782. Most wood charcoal from locus 1782 (see Table 4.2) was identified as *Quercus* (oak), and one fragment as belonging to one of the Salicaceae species (willow) according to Neal Lopinot (personal communication 1997). The oak wood fragments might be part of the ledge that was found near the bottom of the well. Oak is the most commonly used wood for well constructions because of its durability (personal observation).

Crops

Only two *Triticum dicoccum* seeds and a few Cerealia fragments were retrieved from this pit. One *Lens culinaris* seed was identified from one sample.

Fruits

A total of 44 fruit seeds was found, this being the most abundant archaeobotanical category in the feature. Although only a few *Cornus mas* stone fragments were found in the light and heavy fraction samples (Figure 2.13), many more strawberry seeds were recovered (Figure 2.18). All 40 strawberry seeds were found in only two samples in loci 1756 and 1782. In these samples, the seeds of *Rubus* and *Sambucus ebulus* (Figure 2.14–2.15) were also found. One of the two identified *Sambucus ebulus* seeds was mineralized, while the others were carbonized. The dimensions of the carbonized *Sambucus ebulus* seeds were 1.9 mm in length, 1.0 mm in width, and 1.8 mm in thickness.

Weedy Herbaceous Plants

Seeds of only two taxa were identified in this feature: three *Polygonum* seeds and six Solanaceae seeds. Most of these

Figure 3.2. Feature 30: pit/well (Tringham ©Creative Commons License).

seeds came from the same loci where fruit seeds were found.

Unknown
One small unidentified seed was included in this category.

Taphonomy and Interpretation
As indicated above, most of the plant remains were recovered from the western part of the pit in the deeper layers. In the samples from this section, the most abundant seeds were of fruits (78%). The majority of the recovered fruits were actually strawberry seeds (Figure 2.18). Although the 40 strawberry seeds seems like large quantity, it is less than the usual number of seeds per single fruit. One single strawberry fruit can have more than 100 seeds. The presence of the strawberry seeds perhaps indicates deliberate gathering of fruits and may also imply the seasonality of layers 10 and 11, where they were found and dates them to late spring-early summer, although berry fruits could have been dried and used afterwards. *Fragaria vesca* plants are one among pioneer species in cleared woodlands. Commonly, they grow in lighter spots of newly cleared soil that is rich in nitrates. Such land could have been present in the Opovo environs where strawberry plants could have thrived.

It is difficult to interpret the presence of a few *Polygonum* and Solanaceae seeds in this feature where almost no cereal remains were found. Usually *Polygonum convolvulus* seeds are interpreted as weed seeds in cereal crops. Similarly, it is difficult to interpret the presence of Solanaceae seeds. Whether these seeds were separate intrusions, or whether they indicate some specific use of Solanaceae and *Polygonum* plants, I was not able to determine.

On the basis of the presence of fruit seeds and the wooden ledge at the bottom of the feature, one might romantically conclude that someone was washing fruits on the edge of the well that had a ledge made of oak wood and dropped some of the fruits into the well. Afterwards the seeds might have been burnt in situ when the well was filled. But the situation is even more complicated if we consider the mode of arrival of these carbonized seeds, other than burning in situ, for which there is not enough evidence. Then it is possible that the seeds were previously burnt and arrived in carbonized form with pit infill. Only one uncarbonized seed was recovered from these samples. What kind of use or processing of these plants would have led to carbonization of weed and fruit seeds? It is possible that the seeds were included in the garbage infill of the pit. However, in that case, one would expect more diverse plant remains, including cereal or water chestnut fragments, and not just fruit and small weed seeds.

Animal Bone
Regarding the bone analysis, it seems that there were two features within one that was interpreted as Feature 30. Thus, levels 5 to 7 represent a small pit full of animal bones that was perhaps cut through the filled well. The well was filled with house rubble and was much deeper than the pit above. Russell (1993:431) believes that the contents of these two features are separate, although they could have been two components of a single feature. Russell's ascertain would indicate that garbage was added along with house rubble in the same pit. Among the identified bones, the most abundant were of red deer, followed by cattle, pig, roe deer, sheep/goats, dogs, and aurochs (Russell 1993:Figure 8.20).

Feature 31: Pit with Rubble Topping

Archaeological Description
Feature 31 is a pit about 3 m deep located in the northwestern corner of the excavation block in quadrants 58-60/117 (Figures 3.3 and 1.9). It dates to either BH 1 or BH 2. Well-preserved bones, ceramics, stones, charcoal, and lenses of ashes were excavated from the lower levels. In layer 8, flint blades, scrapers, an obsidian fragment, a grindstone, axe fragments, bone tools, and fragments of clay amulets were found. From layer 10, locus 1902, a stone axe, an amulet, and a complete ceramic altar were recovered. In layer 11, loci 1905 and 1906, a large amount of flint including both cores and blades were excavated, in addition to ceramic figurines, pottery, and amulets. Also, bone tools, a red deer skull, and fish bones were retrieved. Above these layers, there were one or more layers of in situ burning which contained chunks of house rubble. Above these layers, the pit was filled with more house rubble containing very few cultural remains, although a copper bead and a fragment of decorated house rubble were found here. In the upper part of the pit, burnt and unburnt materials were excavated, including fragments of a human skeleton. According to Russell (1993), it seems that bones were not cremated, but were a secondary deposition of human and animal bones that had been previously fragmented.

Archaeobotanical Description
From Feature 31, 195 liters of sediment were floated. Each of the light fraction samples was analyzed, and heavy fraction samples from 115 liters, 58% of the total of 195 liters of sediment were also examined (Tables 2.1–2.3). Impressions of cereal chaff were observed on the fragments of the house rubble contained within the heavy fraction samples. This pit contained numerous spikelet forks and glumes of *Triticum monococcum*. Most of the spikelet forks were found in two samples only, one from layer 10 and the other from layer 11. In comparison with the number of spikelet forks, there were far fewer *Triticum dicoccum* and *Triticum monococcum* seeds. Only layers 7–12 and 14 contained archaeobotanical material. The most prolific samples were from loci 1902, 1905, and 1906 from layers 10–12 in quadrants 58–59/117. A sample from locus 1906 was exceptionally burnt and contained relatively few plant remains. The remains of Cornelian cherry stones and water chestnut fragments were distributed mainly in layers 7–8. All of the seeds were carbonized.

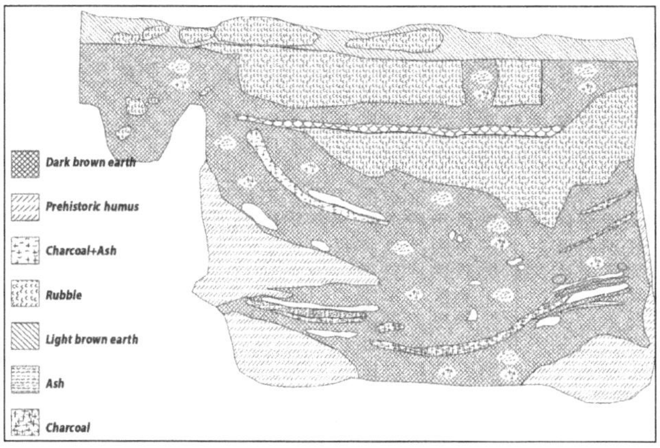

Figure 3.3. Cross section of Feature 31: pit with rubble topping. Scale 1:40. (Tringham ©Creative Commons License).

Wood

All 14 samples from this pit contained wood. Wood abundance was rated 2 on a scale of 1–3, although in the majority of the deeper layers the wood abundance was rated 3.

Crops

This feature, as mentioned above, contained mainly byproducts of crops rather than crop seeds. There were 353 *Triticum monococcum* spikelet forks identified by absolute count, and the density was 181.03 per 100 liters (Tables 2.1 and 2.3). A single spikelet fork was very light, and 155 *Triticum monococcum* spikelet forks (from one sample) weighed 0.13 g. Additionally, there were six *Triticum dicoccum* spikelet forks retrieved. It is possible that some of the *Triticum dicoccum* spikelet forks/glume bases were misidentified as *Triticum monococcum*, but most of the spikelet forks were narrow, which is generally a characteristic of *Triticum monococcum*. Thirteen *Triticum dicoccum* seeds were identified, and all came from layer 8. Measurements of five whole *Triticum dicoccum* seeds were taken. The dimensions were on the average length 5.55 (5.50–6.20) mm, breadth 1.60 (1.50–1.70) mm, and thickness 2.05 (1.80–2.30) mm. These seeds were narrower than the average *Triticum dicoccum* seeds from other Opovo features. Furthermore, there were three wheat seeds from locus 1770, layer 8, and two of the seeds were very small. One of the three seeds resembled *Triticum compactum* type seed, measuring 3.5 mm in length, 2.30 mm in breadth, and 1.7 mm in thickness. The unusual small, narrow seeds may have been byproducts of sieving of the main crop. None of the legume or flax seeds were recovered from this feature.

Fruits

In total, 12 fruit seeds were recovered from this pit. Most of the fruit remains were fragments and whole stones of *Cornus mas*. Altogether, 21 fragments were recovered, which is equivalent to approximately four whole stones or fruit drupes of Cornelian cherry. In addition, three *Rubus* sp. seeds (Figure 2.17) and five *Fragaria* cf. *vesca* seeds were found.

Water Chestnuts

Twenty *Trapa natans* fragments were identified form the light fraction, and one fragment from the heavy fraction samples.

Weedy Herbaceous Plants

Out of the total of 46 weedy herbaceous seeds, 39 were Solanaceae. Most of these seeds were from layer 8, in locus 1770, which is not the same locus where the spikelet forks were found. Five whole Solanaceae seeds (Figure 2.24) were measured and the dimensions ranged between 1–1.5 mm in diameter. Also, four *Polygonum* cf. *convolvulus* and four *Chenopodium* sp. seeds were recovered from this feature.

Unknown

Three specimens were included in this category. One of the seeds was possibly a modern seed of *Sambucus* sp. For the other two specimens, I was not positive whether they were seeds at all. A carbonized pedicel of small berry fruit was recovered and included in this category. It is possible that a pedicel was from the *Sambucus* berry. A rather large fragment of unknown plant was also included in this category.

Taphonomy and Interpretation

Plant remains from Feature 31 consist mainly of einkorn spikelet forks (78%). Weed seeds constitute 10%, fruit seed 3%, and water chestnut fragments 4% of total plant remains (Figure 3.11). Such a combination of various taxa indicates that at least the lower layers consisted primarily of einkorn byproducts. Einkorn spikelets had to be threshed and/or pounded to remove the tightly enclosing glumes. Whether the spikelets were heated by parching or were roasted for some other purposes remains uncertain. Wheat chaff could be used to start fires or to produce smoke for tanning. Because there was in situ burning observed in the pit, it is possible that some plant remains were burnt while they were already in the pit. There is another possibility that after the spikelets were pounded, the glumes were sieved out and thrown into house fires, and than the ashes from those fires were discarded in the pit. Because all the spikelet forks came almost exclusively from two samples, it is quite possible that they represent the results of one short dehusking session. Over 350 spikelet forks were found, equivalent to about 18 einkorn spikes at 20 spikelets per einkorn spike. Twenty spikes are equal to a handful of spikelets. Perhaps hulled spikelets were added, handful by handful, to a mortar for further pounding. If, on the basis of the present einkorn seed weight, we assume that one spike (weighing about 0.4 g) had 20 grains, we can then calculate the weight of grain produced by pounding a handful of einkorn spikelets. Such a handful of spikelets would produce about 8 g of einkorn grain. More than 125 handfuls of einkorn spikelets had to be processed to yield 1 kg of dehusked grain.

Because so few weed taxa were recovered from this feature, I was not able to estimate the sowing time of the einkorn crop, nor the kind of soil on which the einkorn field might have been located.

It is interesting to note that most of the 13 emmer seeds were not found in the same samples where einkorn seeds and einkorn spikelet forks were recovered. This might indicate that the two kinds of wheats are completely separate intrusions, because einkorn seeds and spikelet forks were found in levels 10 and 11, while emmer seeds were retrieved from level 8. Also, in the samples from level 8, Solanaceae seeds, Cornelian cherry stones, and water chestnut fragments were found, where no wheat spikelet forks/glume bases were recovered. This suggests that the Solanaceae plants might not have been weeds in einkorn fields, but rather the seeds arrived in the pit independently from cereal byproducts. This mode of arrival of Solanaceae seeds in this case provides strong evidence that the fruits or the seeds were intentionally gathered for food or medicine. Because none of the recovered Solanaceae seeds resembled seeds of the non-toxic Solanaceae species, I do not know whether the Solanaceae plants and seeds were deliberately collected for medicinal use or food, and if the toxins were removed through some sort of processing.

The presence of strawberry seeds would indicate late-spring early summer use of the pit. Cereal could have been processed year round, and Cornelian cherry stones and water chestnut fragments would indicate autumn gathering, although they could also have also been consumed through the year.

Animal Bone

In this feature, animal bones were abundant, and red deer bones were dominant, followed by cattle, roe deer, wild pig, dog, sheep/goat, aurochs, and hare (Russell 1993:Figure 8.14). According to Russell (1993), level 10 represents remains of animals killed in autumn-winter-spring, level 9 covers all year, level 8 summer-autumn, and level 6 spring and autumn-winter. However, there were no plant remains recovered from level 6.

Feature 34: Oven Floor

Archaeological Description

Feature 34 is described by excavators as an isolated oven that probably belongs to BH 3 (Figure 1.9), although there are no precise stratigraphic data to determine the horizon to which it belongs (BH 2 or BH 3). It is located in the middle of the northern part of the excavation block in quadrants 62/116–117.

Archaeobotanical Description

From this feature, 10 liters of sediment were floated (Table 2.1), but only one carbonized unknown seed was recovered from locus 1446 in layer 6 below the oven floor. The seed was not preserved well enough for precise identification, but it resembled a *Rubus* sp. seed.

Feature 41: Pit or Two Pits

Archaeological Description

Feature 41 is a large pit that is cut by a trench (Feature 42). This pit has a diameter ca. 3 m (Figures 1.9 and 3.4). It was sealed by unfired clay that formed the foundation of a later oven (Feature 13 of BH 2). Still later it was cut by Feature 42, which is a bedding trench of House 1 (BH 1). Feature 41 consisted of numerous layers of ash and garbage fill. It contained an articulated puppy skeleton and a partial dog skull somewhat lower in the same area (Russell 1993:84). Artifacts such as a head of a crude figurine, clay disk with a bull's eye design, and a fragment of seated figurine were also found in the pit. In addition, flint flakes, axe fragments, grindstones, bone tools, spondylus shell beads, ceramic pots, and other artifacts were recovered. From locus 1470, several flint flakes, and a polished stone axe flake were found. From locus 1679, a few flint flakes were retrieved. Excavated material from this pit was divided into west (58/109–110) and east (59–60/109–110) parts, but at first it was assumed that the pit was used and filled in one period: BH 3 (Figure 1.9).

Tringham thought that the pit was originally dug to obtain clay soil for the construction of House 4 or House 3 and later used for the disposal of rubbish (Tringham et al. 1992). Later, the authors supposed that the pit might have had two episodes of use-life (Ruth Tringham, personal communication 1997, 2004). The western part was used earlier (BH 3) and was intruded by a later pit in the eastern part (BH 2).

On the basis of archaeobotanical evidence, I was not able to determine whether there was a difference between the western part of the pit (BH 3), and the eastern part (BH 2). The richest samples came from the northern part of the pit, more specifically from the quadrants 58/110 and 59–60/109 in levels 7 and 8, regardless of their east or west provenience. Quadrants 59–60/109 also had the most diverse plant taxa.

Archaeobotanical Description

Feature 41 was among the first to yield a considerable amount of charred grain from the flotation of a large quantity of sediment. During the excavations, a few charred cereal grains were observed and the pit seemed promising. Therefore, 610 liters of sediment from it were floated. Flotation samples for archaeobotanical analysis were taken from both the western and the eastern parts of the pit from layers 7–12. All the light fraction samples were analyzed (Table 2.1), and heavy fraction samples from 525 liters (86%) of the total 610 liters of sediment were also examined (Table 2.3). The pit contained the richest samples in absolute count and density of all seeds per sediment volume (Tables 2.1–2.3). The most prolific samples were from layers 7 and 8 in quadrants 59–60/109 (locus 1679) and 58/110 (locus 1470), although additional samples were taken from deeper layers (down to level 12). Sample 88063 from locus 1679 yielded the highest frequency of various seeds, although only 5 liters of sediment were floated. Furthermore, in another sample 88078, also from locus 1679, there were many weed seeds but no cereal grains or spikelet forks/glume bases. Only a small quantity (0.09 g) of cereal fragments was retrieved. Both charred seeds and some mineralized seeds of the same species were identified together in the same samples. Nevertheless, most of the seeds were charred, but some mineralized seeds included Solanaceae, *Chenopodium* sp., *Polygonum* sp., *Fragaria* sp., *Sambucus* sp., and *Lithospermum* species. One *Triticum* sp. spikelet fork from a heavy fraction sample was not fully carbonized, either.

Wood

This feature was also rich in wood. Out of 41 samples, only nine did not contain charcoal. Wood abundance for the whole feature was rated 2 on a scale of 1–3. From locus 1470, in one sample (10 liters of sediment) 0.27 g of charcoal were retrieved, and in another sample from locus 1635 (20 liters of sediment), 0.42 g of charcoal were obtained. From heavy fraction samples, 5.95 g of wood were separated.

Crops

Feature 41 is the richest feature in cereal grains, by absolute count (601), density (98.52), absolute weight (13.26 g), and weight density (2.17 g) from the light fraction samples (Tables 2.1 and 2.2). The heavy fraction samples are the richest from this feature (Table 2.3). From this feature, 529 *Triticum dicoccum* seeds were recovered from the light fraction and 96 from the heavy fraction samples (Figure 2.4). Also, 44 *Triticum monococcum* seeds were retrieved from the light fraction samples and nine from the heavy fraction (Figure 2.5). Feature 41 is the only feature where seeds of *Hordeum vulgare* were recovered (Figure 2.10), mostly from layer 8, quadrants 59–60/109. However, only six *Triticum* spp. spikelet forks were found in the light fraction and an additional 16 *Triticum* spp. spikelet forks from the heavy fraction samples. One of the seeds from the heavy fraction sample was not fully carbonized. Several *Triticum* spp. spikelet forks from the heavy fraction resemble the spikelet forks of *Triticum monococcum*, but many of them

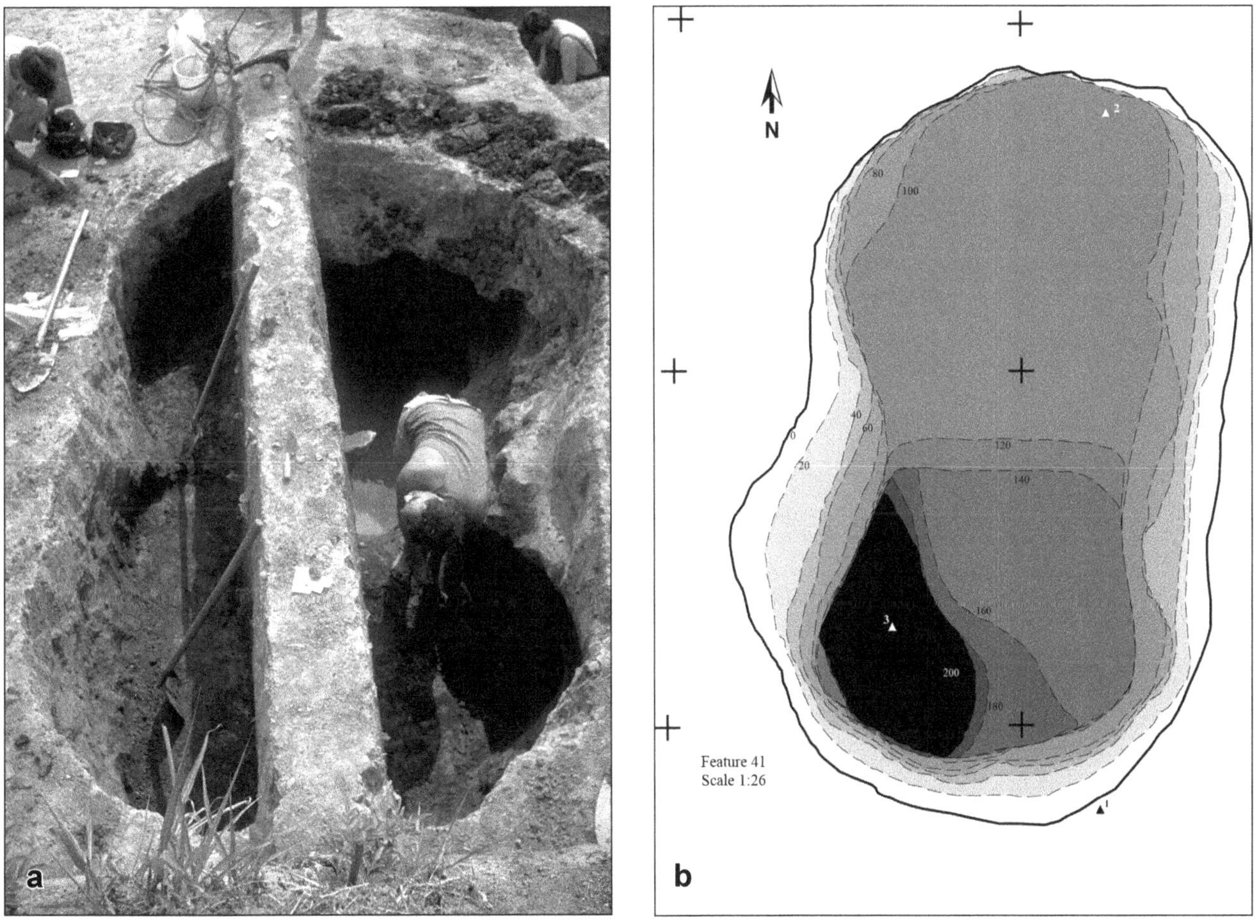

Figure 3.4. Feature 41: **a)** pit or two pits; **b)** diagram showing depths (Tringham ©Creative Commons License).

do not have pronounced second lateral nerves and were hence classified only as *Triticum* spp. Many of the cereal grains were fragmentary and could not be more precisely identified; thus, they were included in the Cerealia category (total weight 4.56 g). Among the Poaceae seeds, there was one small-elongated Poaceae seed probably of some weedy grass, and one rounded seed similar to a *Panic* species. One cereal seed fragment was preserved with tightly enclosing chaff. Despite the existence of chaff, it was not possible to identify the seed more precisely.

Most of the cereal grains were very burnt and some of them badly eroded and exploded. I was able to take dimensions of 56 *Triticum dicoccum* seeds (Figure 2.3). The average dimensions were: length 5.47 (3.10–6.70) mm, breadth 2.44 (1.50–3.10) mm, and thickness 1.87 (1.30–2.50) mm. These dimensions fall in the range of the *Triticum dicoccum* seeds, although their index L/B is rather high, indicating that the seeds were narrower than typical emmer seeds. *Triticum dicoccum* seeds from this feature comprise the majority of the other measurable *Triticum dicoccum* seeds, and they are similar to the dimensions of a few measurable emmer seeds from Feature 26 and 31. Three seeds of *Triticum monococcum* were also measured, and dimensions were very similar to the four measured seeds of *Triticum monococcum* from Features 26 and 46, although they are somewhat wider. An additional six *Triticum monococcum/dicoccum* seeds were measured. The seeds of the latter ambiguous category show characteristics of both *Triticum monococcum* and *Triticum dicoccum*, but when the length of the seeds was plotted against the thickness (Figure 2.3), these seeds clustered more around *Triticum monococcum* then *Triticum dicoccum* seeds, and were hence classified as such. Nine *Hordeum vulgare* seeds were measured, and average dimensions were length 4.98 (4.20–6.00) mm, breadth 2.82 (2.20–3.30) mm, and thickness 1.97 (1.80–2.30) mm. The seeds seemed to be of the *Hordeum nudum* type, although the characteristic wrinkling on the dorsal side was not apparent. Also, since the barley seeds were badly burnt, it was not possible with so few specimens to determine whether the seeds were symmetrical or asymmetrical, and I could not attribute them to either the *distichum* or to the *hexastichum* form.

Among the legume seeds, five *Lens culinaris* seeds were identified, and an additional 16 Fabaceae seeds which could not be more precisely identified. Among the 16 Fabaceae seeds, three were of *Vicia* type and one was possibly a *Pisum sativum* seed. All legume seeds were from the layers 7 and 8, where cereal grains were also found. In addition, two Fabaceae seeds were recovered from the heavy fraction samples; both seeds resembled *Pisum sativum* seeds.

Three *Linum usitatissimum* seeds were identified in layer 7 of the quadrant 58/100.

Fruits

A total of 51 fruit seeds were found in Feature 41. There were 40 *Cornus mas* fragments, out of which two whole stones were found, and from the weight of fragments 16 whole stones were extrapolated. Also, 23 *Fragaria vesca* seeds were found and one among them was not carbonized. Two seeds of *Sambucus* cf. *ebulus*, long-seeded type (Figure 2.15) were carbonized and one was uncarbonized. Seven seeds of carbonized *Rubus* sp. were identified, and one was not carbonized.

Water Chestnuts

In Feature 41, 43 *Trapa natans* fragments, including one diagnostic top part of the fruit, were found.

Weedy Herbaceous Plants

Among the weedy herbaceous plant seeds, the most numerous were Solanaceae seeds; 95 were carbonized and five were mineralized. In addition, 20 Solanaceae seeds were recovered from the heavy fraction samples and they were all mineralized. Most of the carbonized seeds had distinctive reticulation on the surfaces with big cells and thick cell walls. All of the seeds had the hilum placed laterally; in some seeds it was placed in the middle and in others more towards the top. The diameter of several measurable seeds ranged from 1.0 mm to 1.7 mm. The mineralized Solanaceae seeds are flatter and larger than the charred ones, but because the distinctive pattern of the coat was not preserved, it was impossible to identify them to the species level. Twenty-four *Polygonum* sp. seeds were identified and only one was not carbonized. Most of these seeds belonged to the species with three-sided fruits, and there were certainly several *Polygonum convolvulus* seeds among them. There were also very small three-sided seeds that might have belonged to some of the *Rumex* species. One *Polygonum* sp. seed resembled a *Polygonum hydropiper* seed. Two seeds were flat, and one with a diameter ca. 2 mm could be identified as *Polygonum lapathifolium* (Figure 2.27). Another *Polygonum* sp. seed was rather elongated, more than a typical seed of *Polygonum aviculare*. Altogether, 15 *Bromus* sp. seeds were found, and seven most probably *Bromus arvensis* (Figure 2.21). Several seeds were of a long type, and they were *Bromus* cf. *sterilis*. Six *Chenopodium* sp. and two *Galium aparine* seeds were identified.

Unknown

The Unknown category includes 17 specimens from the light fraction and four from the heavy fraction samples. One uncarbonized *Lithospermum arvense* seed and one *Spergula* sp. seed were also found, but because they may have not have been ancient, I included them among the unknown seeds. One whole seed and a half seed similar to the Asteraceae species were also included in this category. Two small seeds resembled the *Polygonum* or *Rumex* species. Two other seeds were preserved well and identifiable, but were unknown to me. The "Unknown" category also included, an uncarbonized fragment of hard shelled fruit is included, and a carbonized fragment of a fruit stone similar in shape to a *Cornus mas* stone, but with an entirely different cross-section. In addition, one carbonized small bud was found.

Taphonomy and Interpretation

In sum, 69% of plant remains from Feature 41 were cereal grains. Among cereals the most abundant were *Triticum dicoccum*, although many specimens were badly burned implying intensive heat. The next most common category was weed seeds, predominantly Solanaceae seeds. In contrast, only a few *Triticum* spp. spikelet forks were identified (1%). Legume seeds constitute 2% of the identified seeds, and fruit seeds 6%. *Trapa natans* fragments compose 5% of the sample. Such a combination of plant taxa indicates that Feature 41 (Figure 3.11) was a pit containing mostly burnt cereals which were probably charred during cooking or roasting, or during the drying. Because so few spikelet forks/glume base fragments were found, it seems that the pit was not used as a refuse pit for byproducts of emmer processing, but rather as a refuse pit for processed burnt grain. Similarly, *Triticum monococcum* and legume seeds could be regarded as contaminants of the main cereal that were not deliberately removed from the emmer crop. However, it is possible that *Lens culinaris* seeds were deposited in the pit, independently. It is also possible that the byproducts (chaff) from cleaning hulled cereals of the grains that arrived in this pit were used for fodder or as temper or else were thrown into another pits (e.g. Features 31, 59 and 61). During the excavations, it was supposed that the western part of the pit, which was clay lined and contained a large quantity of cereal grains, might have been a silo. Perhaps that pit was dug as a silo, but later used as a garbage pit, as indicated by various non-cereal remains, bones, and other artifacts found in it. Some of the very burnt cereal grains might have been left after fire-sterilization of the primary silo, however.

The seeds of *Bromus* sp. were quite large and likely were weed intrusions of an emmer crop that were not removed from the main crop by coarse sieving. *Bromus* species seeds belong to the type called "speirochore", i.e. weed seeds that are not aerodynamic and whose size is generally close to that of cereal grains. Those seeds are intermediate in size between the mesh of the fine sieve and that of the coarse sieve, so it is difficult to remove them (Lundstrom-Baudais and Bailly 1995:177). *Bromus arvensis* together with *Lolium temulentum* and some other weed seeds are indicative of autumn-sown cereals. However, some of the identified *Polygonum* species such as *Polygonum lapathifolium* are more common in spring-sown seeds.

Most of the weedy herbaceous seeds are composed of small Solanaceae seeds (diameter ca 1.3–3.0 mm) which could have been easily separated from the emmer crop by fine sieving, unless whole *Solanum* fruits were harvested still unripe together with the emmer crop, but the fruits could have also been hand-picked from the sieves. Most of the Solanaceae species fruits ripen from late summer to autumn when the cereals were already harvested. Therefore, it is not easy to explain the appearance of such small seeds together with prime grain cereals. Because not all of the Solanaceae seeds were identified to the species level, it is possible that more than one species was represented in this feature. Some of the Solanaceae species can be weeds of arable land, but several such as, *Solanum dulcamara*, are more common in meadow thickets and forest clearings. Often, many Solanaceae species like *Hyoscyamus niger* thrive on fertile refuse sites. Therefore, although Solanaceae seeds could have been part of the emmer crop assemblage and could have arrived with the harvested crop from plants growing as sagetals on the fringes of the fields, they also could have been a separate and rather independent intrusion in this pit. If the Solanaceae seeds are not part of a crop-weed assemblage, then it is difficult to explain how they arrived in mostly carbonized form in the pit. *Physalis alkekengi* has palatable fruits and perhaps the fruits were dried. It is possible that leaves were cooked in soups or stews, but it would be difficult to explain how the seeds occurred in pits. Since there is evidence that the *Hyoscyamus niger* seeds were used medicinally and in alchemical ointments since antiquity (Schoch et al. 1988:179); the preparations might have necessitated heating which accidentally resulted in seed burning. Indeed, burnt seeds are used in folk medicine for severe toothache (Lempaiäinen 1992:238). Or, if Solanaceae plants were undesirable poisonous plants thriving around the house and pit, they may have been burnt in the pit, because both carbonized and some uncarbonized seeds were found in the same samples. In addition, heavy in situ burning occurred in the quadrants where Solanaceae seeds were found.

On the other hand, is highly possible that *Polygonum* seeds were harvested together with the cereal crop, but one would expect to find more *Polygonum* seeds had the main cereal crop not been fully processed.

Similarly, it is difficult to explain the occurrence of fruit seeds as sagetals or intrusions from the fields that were previously left as fallow. One would postulate that berries of *Rubus*, *Sambucus ebulus*, and *Sambucus nigra*, and especially, of *Cornus mas* would have been easily removed from the main cereal, and then deposited in the pit. It is more likely that they were deliberately collected and arrived as separate intrusions in the pit. Such an interpretation seems plausible considering the presence of fruit seeds that were perhaps deposited separately from the cereal crop (although they may have been part of the fire ashes thrown from the house into the pit).

Quite a few of the small seeds were mineralized, and they mostly came from samples that were full of house rubble. All Solanaceae seeds from the heavy fraction were mineralized, thus indicating that they were deposited uncarbonized in the pit. Water chestnuts could have been eaten row, roasted, or cooked and the spiky husks removed by peeling and thrown into the house-fires (and later into the pit) leaving the inside starchy parts to be eaten.

Considering the different numbers of seeds per plant and the fact that the cereals from this feature had been processed before deposition, the number of weedy herbaceous plant taxa cannot be indicative of a weed infestation of the crop fields. However, *Polygonum convolvulus*, *Polygonum lapathifolium*, *Polygonum hydropiper*, possibly *Rumex* species, and *Galium aparine* were in general common weeds in the cereal fields, and some of the *Polygonum* species also could have been ruderals. If we assume that the weeds grew together with emmer in the fields or at their fringes, we might speculate about the ecology of the plot where the wheat was grown. *Polygonum convolvulus* does not have a high demand for humidity and can be a weed in either spring or winter crops, while *Polygonum lapathifolium* does not have a demand for warm places. *Galium aparine* thrives in wetter nitrogen-rich places such as river banks and lake-shores, but it is also a weed of winter crops. If some of the recovered *Chenopodium* sp. seeds belonged to *Chenopodium album* plants, they might have been weeds, ruderals on nitrogenous soil in the settlement, or else were deliberately collected.

According to Ellenberg's (1979) indicator values regarding the preference of the present weeds for nitrogen, acidity, and humidity, most of the identified plant taxa from Feature 41 are cosmopolites. Concerning humidity, there is a larger number of species, such as *Galium aparine*, *Polygonum convolvulus*, and *P. lapathifolium*, all indicative of wetter soils. If we assume that most of the weed seeds were harvested together with the emmer crop, then we can assume that there were dump areas in emmer fields.

Since the richest samples come from levels 7 and 8, it can be postulated that the plant remains ended up in the pit in a rather short period of time. If that is true, then the dominance of emmer seeds may be indicative of emmer cultivation and some of the present weed seeds can be explained as infestations in emmer fields. Several weed taxa can grow in either winter or summer crops, but on the basis of the presence of *Galium aparine* and *Bromus arvensis* seeds, which most commonly infest winter crops, I would suggest that the emmer was winter sown.

Lentils were probably cultivated and consumed, although represented by small numbers in this pit. Because lentils are usually spring-sown, it can be postulated that crop rotation was practiced, albeit difficult to prove. It is possible that the Opovo inhabitants rotated emmer with legumes. However, *Lens culinaris* is a crop that does not like soil with high humidity. The presence of some of the hygrophytic weed species among emmer seeds might indicate that legumes were not grown on the same plot where emmer was previously harvested. Thus, we can assume that legumes whose seeds were found in this pit were grown in different fields from the emmer crop.

Regarding the use life of Feature 41 on the basis of plant remains, the pit was clearly used most intensively used during the time of the site when layers 7 and 8 accumulated. In those layers, seeds of cereals, weeds, legumes, flax, as well as water chestnut fragments were all found together. In the deeper layers, there were fewer seeds, mostly cereal fragments. Therefore, it is difficult to establish the seasonal use of this pit, or which seasonal activities it preserved. The plant indicating the earliest seasonal use of the pit is strawberry. Strawberry fruits ripen in May. The seeds of elderberries, the stones of Cornelian cherry, and the water chestnut fragments would be indicative of fruit collecting in autumn. Some of the fruits like Cornelian cherry could have been dried and consumed during the winter. Cereals were harvested in summer, but could have been consumed and deposited throughout the year. The majority of the identified taxa including cereal, weed, and fruit seeds, and water chestnut fragments were found together in the same samples; thus, it would be difficult to determine different seasonality for separate layers, and they could easily represent a full year.

Animal Bone

Most of the bone specimens belong to red deer, roe deer, and wild pig. Among the domestic animals, the most abundant bones are of cattle (Russell 1993:Figure 8.21). Among the deer bones, there is a high number of red deer mandibles. Also, there is a large number of bones of immature deer in this pit. According to Russell (1993:432), levels 11 through 7 each appear to represent the full year. She determined that level 14 could be assigned to autumn, level 13 to autumn-winter-spring, and level 12 to summer. Because only a very few plant remains were discovered in these deeper layers, I was not able to infer their seasonality of use, but would agree with Russell that at least layers 7–11 represent a full year's deposition.

Feature 42: Bedding Trench for the Foundation of the South Wall of House 1

Archaeological Description

Feature 42 is a trench that is, according to the excavators, a foundation trench of the southern wall of House 1 (Figure 1.9). The house is dated to BH 1, and hence the foundation trench should be contemporaneous. The trench did not contain much archaeological material. It transects diagonally Feature 41 (Figure 3.4) which is dated either to BH 3 and BH 2. The foundation trench that cuts the pit stretches narrowly through quadrants 59/109–110. There are no results of faunal analysis reported separately for this feature.

Archaeobotanical Description

Feature 42 was among the first features to yield considerable amounts of visible plant material. Thus, 80 liters of sediment were floated, and all the light and heavy fraction samples were analyzed (Tables 2.1–2.3). The samples were mostly from three loci: 1493 and 1494, both from layer 7, and locus 1621 from layer 8. In this feature the most numerous remains were cereal fragments. Almost all seeds were charred except for a few that might be of modern origin and were classified as "Unknown". To facilitate flotation, some of the samples were presieved to remove large fragments of house rubble included within the sediment.

Wood

This feature did not contain large quantities of wood. The wood abundance was rated 1 on a scale from 1–3. In the heavy fraction samples, a total of 0.58 g of wood was recovered. In the light fraction, in only one sample from locus 1621, in layer 8, the wood abundance was rated 2.

Crops

From this feature, 15 *Triticum dicoccum* and three *Triticum monococcum* seeds were recovered. In heavy fraction samples, two *Triticum dicoccum* and *Triticum* sp. spikelet forks were identified. Most of the cereals were found in the samples from layer 7. Among the *Triticum dicoccum* seeds, one large seed was recovered: 6.7 mm in length×1.8 mm in breadth×2.8 mm in thickness. On the other hand, one small *Triticum monococcum* seed was found in the same sample. Every sample contained Cerealia fragments that could not be identified more precisely. Also one Poaceae seed was recovered from layer 8; it had a truncated top and resembles an *Avena*-type seed in gross morphology. No *Avena* species were domesticated by the Neolithic period, and it is possible that the seed belonged to some of the wild *Avena* species present in the local flora, such as *Avena sterilis*, *Avena fatua*, or others. In addition to cereals, one Fabaceae seed was recovered from both light fraction and heavy fraction samples. A Fabaceae seed from the light fraction samples resembled *Lens culinaris*; the one from the heavy fraction was probably a type of Fabaceae seed, but was classified as "Unknown".

Fruits

Three *Cornus mas* fragments were retrieved from the light fraction samples and one fragment from a heavy fraction sample, all from the upper layer of this trench. When extrapolated to the number of whole stones, only one Cornelian cherry stone is represented.

Water Chestnuts

Five *Trapa natans* fragments were found in the light fraction samples. No other fruit seeds were recovered.

Weedy Herbaceous Plants

Altogether six weed seeds were retrieved from this feature, including four Solanaceae seeds, one *Polygonum convolvulus* seed, and one *Bromus* long-type seed. The long-type *Bromus* sp. seed is tentatively identified as *Bromus sterilis*.

Unknown

In this category, four specimens were included for the light fraction and one specimen from the heavy fraction samples. Some of the unknown seeds were only partially preserved, and could not be identified more precisely.

Taphonomy and Interpretation

In sum, plant remains from Feature 42 consisted 60% of cereals and 19% of weed seeds, followed by water chestnut fragments and legume seeds (Figure 3.11). This feature contains all the same taxa as Feature 41, the pit it intersects. Taxa composition is also very similar when densities are compared. Thus, we may conclude that the contents of Feature 42, a foundation trench, are actually the same as the contents of Feature 41, a pit. It seems that the trench intersected the pit, Feature 41, only in the upper layers, including layers 7 and 8, and that the lower layers of the pit remained undisturbed. Although this trench is the only feature from BH 1 that yielded plant remains, those remains can be assigned to Feature 41.

Feature 45: Ash Pit

Archaeological Description

Feature 45 is a shallow ash-filled pit on the west side of House 4 (Figure 3.5). It lies below House 1 from BH 1, in quadrants 59/112–113 (Figure 1.9). The pit was assigned to BH 3. It is a medium-sized pit with a diameter less than 2 m. In the lower levels of the pit, a whole pot and an aurochs horn core were found mixed together with many layers of ash. Large quantities of charcoal were also recovered. Above the ash layers was a garbage layer without ash overlain by a small area of oven floor, which was disturbed. This layer was covered by a layer of in situ burning of soil that capped the pit. The capped layer was pierced by a postmold that is probably contemporaneous with House

Figure 3.5. Feature 45: ash pit (Tringham ©Creative Commons License).

1 from BH 1. It seems possible that in some stages of its use the pit was an oven floor or hearth floor.

Archaeobotanical Description

From this feature, 20 liters of sediment were floated, and all light fraction samples were analyzed (Tables 2.1–2.3). Two samples were taken from loci 1868 and 1882, in layers 9 and 12 respectively. Only the sample from locus 1868, layer 9, yielded archaeobotanical material.

Wood

Wood was recovered from one sample only, from locus 1868. The wood abundance was rated 3, on the scale 1–3. The other samples did not contain any charcoal.

Crops

Only one seed *Triticum dicoccum* and two spikelet forks of *Triticum monococcum* were recovered from the one locus 1868 sample.

Water chestnuts

Only two fragments of *Trapa natans* were identified.

Taphonomy and Interpretation

Plant remains from Feature 45 consist almost entirely of wood charcoal. I was not able to determine whether the ash in the deeper layers was a result of throwing ash discarded from house fires, or whether the pit was made especially for roasting and baking. The few cereal and water chestnut remains may be considered incidental intrusions and could point to the intentional use of this pit for roasting.

Animal Bone

In Feature 45, the most abundant remains were the bones of red deer, closely followed by roe deer, then by pig and cattle. Because of the small sample size, the seasonality pattern was uncertain; however, it is possible that there was only a brief deposition of bones in deeper levels dating from summer to early winter (Russell 1993).

Feature 46: Pit

Archaeological Description

Feature 46 is a medium-sized filled pit that is bell-shaped in the lower levels (Figure 3.6). It is located in the central southern section of the excavation block, and is dated to BH 3 (Figure 1.9). In the lower levels, it was filled with garbage mixed with much wood charcoal and ash. The pit contained ash lenses. Above this infill, there were two layers of wood charcoal mixed with ashes including chunks of yellow clay; each layer was covered with in situ burnt clay. The ash layers seemed to be without artifacts. Unburnt clay was on top of these layers and it may also have been part of the pit. It seems possible that during its life the pit was used as an oven or hearth floor. Bones of mammals and fish were found in locus 1724, layer 9. A human vertebra, an antler, a bone spatula, a clay amulet, a few flints, and fragments of ornamented pottery were found in locus 1724. Similar artifacts were found in locus 1753, in addition to a shell bead and a grindstone.

The pit was assigned to BH 3 according to the excavators, but according to Russell (1993:85), it might be contemporaneous with Feature 31, a pit in the northern part of the block that is dated to BH 1. Two parts of a composite antler tool, made from the same piece of antler and fitting together were found separately in these two pits (Feature 46 and 31). Russell believes it is unlikely that part of the tool from BH 3 was kept as an heirloom until the later occupation during BH 1 (Russell 1993). Regarding stratigraphic analysis of Feature 46, this pit starts too low in the profile to be assigned to BH 1; thus, both pits might belong to BH 3, and not to BH 1 (Feature 31) and BH 3 (Feature 46).

Archaeobotanical Description

From this feature, 105 liters of sediment were floated (Tables 2.1–2.3). All the light fraction samples were analyzed and almost all heavy fraction samples (90%) were analyzed as well (95 liters). The most prolific samples were from layers 9 and 10 of the quadrants 61/108–109. Feature 46 is the pit that yielded the largest quantity of water chestnut remains. Also, cereal, legume, fruit, and weed seeds were found. All the seeds were carbonized except mineralized grape seeds that were recovered from locus 1754, which was described as a small pit that was included in this larger pit. Most cereal and legume seeds were retrieved from loci 1724 and 1753, levels 9 and 10.

Wood

This feature was rich in wood, almost all samples containing some wood charcoal. Especially in the heavy fraction, there were samples in which large quantities of wood were found. The wood abundance is rated 2, on a scale from 1–3.

Crops

Ten *Triticum dicoccum* and 27 *Triticum monococcum* seeds were recovered from this feature, but no wheat spikelet forks were retrieved. Several *Triticum dicoccum* seeds are very narrow, and in several instances it was difficult to distinguish between the two wheat species. Two seeds were classified provisionally in the arbitrary category *Triticum monococcum/dicoccum*, and the average dimensions are: length 5.55 mm; breadth 1.70 mm and thickness 2.00 mm. When the lengths of these seeds were plotted against thickness, the values clustered among other *Triticum monococcum* seeds, and therefore they were classified as *Triticum monococcum* (Figure 2.3). Measurements of 14 whole *Triticum monococcum* seeds were taken. The average dimensions are: length 5.14 (3.20–6.70) mm; breath 1.66 mm (1.10–2.20) mm, and thickness 1.75 (1.00–2.20) mm. These *Triticum monococcum* seeds are rather slim, and are the narrowest in comparison to other *Triticum monococcum* seeds from Features 26 and 46. One small seed of Poaceae (3.5 mm long) with an embryo placed on its lateral side was also recovered.

This feature contained the highest number of lentil seeds in all the Opovo features. In total, 13 *Lens culinaris* (Figure 2.11) and one Fabaceae seed were found there. The diameters of measured lentil seeds ranged between 2–3 mm.

Fruits

Only three *Cornus mas* stone fragments were recovered from Feature 46, which equals approximately one stone when extrapolated to the number of whole stones. However, it is interesting to note that in a small pit in layer 10 of quadrant 61/108, six whole mineralized seeds of *Vitis vinifera* s.l. were found (Figure 2.16). These seeds by size and shape resembled the *sylvestris* subspecies, but since the outer coat was not preserved the identification to the exact subspecies was impossible.

Water Chestnuts

The best represented plant taxon from this feature is *Trapa natans*. There were 272 water chestnut fragments (259.05 density per 100 liters) found in this pit (Tables 2.1 and 2.2). All fragments were recovered from two samples from level 9 in a single locus 1724. In addition, 14 fragments were retrieved from the heavy fraction (Table 2.3). The weight of *Trapa natans* fragments from both light and heavy fractions was 1.31 g. Many of the fragments consisted of actual spikes of the endocarp (Figure 2.19). The spikes were not very elongated and resembled locally found fruits classified by Janković (1958) as a separate species, *Trapa brevicarpa*.

Weedy Herbaceous Plants

In this category, the most abundant are the seeds of *Polygonum* sp. Twenty-one seeds of *Polygonum* cf. *convolvulus* were found (Figure 2.26). Also, 16 Solanaceae and two *Galium aparine* seeds (Figure 2.28) were recovered from the feature. Most of the Solanaceae seeds came from the same samples in which water chestnuts were found, i.e. from locus 1724, level 9, while *Polygonum* and *Galium* seeds were from the samples where cereal and legume seeds were recovered: level 10. Four *Chenopodium* seeds were also identified, and one had a truncated margin (Figure 2.29).

Unknown

Five seeds were included in this category. One of the seeds was of the *Fumaria* species, but since it was not carbonized and could be of more recent origin, I classified it as "Unknown". Furthermore, there were two types of possible fruit fragments, one of a drupe-stone, and one of the pericarp from a nut-type fruit.

Taphonomy and Interpretation

In sum, 73% of plant remains from Feature 46 (Figure 3.11) consisted of water chestnut remains. However, not a single whole fruit was recovered at the Opovo site. Such an abundance of water chestnut fragments does imply deliberate collecting and processing of water chestnut fruits at the site. Water chestnuts might have been first roasted and the burnt spiky endocarp peelings thrown in the pit. Another possibility was that the fruits were peeled prior to eating or after cooking and the spikes were thrown into fires and discarded together with the ashes into pits. The dense endocarp fragments were likely to get preserved even when thrown into fire. Water chestnut fragments were found in only two samples from locus 1724, layer 9; they may indicate a late summer-fall use of the pit. Cereal seeds were found in the same locus as water chestnuts but in layer 10, suggesting that the cereal remains were discarded prior to the water chestnut fragments. It is uncertain, whether the cereals were deposited shortly before the water chestnuts or a year earlier. Cereals can be stored and processed year round. In the same samples where cereals were found, *Polygonum convolvulus* and *Galium aparine* seeds were also recovered. These seeds could have been weedy infestations of cereal crops and sieved out from the cereal crop. In this pit no wheat spikelet forks were found, implying the cereals were dehusked before the grains ended up in the pit. The

Figure 3.6. Feature 46: **a)** pit; **b)** cross section (Tringham ©Creative Commons License).

presence of *Polygonum* seeds cannot indicate the time of the cereal sowing because *Polygonum convolvulus* plants can accompany both winter and summer crops; *Galium aparine* usually indicates winter-sown cereal. Therefore, we can suppose that some of the cereals from this pit were winter-sown. The appearance of Solanaceae seeds is again ambiguous, since most were found in samples with water chestnuts and not cereals, implying that they might not have been weedy intrusions in crops but rather arrived in the pit as separate intrusions. Certainly Solanaceae as terrestrial plants could not have been collected along with fruits of aquatic plants such as water chestnuts. Also, it is difficult to conceive what type of processing would result in a combination of Solanaceae seeds and water chestnuts.

Lentil seeds were recovered from the samples where cereal grains were found. In the sample where 18 *Triticum monococcum* seeds were recovered, most of the *Lens culinaris* seeds were also found. Lentils are spring sown crops, whereas the presence of *Galium aparine*, as mentioned above, is usually indicative of a winter crop. Thus, we may conclude that lentils were not an intrusion in the cereal crop and were grown separately.

Mineralized *Vitis vinifera* cf. *sylvestris* seeds were found in locus 1754, described as a small pit adjacent to the one containing cereals. However, this locus is from the same layer 10 where cereals were found, suggesting the grape pips were discarded at approximately the same time as the rest of the filling of Feature 46. Six uncarbonized grape pips were found in this pit with a few other carbonized seeds, e.g., *Triticum monococcum*, *Lens culinaris*, *Polygonum convolvulus*, and *Cornus mas* fragments. Thus, the composition of taxa found in locus 1754, a small pit within Feature 46, was similar to the taxa composition of other loci from this feature, except for the grape pips, which were unique.

Animal Bone

In this feature, the most abundant remains are the bones of roe deer followed by red deer, and pigs. A few bones of cattle were also found (Russell 1993:Figure 8.23). From this pit a small sample size of bones was recovered, thus the pit could represent as little as six months of accumulation in levels 12 through 8, from summer to early winter.

Feature 48: Pit

Archaeological Description

Feature 48 is a bell-shaped, medium sized but deep pit located in quadrants 61–62/100 (Figure 1.9). On top was Feature 18, a patch of secondarily deposited rubble (Figures 1.11 and 3.7). In layer 9, several pieces of oven floor with black surface were found. It is unclear to which building horizon the pit belongs. Because the pit is located at the southeast side of House 1, it may have been dug and

Figure 3.7. Feature 48: pit (Tringham ©Creative Commons License).

used in BH 2 and/or BH 3 according to more recent interpretation (Ruth Tringham, personal communication 2004). The bottom of the pit has an area that is red from crumbled house rubble.

Archaeobotanical Description

From this feature, 41 liters of sediment were floated. All the light fraction samples were analyzed (Table 2.1), and heavy samples from 35 liters (85%) of sediment were also examined (Table 2.3). In general, this pit was poor in plant remains and only samples from layers 8 and 10 yielded a small amount of archaeobotanical material; however, no plant remains were found in the samples from layer 9. The density of all plant remains was rather low, except for Solanaceae seeds (Table 2.2). All plant remains were charred.

Wood

Wood was not abundant from the few samples taken from this feature, which was rated 2 on the scale from 1–3. However, one small sample (1 liter) from locus 1689, layer 8, consisted entirely of wood charcoal and ash.

Crops

Altogether only five cereal grains were recovered from this feature, and were identified as *Triticum monococcum* and *Triticum dicoccum*. No spikelet forks were found. Only one Fabaceae seed was retrieved.

Fruits

Only one *Cornus mas* stone fragment from the light fraction and one fragment from the heavy fraction sample were identified. No other fruit seeds nor water chestnut fragments were found.

Weedy Herbaceous Plants

Among the weedy herbaceous plant seeds, only 18 Solanaceae seeds were recovered from one sample (15 liters) from locus 1775, layer 10. It was possible to measure 14 whole seeds. The dimensions averaged 1.29 mm long

(0.80–1.50 mm), 1.09 mm wide (0.70–1.40 mm), and 0.80 mm thick. The Solanaceae seeds from this feature were generally smaller but plumper than the other 10 measured seeds from Features 31 and 41. Many of these seeds were rounder with pronounced reticulation, and thus resembled *Hyoscyamus niger* seeds; however, the identification remains uncertain (Figure 2.23).

Unknown

In this category only one-half of an unknown seed was included.

Taphonomy and Interpretation

In sum, very few plant remains were recovered from this feature. A few wheat seeds of *Triticum monococcum* and *Triticum dicoccum* and one legume seed could be explained as incidental intrusions or as food refuse. The absence of spikelet forks/ glume bases and seeds other than Solanaceae weed seeds may indicate that cereal byproducts were not discarded in this pit. The high number of Solanaceae seeds (74%) could not be regarded as possible infestations of cereal crops, otherwise, other cereal byproducts would have accompanied them (Figure 3.11). Therefore, it is possible that Solanaceae seeds were separately and deliberately discarded in the pit. As mentioned, the fruits of most Solanaceae species ripen in late summer and autumn. If most of the seeds were actually from *Hyoscyamus niger* berries, they could have been collected for medicinal and/or ritual purposes. These plants thrive on well lighted, warm, moderately dry, not acidic, and nitrogen-rich soils (Ellenberg 1979:74). Nitrogen-rich soils are often encountered in manure deposits. These highly poisonous plants are avoided by animals and thus could not have been contained in animal dung if it were deposited in this pit.

On the basis of plant remains, this pit differs from the typical refuse pits found in the Opovo site. If the Solanaceae seeds are actually *Hyoscyamus niger* and thrived near houses on dump heaps, then they may indicate that the settlement was located at a light and well-dried place. If, however, the plants were gathered or weeded out from elsewhere, then some of the fields in the vicinity of the settlement were dry, not acid and probably manured. However, the most challenging problem to explain is how *Hyoscyamus* fruits could have been used or discarded so that their seeds were charred, unless they were used medicinally. If they were undesirable plants, they could have been weeded out and thrown into a fire and then discarded with ashes into the pit. If they were intentionally gathered, then the fruits could have been dried and then burned in a house fire, or else heated for making medicine, or some kind of poison.

Animal Bone

Although not many bones were recovered from this feature, the most abundant were of roe deer, followed by red deer, cattle, pig, and sheep/goat (Russell 1993:Figure 8.16). No seasonality patterns were reconstructed from this pit. Many fish bones were recovered from layer 8.

Feature 51: Posthole/Pit

Archaeological Description

Feature 51 is possibly a small pit or posthole adjoining or underneath House 4 and belongs to BH 3 (Figure 1.9). It is less than 50 cm wide in diameter. It is on the edge of the southern wall of House 4 and might be one of the foundations of the house; but since it contained a lot of archaeological material, the excavators thought it might actually represented a pit.

Archaeobotanical Description

From this feature, 10 liters of sediment were floated from locus 1730, layer 8, quadrants 61/112. In a light fraction sample, only one seed of *Polygonum* sp. was found. There were no other seeds or water chestnut fragments recovered. Also, the one analyzed sample contained no wood.

Taphonomy and Interpretation

With only one seed recovered, it is difficult to judge the function of this feature. If it were a rubbish pit, there would have been more diverse plant remains. If it were a posthole and contained more plant material, it would be possible to interpret it more precisely. When spilled, remains may end up in the postholes as a result of being swept aside. However, flotation samples taken from the floor of House 4 contained very few plant remains, indicating crop processing took place outside House 4. There were no ovens found within House 4, nor many cereal remains; therefore, there was little chance that some of the seeds or byproducts ended up in the posthole.

Animal Bone

There are no separate results of bone analyses from this feature.

Feature 52: Pit

Archaeological Description

Feature 52 is a large pit that cuts into the western profile of the excavation block, quadrants 58/113 (Figure 1.9). The pit has a rubble topping that was described as Feature 37. On the top of this house rubble was another layer of secondarily deposited rubble from Feature 2 (House 1), or possibly this rubble was part of Feature 37 (Figure 3.8). The lower levels of the pit were filled with ash and garbage and much of the sediment was a dense clay matrix. Layer 8 consisted of clay matrix and was full of fish bones. A dog skull, a flint core, and fine pottery fragments were found in layers 9 and 10. From the part of the pit designated as Feature 52, much flint debitage was excavated in the deeper layers. Also, archaeobotanical remains were found only in the part of the feature designated as Feature 52, which probably dates to BH 2.

Figure 3.8. Feature 52: pit (Tringham ©Creative Commons License).

Archaeobotanical Description

The largest number of wheat spikelets and glumes were found in this feature, although not more than 90 liters of sediment were floated. All the light fraction samples were analyzed (Table 2.1), but only one sample from the heavy fraction (10%) was examined (Table 2.3). Not a single archaeobotanical item was recovered from this heavy fraction sample. The richest light fraction samples were from locus 1886, from layer 10, described as a clayish and ashy pit in the south part of the feature. All plant remains were charred and most were well burnt.

Wood

The wood charcoal was found in the same samples in which wheat spikelet forks were the dominant taxon. The average wood abundance for the whole feature was rated 2 on the scale of 1–3. However, in only one sample (10 liters) from locus 1886, 7.12 g of wood charcoal was found.

Crops

In the cereal category, mostly byproducts of cereals and not many whole cereal grains were recovered. In comparison to 490 *Triticum monococcum* spikelet forks, only 13 whole *Triticum dicoccum* and 11 *Triticum monococcum* seeds were found. Although there were few whole cereal grains, many cereal fragments were recovered (0.92 g), representing the highest density of cereal fragments by weight (Table 2.2). This feature contained the largest number of spikelets found by absolute count and density (544.44 spikelets per 100 liters) in all of the Opovo features (Tables 2.1 and 2.2). Because so many spikelet forks/glume bases were found together, I was able to weigh them. Thus, 190 *Triticum monococcum* spikelet forks weighed 0.11 g, 112 spikelet forks weighed 0.08 g, and 114 spikelet forks weighed 0.07 g.

Additionally, two *Lens culinaris* and two Fabaceae seeds were recovered. One *Linum usitatissimum* seed was retrieved from the same sample in which the majority of the spikelet forks were found.

Fruits

Only seven fragments of *Cornus mas* stones were found, which is approximately equal to one drupe-stone. No other fruit seeds were found.

Water Chestnuts

Four *Trapa natans* fragments were recovered from locus 1797 in level 8.

Weedy Herbaceous Plants

A total of 15 weedy herbaceous plant seeds was recovered from this feature, out of which nine were identified as Solanaceae seeds. Three *Polygonum* seeds and one seed each of *Bromus*, *Lolium*, and *Chenopodium* species were identified.

Unknown

In this category, six specimens were included. One uncarbonized seed of cf. Boraginaceae, which might have been of modern origin, was included among the unknown seeds. Also, one seed of cf. Astereceae, and one seed of cf. Poaceae, resembling a *Panicum*-type, and a small rounded seed were included in the same category (Figure 2.31). Among the unidentified fragments was a pedicel of a small berry fruit.

Taphonomy and Interpretation

Plant remains from Feature 52 consisted mostly of wheat spikelet forks (91%). All other cereal, legume, weed seed, and water chestnut fragments comprised 9% of the total plant taxa found in this feature (Figure 3.11). Such a composition of taxa indicates that wheat byproducts resulting from dehusking einkorn spikelets were primarily deposited in layer 10. It is most probable that after spikelets were pounded, the glumes were sieved out and thrown into a fire; later, the ashes were discarded into the pit along with the spikelet forks. Because all the spikelet forks came almost exclusively from three samples, it is quite likely that they represent refuse from one short dehusking session. Over 490 spikelet forks were found equating to about 24 einkorn spikes with around 20 spikelets per spike. Twenty-four spikes is one large handful of spikelets. If, on the basis of the present einkorn seed weight, we assume that 24 spikelets had 20 grains each (0.4 g per spike), we can then calculate the weight of grain produced after pounding a handful of einkorn spikelets. A handful of 24 spikes would produce ca. 10 g of einkorn grain. Then, to obtain ca. 1 kg of grain, about 120 handfuls of einkorn spikelets had to be processed. To reconstruct the time necessary for dehusking ca. 1 kg of einkorn, I relied on experimental data on the time invested in pounding glumed wheats. One of the strictly controlled experiments carried out by Meurers-Balke and Lüning (1992), although with a group of inexperienced students, showed that 1 kg of unheated emmer can be dehusked in a wooden mortar with a wooden pestle in 40 minutes. They found that dehusking with a saddle-quern is much less efficient than is a vertical wooden mortar hollowed out from a tree trunk. Therefore, it is reasonable to assume that experienced persons in the Opovo settlement could have pounded one kilogram of glumed wheat in approximately 30 minutes.

In the same sample in which spikelet forks were found, the majority of the weed seeds were identified. Because the most prolific samples were from the ashy part of the pit, where the largest quantities of wood charcoal were found, it seems that ashes from the house fire were thrown into this pit. *Bromus*, *Polygonum* and *Chenopodium* seeds could easily be weedy intrusions in the cereal crop. Although *Polygonum* and *Chenopodium* could have been sieved out, *Bromus* seeds were either picked out manually, or else the sieve was coarse enough to allow brome seeds to pass through. The presence of Solanaceae seeds is perplexing. The seeds were found together with cereal by products, and hence their presence in charred form in this pit could be explained as weedy intrusions in cereal crops along with other weedy seeds. Nevertheless, Solanaceae seeds were present in other Opovo pits, regardless whether cereal remains were found in the same samples or not.

On the basis of identified weeds from these samples, it is difficult to say whether the einkorn crop was sown in autumn or spring. Short-seeded *Bromus* cf. *arvensis* species would point to winter cereals. However, both *Polygonum convolvulus* and *Chenopodium* species can be weeds of spring or winter crops. They can also be ruderal plants, or intentionally gathered. The identification of some weeds to the exact taxon was not possible, and since most of these weeds are cosmopolites, the location of cereal fields was uncertain.

Trapa natans fragments were found in layer 8, in a layer above cereal remains. This indicates that water chestnut remains were deposited later than wheat byproducts. However, it is interesting to note that Cornelian cherry stone fragments were found mixed with wheat byproducts. Plant remains were mostly found in layers 8–10 with cereal, lentils, and flax seeds, and not from deeper layers. These plants could have been processed all year round, and hence their presence does not indicate of any specific season. Cornelian cherry and water chestnut fragments would point to autumn gathering, although the fruits of both species could have been saved for later use.

Animal Bone

Small bone samples were found in this feature. Regarding bone remains, the most abundant were of cattle, followed by red deer, pig, roe deer, dog, sheep/goat, and hare. This pit has a proportionally large number of cattle bones compared with other features (Russell 1993:Figure 8.15). Level 14 could be assigned to autumn, levels 13 and 12 to winter-spring, level 11 to autumn and perhaps to winter-spring, and the levels from 8–10 to autumn-winter-spring, according to Russell (1993:432).

Feature 59: Pit

Archaeological Description

Feature 59 is a pit with horizontal layers of fill (Figure 3.9). It is located in the northeast part of the excavation block

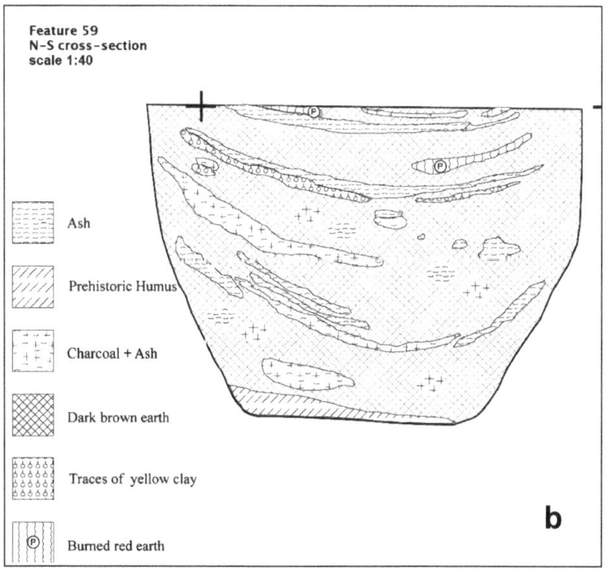

Figure 3.9. Feature 59: **a)** pit; **b)** cross section (Tringham ©Creative Commons License).

(Figure 1.9), abutting the eastern block profile (quadrants 65/116). It has a rubble topping that is Feature 33. Feature 59 certainly belongs to BH 3. It contained much wood charcoal and ash in layer 10 and in the deeper layers. There was an animal disturbance observed in the south side of the pit in layers 9–12. In layer 10, a flint core was found, and in layer 11, an altar fragment, bone tools, bones, and pottery sherds were discovered.

Archaeobotanical Description

Feature 59 is one of the pits that yielded a considerable number (76) of wheat spikelet forks. From this feature, 25 liters of sediment were floated, and all the light fraction samples were analyzed. One small sample taken from layer 8, locus 1825, yielded no archaeobotanical remains, but the two samples from layers 10–11 had a considerable quantity of plant remains, especially spikelet forks and large quantities of wood charcoal.

Wood

Deeper layers of this feature contained large quantities of wood, rated 3, on a scale of 1–3. There were several large fragments of wood and five types of wood present in one sample from locus 1898. Among 20 identified wood charcoal fragments, there were six *Quercus*, eight *Cornus*, three Salicaceae, and one Ulmaceae specimens (see Table 4.2). In addition, another type of wood was identified and described as finely porous early wood zone with a rather abrupt dense late wood (Neal Lopinot, personal communication 1997).

Crops

Among the crop remains, the most numerous were *Triticum monococcum* spikelet forks. There were 76 spikelet forks identified from this feature. However, not many cereal grains were found. Only eight *Triticum dicoccum* and five *Triticum monococcum* seeds were recovered. In addition, one *Lens culinaris* seed and one large type Fabaceae seed were retrieved.

Weedy Herbaceous Plants

In total, seven weed seeds were recovered: one *Bromus* seed, three Solanaceae, and three *Chenopodium* sp. seeds. One among the *Chenopodium* seeds had a truncated margin, a characteristic of domesticated morphs of chenopods; the others did not have truncated margins and were possibly another morph of *Chenopodium* plants. Although only a few chenopod seeds were recovered from this feature, the density (12.00) was rather high considering the amount (25 liters) of floated sediment (Table 2.2).

Unknown

Among the three fragments included in this category, one was a fragment of an unknown plant that had an inner structure similar to a *Cornus mas* stone. The fragment was not circular, but was flatter and larger than typical *Cornus mas* stones.

Taphonomy and Interpretation

In this pit, there was no evidence of in situ burning. Similar to other pits where in situ burning was absent, charred plant remains provide evidence the plants were burnt before they were deposited in the pit. The presence of large quantities of wood charcoal in the same samples where charred spikelet forks were found also favors such an explanation. The appearance of a large number of wheat spikelet forks (78%) is evidence of daily cereal processing (Figure 3.11). These byproducts were probably thrown into house fires and the ashes were later dumped into the pit. *Chenopodium* plants could have been weedy infestations of cereal crops, but *Chenopodium* plants are also ruderal plants that grow on nitrogenous soils. However, as with the identified charred Solanaceae seeds, it is difficult to conceive how the seeds were charred if they were not thrown into fires. If they were not thrown into fires as byproducts of cereal processing, then perhaps they were intentionally gathered and exposed to house fires because there was no evidence of in situ burning. Again, since cereal remains were the dominant taxa, and cereals could have been stored and processed year round, it was difficult to establish seasonal accumulations in the pit.

Animal Bone

A considerable number of bones were recovered from this feature. The most abundant were of red deer, followed by roe deer, pig, and cattle (Russell 1993:Figure 8.25). Seasonality reconstruction was not possible, however.

Feature 61: Pit

Archaeological Description

This feature is a pit abutting the northern edge of the excavation block (quadrants 60/117) and it is dated to BH 3 (Figure 1.9). It is a rather small, bell-shaped pit less than 1 m in diameter (Figure 3.10). Layers 10–13 consisted of dark brown sediment with lenses of ash and wood char-

Figure 3.10. Feature 61: pit (Tringham ©Creative Commons License).

coal. There was no evidence of in situ burning. During the excavations, it was observed that a fair number of bones was present, especially in layer 8, and less in layer 10. This pit also contained other archaeological material. From layer 13, much charcoal, bone, and ceramics were recovered.

Archaeobotanical Description

From this feature, 30 liters of sediment were floated. All the light fraction samples were analyzed. Two floated samples were from the same locus 1887, one from layer 10 and the other was from layer 13. All the seeds were carbonized except one Solanaceae seed and one *Galium aparine* seed.

Wood

A large quantity (6.25 g) of wood charcoal was recovered from the deeper layer of this feature. Among 20 wood charcoal specimens, there were at least three types of wood: twelve *Quercus* (oak), four Salicaceae (willow/poplar), and three *Ulmus*/Ulmaceae (elms) fragments. Some of the Ulmaceae wood charcoal resembled to *Celtis* (hackberries) species. The wavy banding typical of the late wood elements in elms and hackberries was relatively limited, indicating that that there is relatively little late wood. Another type of wood was identified as finely diffuse, porous wood whose vessels are much more scattered than in *Salix* (Neal Lopinot, personal communication 1997).

Crops

Among the crops, cereal byproducts were the most abundant. There were 20 *Triticum monococcum* spikelet forks and one *Triticum dicoccum* spikelet fork. It is possible that more *Triticum dicoccum* spikelet forks were present among the identified *Triticum monococcum* spikelet forks, but since most of the identified spikelet forks were narrow and small, they were classified as *Triticum monococcum*. Not a single *Triticum monococcum* seed was recovered from this feature. I expected to find at least some *Triticum monococcum* seeds with the *Triticum monococcum* spikelet bases, but only one emmer seed was found.

Fruits

Only two fragments of *Cornus mas* stones were recovered.

Water Chestnuts

One fragment of *Trapa natans* was found in this feature.

Weedy Herbaceous Plants

Three Solanaceae, one *Galium aparine*, and one *Chenopodium* sp. seed were recovered from this feature.

Unknown

In this category, one specimen was included that could not be more precisely identified.

Taphonomy and Interpretation

In sum, this pit is dominated by wood charcoal and wheat spikelet forks. Spikelet forks comprised 77% of plant taxa, followed by 15% weed seeds (Figure 3.11). The paucity of *Trapa natans* and *Cornus mas* stone fragments may indicate a late autumn accumulation in level 10. *Galium aparine* seeds could point to winter-sown cereals, while a presence of *Chenopodium* sp. seed could indicate both summer and winter crops.

Animal Bone

Red deer bones were the most abundant, followed by pig, roe deer, dog, cattle and sheep/goat. Not enough diagnostic bones were recovered to draw conclusions about seasonal accumulation in this feature (Russell 1993).

Discussion of Context

Features

To make the results comparable among features, I computed the density of taxa found, expressed as the number of specimens per 100 liters of sediment volume of 15 features and non-feature samples. For many features, more than 100 liters of sediment were floated and the density numbers represent real values; for the smaller pits, where less than 100 liters sediment was floated, the densities represent extrapolated numbers (Tables 2.1–2.3).

The features in the bar graph (Figure 3.12) are ordered according to the excavators' feature assignment to the three building horizons, BH 1–3 (Ruth Tringham, personal communication 2004). It was unclear for several pits whether they belong to the second or third building horizon. From BH 1, only one pit–Feature 42, Bedding trench of House 1, yielded plant remains.

The highest density of cereal grains was found in Feature 41, two joined large pits attributed to BH 2 and BH 3. The highest density of einkorn spikelet forks was found in Features 52 and 59, which were assigned to BH 2 and BH 3, respectively. Water chestnut density was highest in Feature 46, a pit that was assigned to BH 3. The highest density of fruit seeds was in a circular pit, Feature 30, possibly a well and dated also to BH 3. Weed seed density was highest in Feature 48, a pit that may belong to either the BH 2 or BH 3 (Table 2.2; Figures 3.11–3.12).

No clear patterning is demonstrable either in density distribution or in combination of plant taxa from different proveniences (Figure 3.12). When comparing the density values from pits and houses, the density values of all the samples from the houses were so low that only pit could be compared (Table 2.2; Figure 3.11).

Cereal remains were found in almost all features. Also Solanaceae seeds were present in all pits, except Feature 45. Furthermore, water chestnuts were very common, absent from only three pits.

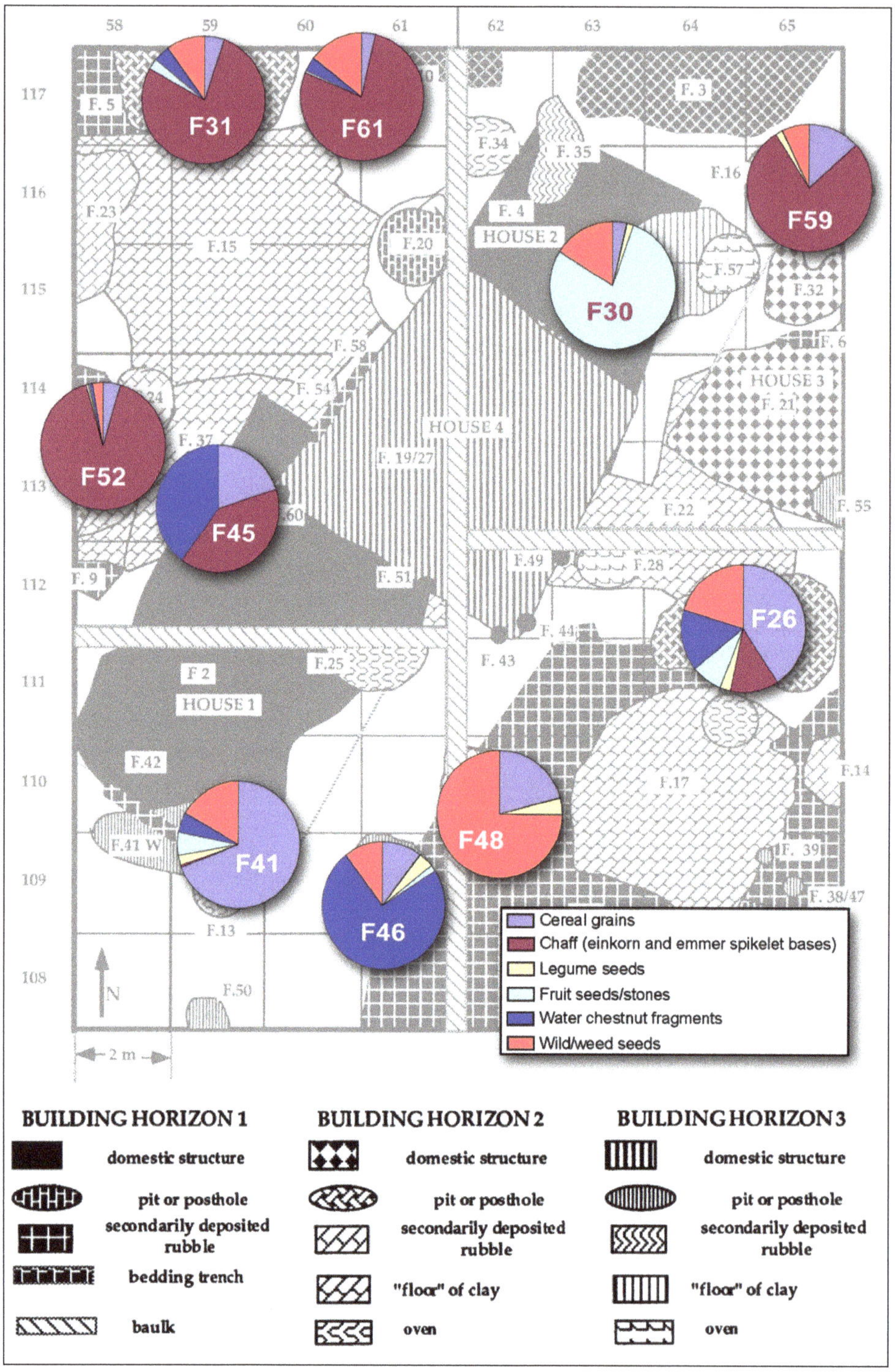

Figure 3.11. Pie charts overlying the pits indicate the distribution of various plant remains by density.

However, if the density distribution of plant remains in various pits is grouped according to the provenience within the site, regardless of building horizon, than a pattern begins to emerge (Figure 3.11). The pits in the northern part of the excavation block contain the highest number of spikelet forks in absolute count and density (Tables 2.1 and 2.2). In Features 31, 51, 59, and 61, einkorn spikelet forks dominate the samples from the deeper layers. Pit, Feature 31 is attributed to BH 2/3, while pits, Features 51, 59, and 61 are attributed to BH 3 (the earliest phase).

It appears that byproducts of cereal processing were discarded into the pits in the northern part of the site (Figures 3.11 and 3.13). Perhaps dehusking the hulled cereals such as einkorn took place near the pits either in the summer months or inside the structures during winter months. In addition to the wheat spikelet forks found in the pits, in the northeastern part of the excavation block many grinding stones were recovered from the upper layers (Feature 3). According to lithic analysis, grindstones from Opovo show little evidence of having been used in grinding wheat

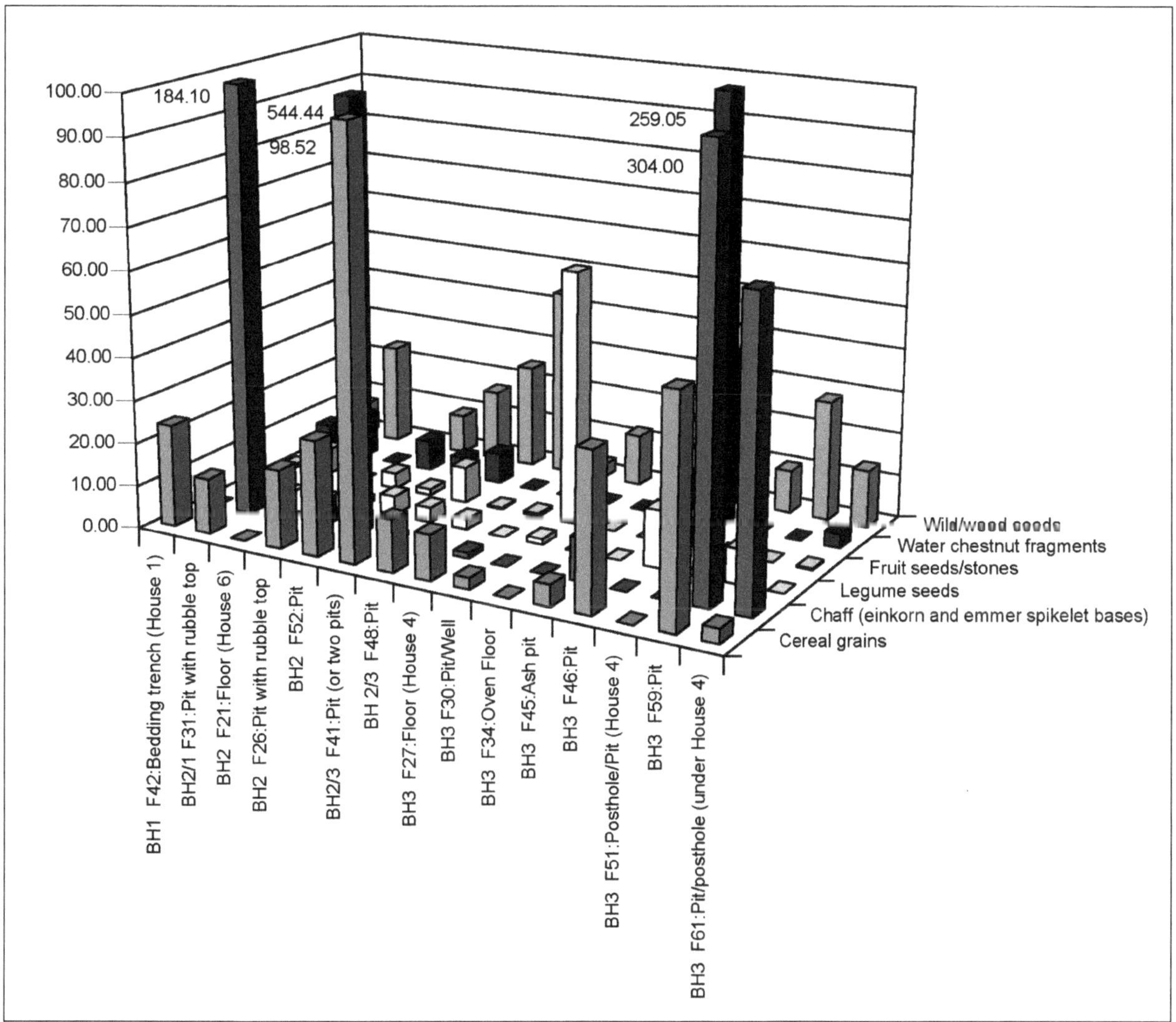

Figure 3.12. Distribution of various plant remains by density (count per 100 liters of sediment). The features are ordered according to the excavators' feature assignment to the three Building Horizons (BH1–BH3).

or other grains, although the hard compact sandstone and gabbro grindstones were suitable for such grinding (Voytek in Tringham et al. 1985:440).

In contrast to the northern part, most of the cereal grains and water chestnut fragments were found in pits in the central and southern parts of the excavation block (Figures 3.11–3.13). Water chestnut fragments were recovered from layer 8 in many features. Weed seeds dominate in pit Feature 48 (due to the presence of 18 Solanaceae seeds recovered from this pit, and a low number of cereal grains). In general, not many spikelet forks in proportion to other plant taxa were recovered from the pits in the southern part of the excavation block. Thus, the refuse from cooking and eating was likely dumped into the pits in the southern part of the site in Features 26, 41, 48, and 46. (Feature 26 is attributed to BH 2, both Features 41 and 48 to either BH 2 or 3, and Feature 46 is attributed to BH 3.)

Cereal grains were, in most cases, recovered from the upper layers 7, 8, and 9, while the cereal spikelet forks were found in the deeper layers 9 and 11 of the pits in the northern part of the site. The absence of more precise chronological dating and the ambiguities of the pits' assignment to the three building horizons (or layers within the pits) made it unclear whether the pits in the northern part containing discard of the cereal processing were used earlier than the pits containing discard of cooking in the southern part.

I also computed correlation coefficients (*Pearson's r* or the *product-moment correlation coefficient*) to determine whether significant relationships can be delineated among various plant taxa, on the one hand, and between animal and other plant taxa, on the other. The correlations were computed using absolute counts of seeds, spikelet forks, water chestnut fragments, and Cornelian cherry drupestones recovered from pits. Absolute counts are the only raw data I could use to compute correlation coefficients, although the numbers of various seeds or fruit fragments are quite variable among different taxa. Plant remains from the floors of House 3 and House 4, as well as from the oven

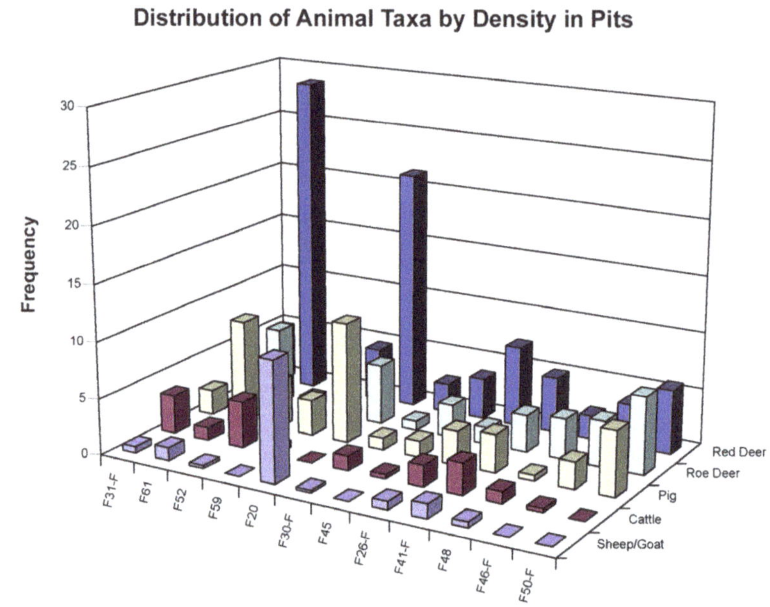

a

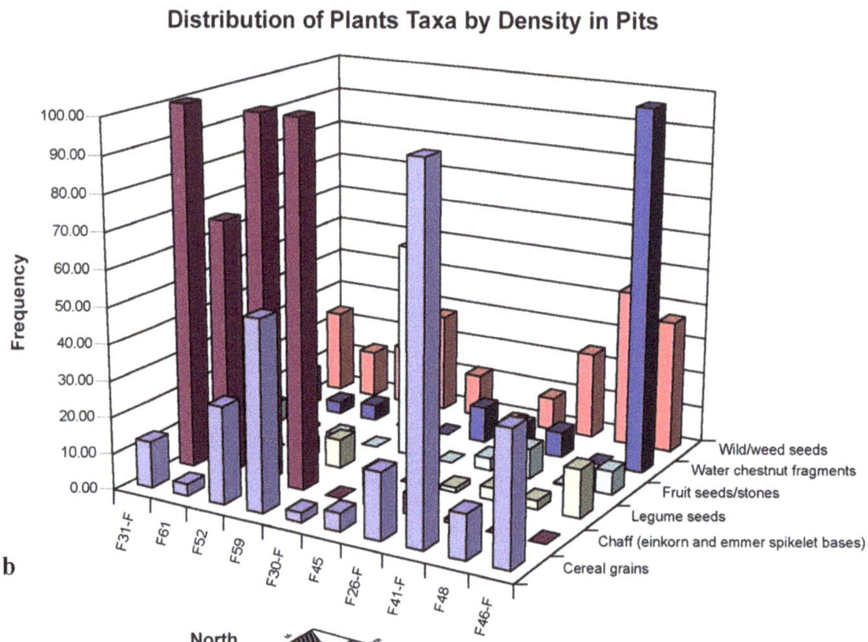

b

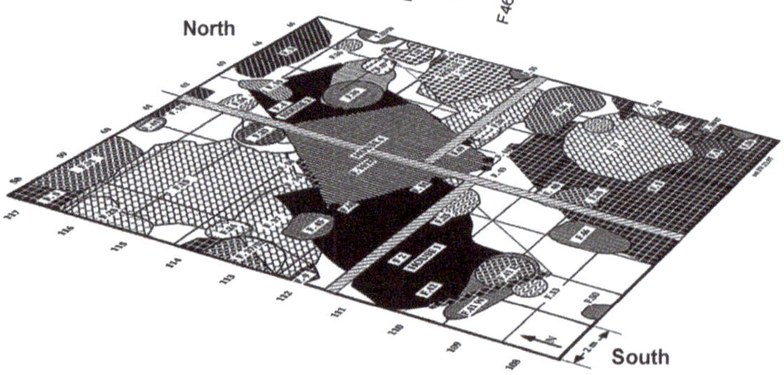

c

Figure 3.13. Distribution of various animal and plant taxa in pits:
a) Distribution of animal taxa by density in pits, count of diagnostic zones per cubic meter of sediment (data source: Russell 1993);
b) Distribution of plant remains by density in pits, count per 100 liters of sediment;
c) Plan of the excavation block.
Pits in a) and b) graphs are grouped according to their north-south provenience within the excavation block.

floor, were not considered because very few plant remains were found in those features and because I compared similar contexts. As expected, the strongest correlations were found within various kinds of animal taxa and within various plant taxa, respectively.

For the plant taxa, I intended to determine whether there are significant correlations between cereal grains and certain weed seeds, and whether water chestnut fragments are accompanied by other plants in a consistent way. Since only 12 pits were considered, the total number of samples was rather low. I could have used a flotation sample as a unit, but because there were many floated samples in several features and just a few in others, when assessing context, I decided to take each pit as one sample regardless of the number of floated samples per feature or the building horizon to which the pit was assigned. Taking each feature as a unit, I was also able to correlate the results of plant analyses with Russell's (1993) results of bone analyses. I considered only correlation coefficients that had low significance probability values (p not higher than .1), so that confidence would be not less than 90%. Also, I did not consider significant correlations if the number (N) of samples was smaller than five features. Furthermore, I did not discuss at length the cases in which the correlation coefficient (r) was smaller than .6000. Among various plant taxa there were strong correlations, for example between *Triticum dicoccum* seeds and Solanaceae seeds (r=.9328; p=.000; N=11). Slightly weaker correlations were between *Triticum monococcum* seeds and Solanaceae seeds (r=.8156; p=.007; N=9). This might suggest that Solanaceae seeds were less consistently present with einkorn crops, than with emmer crops assuming that same post-harvest processing was applied to both types of glumed wheats. In contrast, *Triticum dicoccum* and *Triticum monococcum* spikelet forks showed negative correlations with Solanaceae seeds, although the number of samples was large. The significance probability values were very high, however, p ranging between .653 and .718 respectively; the confidence interval was therefore below reliance. Likewise, neither *Triticum dicoccum* and *Triticum monococcum* spikelet forks show any significant correlations with other weed or cereal taxa, and significance probability values remain very high and below confidence.

In the correlation coefficient analysis only pits as whole units, without internal stratification, were considered; therefore, the significant correlations did not always reflect the possible variations within each feature. In Feature 26, einkorn, emmer, and Solanaceae seeds were found, but Solanaceae seeds were not found in the same samples with cereal grains. This might be explained as a result of different depositional processes occurring in this pit, or simply as a vagary of the sampling, although similar inconsistency was present in other features. Likewise, in Feature 46, Solanaceae seeds were recovered mostly from the samples with numerous water chestnut fragments and not cereal grains. In Feature 48, only a few cereal grains were found and no cereal spikelet forks, but 18 Solanaceae seeds were recovered from the same feature. In contrast, in Feature 52, Solanaceae seeds were found together with cereal byproducts. It is difficult to interpret the strong correlation coefficients between *Trapa natans* fragments and *Chenopodium* sp. seeds (r=.9528; p=.003; N=6) on the one hand, and the less strong correlations between *Triticum monococcum* and *Chenopodium* sp. seeds (r=.8339; p=.039; N=6) and *Triticum dicoccum* and *Chenopodium* sp. seeds (r=.7986; p=.032; N=7), on the other. It is certain that *Chenopodium* fruits of terrestrial plants could not be gathered together with fruits of aquatic plants such as water chestnuts. I was not able to determine, whether the differences in correlation coefficients reflect different culinary practices, resulting in deposition of water chestnut fragments together with *Chenopodium* seeds rather than in deposition of cereal grains with chenopod seeds. Legume seeds showed rather strong correlations with cereal and also with water chestnut fragments (*Trapa natans* fragments vs. *Lens culinaris* seeds: r=.9876; p=.002; N=5). Furthermore, neither emmer nor einkorn spikelet forks show a significant relationship with any of the plant taxa. In most cases the correlations were negative, but in many cases not strongly so, although the confidence intervals were below acceptability. In sum, although the correlation coefficients among various plant taxa seem significant, there are still many exceptions when samples within features are analyzed in greater detail.

I also tried to determine the relationship between deposition of identified animal bones and plants in the pits. Correlation coefficients (*Pearson's r*) were again computed using absolute counts of seeds, spikelet forks, water chestnut fragments, and Cornelian cherry drupe-stones and counts of bone diagnostic zones in pits. For the animal bone counts, I used Russell's data (1993:428–446) about taxa distribution of diagnostic zones (DZ) in pits. Red deer is the dominant animal in most features, regardless of archaeobotanical content. Generally, there were no strong correlations between animal and plant taxa. There were a few moderately strong correlations, however, between total number of cereal grains and number of red deer and roe deer bones, which could be interpreted as cereals being consumed with wild animal meat (red deer vs. cereals: r=.6698, p=.024; N=11; roe deer vs. cereals r= .7880, p=.007, N=6). Similarly, there is some correlation, although less strong, between the number of cattle bones and total cereal grains found (r=.5629, p=.090, N=10). The weaker correlation between cereal grains and cattle bones might indicate that cereals typically were not eaten with beef or were discarded separately. However, there were strong correlations between Solanaceae seeds and cattle bones (r=.7182, p=.029, N=9), and even stronger correlations between roe or red deer and Solanaceae seeds. These correlations between animals and Solanaceae seeds might indicate that Solanaceae plants were somehow included in meat preservation, consumption, or deposition (red deer vs. Solanaceae seeds: r=.7855, p=0.12, N=9; roe deer vs. Solanaceae seeds: r=.9140, p=.001, N=9). There were no significant correlations between water chestnuts and animals, except in the case of the sheep/goat category (r=.9678, p=.007, N=5); these correlations were high but occurred in only five features.

Russell (1993) tried to determine seasonality estimates for animals killed from various layers of several pits, as mentioned in the contextual analysis of plant and animal remains for each feature. She was able to show that the site was occupied year round. Such a conclusion was corroborated on the basis of the evidence of cereal grains and the use of glumed wheat chaff to temper daub, indicating crop cultivation. Russell has suggested that some crop raiding did occur, although most hunting seems to have been done in non-summer months. When the seasonality estimates for the killed animals are compared with the recovered plant taxa from the same feature layers, only in a few instances is there evidence of crop raiding. Crop raiding might be suggested for Feature 41, level 9, in which both cereal and lentil seeds were recovered with bones from deer and wild boar that were probably killed during the summer-autumn months. But in the same layer, there were the bones of domestic pigs, killed during the spring-summer. In the two pits where wheat spikelet forks (Feature 31: layers 10 and 11; Feature 51: layer 10) are the most abundant plant category, there were no bones of animals killed during the summer. This might indicate that bones from consumed animals were not thrown into the same pits with cereal byproducts. Water chestnut fragments were found in layers where no animals were killed during the summer (Feature 45: layer 9; Feature 52: layer 8), or where animals where killed during summer-autumn (Feature 31: layer 8; Feature 26: layer 8; Feature 41: layer 8; Feature 46: layer 9). This does not mean, however, that crop raiding did not occur, since the Opovo settlement was probably surrounded by mixed-oak forest, a natural habitat for deer and boars.

In summary, it seems impossible to establish any firm pattern between animal and plant remains deposited in the pits (Figure 3.13). If any generalizations could be made, the pits rich in bones were also rich in plant remains, regardless whether the plant remains were byproducts of cereal processing or food refuse.

Non-features

In addition to taking samples from various features up to level 3, the sampling strategy was changed to randomly selecting 1 m^2 out of every 2 m^2 quadrant throughout Block I, thus giving a systematic 25% sample. Therefore, this sampling would ideally result in 20 routine samples taken for every level in addition to samples taken from the features. In total, 1235 liters of sediment taken as the routine samples were floated from levels 6, 7, 8, and 9. All 132 light fraction samples were analyzed. Most of the plant taxa that were identified in the light fraction samples from the features were also recovered from the non-feature samples, but in smaller quantities; however, fewer species were represented in the non-feature samples (Table 2.1). Wood charcoal was recorded in 36 of the total 132 samples. While in some samples the wood charcoal was rated 3 on a scale 1–3, the most common abundance was 1. Wood charcoal was identified from two routine samples (Table 4.2). In locus 1418 (quadrants 57/109, layer 7), the entire sample appeared to represent a fragmented stem of extremely knotty or gnarled wood, probably diffuse or semi-diffuse porous wood with numerous and variable rays (1–5 cells wide). In sample from locus 1574 (quadrants 65/115, layer 7) of 20 identified specimens there were two *Quercus* and 18 *Cornus* charcoal fragments. *Cornus* wood appeared to come from a fragmented 9–10+ year-old stem. In addition, one sample of wood charcoal collected during the excavations of deeper layers and without exact provenience consisted entirely of *Quercus* charcoal (Neal Lopinot, personal communication 1997).

A few cereal grains were found but only one einkorn spikelet fork, in contrast to the numerous spikelet forks found in the light fraction samples from features. Likewise, no Fabaceae seeds were found. Nevertheless, some of the fruit and weed seeds and drupe-stones, and water chestnut fragments were retrieved from the non-feature samples, mostly from layer 7. From the same layer, the largest quantity of sediment was floated (660 liters), yielding most of the plant remains. In total, only 55 seeds and 12 water chestnut fragments were found in all non-feature samples.

Although a large volume of sediment was floated from non-feature context, no information was provided about additional plant taxa. Also, there was no single concentration of any of the plant remains, indicating that the random sampling accurately represented plant remains preserved.

CHAPTER 4. VEGETATION RECONSTRUCTION FOR THE OPOVO SITE DURING THE LATE VINČA PERIOD IN THE FIFTH MILLENNIUM B.C.

The Opovo site is located in the southern part of the Pannonain Plain in Voyvodina. I describe the present-day setting of the Opovo site using climatological, pedological, and botanical data, and I evaluate the extent to which this evidence can be used to reconstruct the environment of Opovo during the fifth and fourth millennia B.C. On the basis of the identified plant remains from the Opovo archaeobotanical samples and the requirements of the modern-day plant equivalents, I have reconstructed the potential vegetation for the Opovo site (Figure 4.1).

The idea that steppe was the dominant vegetation type in the southern part of the Pannonian Plain is sustained in descriptions of the past environment, including Opovo (e.g., van der Plas et al. 1985). Similarly, the first Neolithic communities in central Europe were associated with steppe regions in the past. Gradmann's *Steppenheidetheorie* (1901, 1933) has been thoroughly revised, due to multiple lines of evidence indicating the existence of forests instead of steppes in these regions during the Neolithic period (e.g., Chamapier 1952). The Opovo site is surrounded by chernozem soil. The question whether the chernozem soil is truly a consequence of steppe-vegetation or not in the Pannonian Plain has been disputed (e.g., Havinga 1972). The present climate of Voyvodina is wetter and could have been so in the past, especially during the Atlantic period in comparison to drier climates further east where there are typical steppes.

I examine the claims for steppe vegetation around the Opovo site, and provide archaeobotanical, archaeozoological, and archaeological evidence for mixed-oak forests and marshland in the Opovo environs during the Vinča occupation of the site.

Climate and Weather

The modern climate of Voyvodina is moderately continental. Voyvodina has milder climate due to the influence of the Atlantic and the Adriatic (Mediterranean) climate. The average annual rainfall varies between 550–740 mm. The highest rainfall is May and June, and the lowest during January and February. The average annual temperature is 11.2°C, and 18.1°C during the vegetational period (April–October). Relative humidity is 85–95% in winter and 60–70% in summer. High summer temperatures and strong winds contribute to the loss of humidity and the evaporation in the present conditions is high (Katić et al. 1979).

Information about climate in the Opovo region is from the meteorological station in Pančevo, the county seat 20 km south of the Opovo site. The climatic data were taken from the 1985 publication of the *Hydrometeorological Institute Annual* for the period 1948–1984 for the meteorological station located at 44° 53' N and 20° 40' W at an elevation 78 m above sea level (Savezni Hidrometeorološki Zavod, 1985). The Opovo site is located on a hillock with very similar elevation, and both Pančevo and Opovo near the Tamis River in southwest Banat in Voyvodina. The closest long-term climatological data is available from Pančevo, located 20 km south of the Opovo site with same elevation asl (Table 4.1).

Figure 4.1. Vegetation reconstruction for the Opovo site.
Legend: 1-*Trapa natans* (water chestnuts); 2-*Phragmites communis* (reeds); 3-Salicacacae (willow/poplars); 4-Ulmaceae (elms); 5-*Quercus* sp. (oaks); 6-*Cornus mas* (Cornelian cherry trees); 7-*Sambucus ebulus* (elderberry trees); 8-Crops.

The average rainfall for the Pančevo region was 564 mm for the period 1924–1938. The average rainfall from 1948–1984 was 650 mm for the Pančevo station and 629 mm for the Opovo station (Katić 1972). It is possible that precipitation had increased in the later period, but perhaps the average for the 35-year period is more reliable than the average for just 14 years. Precipitation is crucial for vegetation growth from April–October, and an average precipitation in this period was 377.3 mm for the Pančevo station and 358 mm for the Opovo station. The month with the most precipitation was June, which averaged 91.3 mm. In the 35-year period, the maximum rainfall was 981 mm in 1954. In addition, there were five years with rainfall over 800 mm. It is interesting that above-average rainfall often appears in consecutive years. However, there were years in which rainfall was below 500 mm. The minimum rainfall was 464 mm in 1976; however, drought was less severe than in 1950, when the precipitation was 495 mm. (Note the rainfall during the 1950 growing season was only 190 mm in comparison to 297 mm in 1976.) Similarly, there were three years in which the rainfall was at average levels or higher, while the average rainfall in the vegetation period was below 300 mm.

The average temperature in Pančevo was 11.1° C during 1948–1984 (Table 4.1). The hottest months were July and August, with an average temperature of 21° C. The coldest month was January, which averaged temperature -0.7° C. However, temperatures during the winter and spring months are more variable and average temperatures can oscillate between -7° C up to +4° C in January.

The climate during the Atlantic period (5500–2800 B.C.) when the Opovo site was occupied was perhaps somewhat different than the modern one. The climate during the Atlantic phase was warmer and wetter than in modern times, based on the pollen analyses from peats in Serbia south of the Danube (Pantić 1960). The Atlantic period in the Great Hungarian Plain (north of the Opovo region) is also characterized by a warm and humid, but temperate, nearly Submedittereanen climate (Járai-Komlódi 1966). Pollen investigations from the tell site Gomolava, on the Sava River, did not reveal important changes in vegetation during last 7000 years, and thus there was no discussion of climatic change (Bottema and Ottaway 1982).[1]

Table 4.1. Climatic data for Pančevo (1948–1984).

Average temperature	t° C	Average precipitation	(mm)
Annual	11.1	Annual	650
January	-0.7	April-October	377
July and August	21	June	91.3

Setting

Pedological and geological sampling of the Opovo site was performed by a team of Dutch soil scientists from the University of Wageningen, under the direction of L. van der Plas and L. Pons, to obtain information about the potential for agriculture and other resources in the region, the formation of the palaeosoils, and the depth of the cultural layers (Tringham et al. 1985:425–444; 1992:354–356).

Van der Plas et al. (1985) estimated relative age of topographic elements by examining the meander system through an assessment of the altitudinal differences between old river courses, levees, and the site location. Change appears to have been a significant aspect of the local geomorphological history. The principal cause of this change was the Tamis River, which radically altered the loess landscape that had formed much earlier than the occupation of the Opovo site. Levees, gullies, and backwater sloughs created by the Tamis are the most significant features of the immediate topography. The development of Opovo's fluvial landscape appears to have taken place over a short period of time. The abandoned and unfilled depressions and the unfinished levees all point to rapid changes in the course of the river.

The highest elevations of the hillocks in the vicinity of the site are between 81 m and 83 m above sea level. Some of the levees around the Opovo site rise up to 2 m above the highest elevation of the site, indicating the site was once under water, though it was probably occupied before inundation took place. This conclusion is based on the following considerations. In the Opovo region there are two clearly demarcated landscapes: one is characterized by chernozem soils on a higher topography, dating to the early Holocene; the other is characterized by more recent solonetz soils and a lower, fluvial topography (Figures 1.8 and 4.2). The boundary between the two zones is clear in the Opovo area, and follows a roughly N-S line parallel to the Tamis. The prehistoric settlement is located on the older, chernozem soils. The pedological profiles taken in the high ground immediately around the site show that the soil there is a lowland chernozem, highly degraded. These chernozems could not have developed in a permanently waterlogged floodplain such as the one that existed until a modern system of water management was introduced. The depth of the soil suggests a rather long period of soil formation, one that was not finished when the site was occupied and that may have ceased after Opovo was abandoned. The formation of these chernozems was arrested when the conditions of water logging began.

Further evidence that the river shifted its course away from the site after the Vinča period is provided by an auger probe (Probe No. 12) from one of the lowest gullies, which yielded Vinča pottery and house rubble 50–80 cm below the modern surface (Tringham et al. 1992:Figure 2).

These remains tend to support the notion of a lower water table during the Vinča period. To the north, northeast, and south, however, the site was surrounded by depressions in the landscape that were still waterlogged, but no longer part of an active river course (Figure 4.2).

Study of the microscopic structure and the chemistry of soils from the latex profile molds taken from both the excavation blocks (Sonda I) and another 1×1 m small test unit in the immediate vicinity (Sonda II) was performed in the laboratory in Wageningen. Accepting provided pedological information, Tringham et al. (1985:435) conclude: "There is no doubt on the basis of the preliminary observations made in these soundings that the area around the site was marshy, with a great deal of water (either running or stagnant) and a high water table. It is supposed, on this basis, that the area actually available for cultivation would have been limited although very fertile. It is also supposed (but by no means proven) that the area of wooded country around the site would have been limited and that the natural vegetation on the non-marshy area would have been steppe grassland." Similar fluvial landscape with frequent flooding and high underground water was reconstructed for the Vinča culture site Uivar in Romania, also in the Pannonian Plain (Sponholz 2004). Pollen analyses from the site Uivar were unfruitful due to poor pollen preservation, but pollen analysis from the nearby locality Semenec in the Banat Mountains showed the existence of the mixed-oak forest (Rösch et al. 2002, see below).

Taking into consideration the Tringham et al. (1985) account, historical and modern evidence, the rivers Tisa, Tamis, and their branches were much more likely to flood and create numerous lakes, ponds, and marshes in the cut-off meanders around the site in the past. However, the height of underground water seems very debatable. In their first article, Tringham et al. (1985:426) suggest that the water table was higher than today, citing Nandris (1970) and Kosse (1979), and noting that much of the land was seasonally waterlogged. Groundwater was found at 4.5 m below surface while digging the test trench (Sonda II) in 1984, immediately to the east of the excavation block. Tringham et al. (1992) state that Vinča pottery was present in an auger probe (No. 12) at the depth of 50–80 cm below the surface, in one of the lowest lying gullies next to the site. This indicates the water table was lower by ca. 1 m than during the Vinča period. It is likely that the site

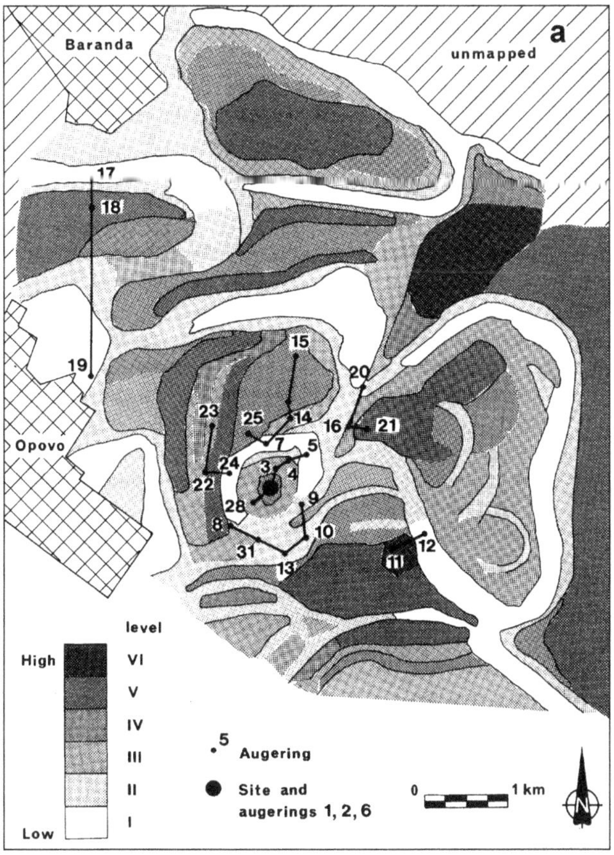

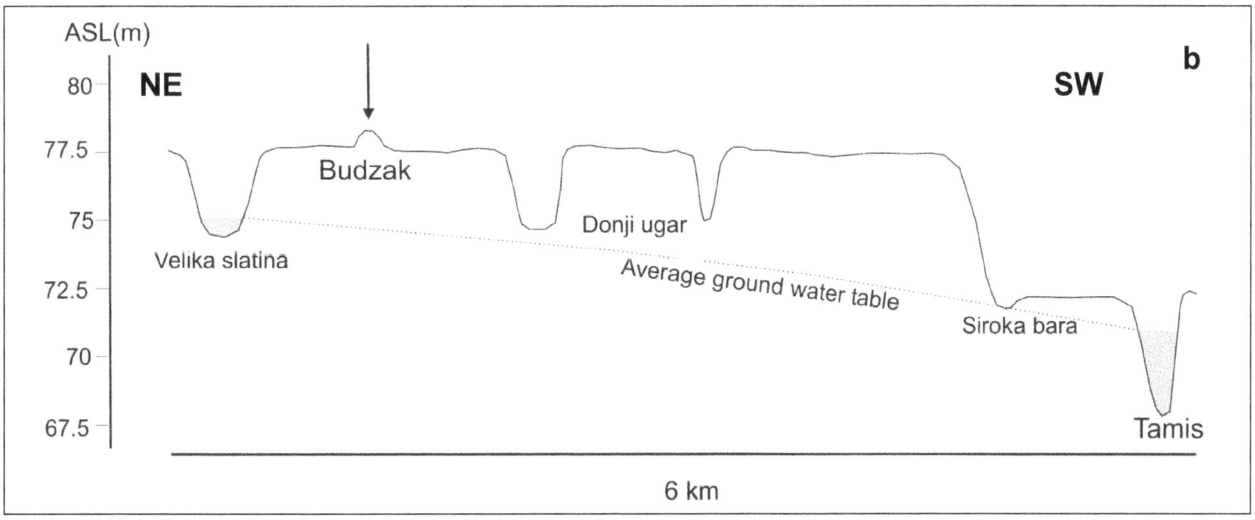

Figure 4.2. Topographic map of Opovo: **a)** provisional geomorphology (after Tringham et al. 1992); **b)** profile – northeast-southwest cross-section (Map of Voyvodina, Section Zrenjanin 4-4, 1969). Arrow indicates location of prehistoric site.

was occupied before inundation took place (Tringham et al. 1992:354), and that during 200 years of occupation, the water table at the Opovo site oscillated during drier and wetter episodes.

Since no pollen analyses are available from the region, detailed pollen and plant macro remains investigations from Linearbandkeramik (LBK) settlements dated to the Early Neolithic Period (fifth millennium B.C.) in central Germany provide data for the reconstruction of the general ecology and plant ecology of the period. Some of these results may be used as analogs in reconstructing the Opovo ecological trends during the Late Vinča occupation, despite geographic and chronological distances. For example, Kreuz (1995:130) in the article about ecological zones of ten early LBK sites in central Germany describes the effects of floodplains on the ecology of each site. Floodplains within river valleys of either small or large rivers could have caused different fluvial deposits. Depending on these sediments and the groundwater table, different soil types would have developed. Hence, there would have been a variety of ecological conditions for plant growth depending on the water regime. Kreuz (1995) has found that some territories (1 km radius) of the three sites appear to include more floodplain terrain than others. She postulates that people selected these places for some specific reason such as hunting or stockbreeding, as suggested by the unusual quantity of pig bones on one of the sites (Angela Kreuz, personal communication with Uerpmann 1990 and 1992, cited in Kreuz 1995:130). According to Kreuz, it is possible that dry and sandy soils existed seasonally on which light-demanding forests with rich undergrowth of herbs and grasses could have grown. In addition, the more prominent (Kreuz uses the term "marked") landscape of the past allowed more space for development of a greater number of light-demanding species. Similarly, the pedological investigations of the Opovo site showed that the original landscape on which the Neolithic settlement was located would have been buried or eroded before the formation of the high natural levees (Tringham et al. 1985:436). If indeed the landscape of the Opovo site was more elevated in the past than today and if there were erosion, then it is possible there were areas suitable for the development of light-demanding vegetation. Therefore, despite seasonally waterlogged areas, it seems possible that dry and non-marshy areas were also present around the Opovo site, especially during the drier years when more land was available for agriculture.

Soils and Vegetation Communities

Today the Opovo site lies in an area of calcareous chernozem on the loess terrace (*Pedološka Karta Vojvodine*, Nejgebauer et al. 1971). In a 2.5 km radius around the site, to the east there is an area of chernozem-like meadow soil with signs of salinization (Figure 1.8). In the north, there are lower-lying areas of solonez-like soils For reconstruction of the past ecology, Tringham et al. (1992) use the formation of *Schwarzerden* in central Germany as an analogy for the Opovo site. The Schwarzerden[2] or chernozems[3] (black soil) as they are called in eastern Europe, form in loess regions. According to the authors' personal communication with van der Plas (1988, cited in Tringham et al. 1992), the formation of Schwarzerden appears to have taken place at the same time as a local formation of a steppe-like environment in Germany, just before or during the occupation of the Neolithic LBK. This suggests to Tringham et al. (1992) that a similar environmental process took place at Opovo a thousand years later. The development of lowland chernozems like those seen at Opovo is described by Tringham et al. (1992) in the following manner: "Chernozem formation is a function of a lime-rich, loess-like sediment under the influence of a climate with no more than 600 mm of annual rainfall, and a mean annual temperature of 6–10° C. The corresponding vegetation in this soil/climate situation is steppe-like, with grass and isolated trees. Steppe grasses support numerous rodents, many of which live in burrows and such burrows were frequently observed in the site's stratigraphy."

There is debate about the formation of the chernozem soils and the spread of steppe vegetation, thought to be responsible for the formation of chernozems. As for the chronology of chernozem formation in eastern Europe, several international pedologists believe that chernozems developed during the Boreal period, (cf. Laatsch 1957; Scheffer and Schchtschabel 1966; and others cited in Havinga 1972). Local pedologists and botanists in the former Yugoslavia have been the proponents of the same hypothesis (e.g., Nejgebauer 1952; Slavnić 1952). According to them, chernozem soils developed before the Neolithic period and therefore before the occupation of the Opovo settlement. Similarly, according to the botanist Slavnić (1952), it seems that forests were absent from Voyvodina during the Boreal period only, when drier and colder climate steppe vegetation had developed that was responsible for the creation of chernozems. Similar evidence for steppe vegetation during the Boreal period was found by Járai-Komlódi (1968) for the areas surrounding the former backwaters of the Danube and the Tisa in Hungary. There, open forests were succeeded by dense forests in later periods. Therefore, chernozem soil could be considered a relic soil, because the postglacial climate of Voyvodina is not characteristic of steppes but rather of forests. These scientists are advocates of the hypothesis that chernozem soils should be regarded as a primary soil development. However, others like Fink (1964) and Havinga (1972) argue that chernozems represent secondary soil development at least for the Pannonian part of Austria and Hungary. According to these researchers, forests once covered the loess where chernozem profiles are now found. Pollen analysis of two peat profiles surrounded by chernozem and Schwarzerde soils in the Austrian part of dry and warm Pannonian climate has shown that open forests preceded steppe vegetation. Apparently,

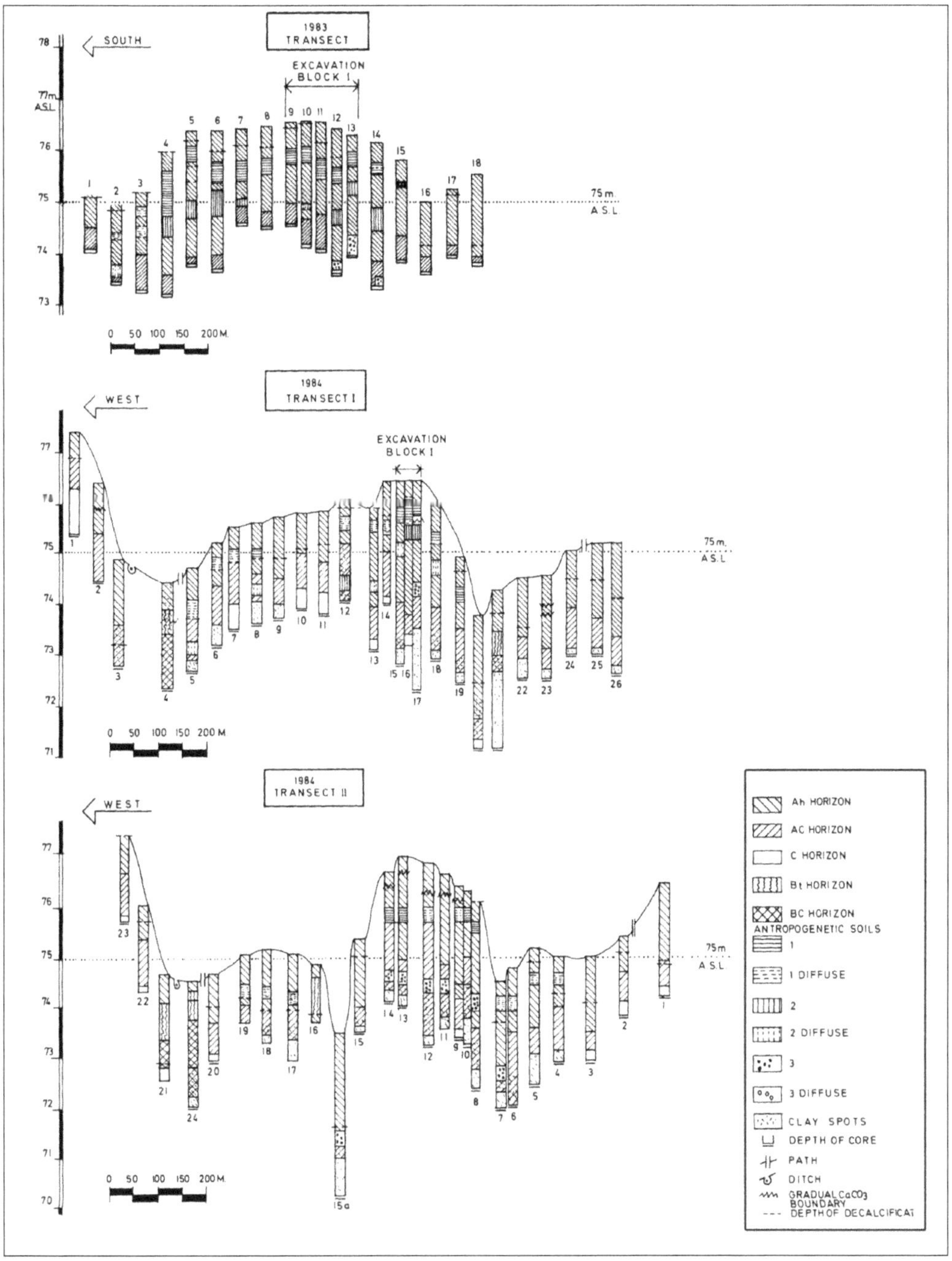

Figure 4.3. Soil probe transects (after Tringham et al. 1985).

steppe vegetation and the corresponding chernozem soil developed later, during the last part of the Subboreal and Subatlantic periods, when humans had greater impact on the vegetation had considerably increased (Havinga 1972). More recent palaeoeclogical studies in northeastern Hungary suggest that postglacial increase in deciduous forests was responsible for the development of brown-earth soils (Willis et al. 1998).

However in the case of Opovo, soil stratigraphy regarding chernozem formation seems to be different from the one described above for Austrian and Hungarian parts of the Pannonian Plain, as provided by the Dutch team of pedologists who analyzed the soil formation at Opovo and gave the following information (Tringham et al. 1985:433–435, Figure 9): The Ah horizon is described as humic chernozem soil, 170 cm thick; and the AC horizon is described as a transition between upper humus layer and the soil parent material (depth 170–280 cm). Below it is the C horizon which is the raw soil material which consists of so-called lowland loess deposited by river like currents. Around the archaeological site, a cultural layer coincides with Ah and AC horizons and partially with the C horizon (Figure 4.3). Hence, the authors conclude that on the basis of palaeosoil investigations, the prehistoric settlement was on the older chernozem soils and that "these chernozems could not have developed in a permanently waterlogged floodplain such as the one that existed until a modern system of water management was introduced" (Tringham et al. 1992:354). Therefore, it seems that there is no indication of a brown-forest layer beneath the chernozem layer at the Opovo site. It remains uncertain, whether such data provide evidence for the primary or secondary development of chernozem soils.

Unfortunately, little pollen sampling has been done in Voyvodina, either to establish vegetation succession through time or to determine the chronology of the corresponding soil formation. Willis and Bennett (1994:326–330; Willis 1995:9–25) surveyed paleoecological evidence from the Balkans and concluded that agriculture had very little impact upon the landscape until 6000 B.P., which is considerably later than prevailing theories about the spread of Neolithic cultures. The authors argue that on the basis of palynological data, there is no evidence for the reduction of woodland associated with anthropogenic disturbance before the suggested 6000 B.P., even for large archaeological sites (e.g., in Thessaly) that are well within the pollen catchment area. They suggest that the population densities must have been low and food production was on a small scale. While this work based on palynological evidence mostly from the southern Balkans sites showed negligible impact of the first farmers on the landscape around 6600 B.C., the new palaeoecological evidence from Hungary provides evidence for widespread burning of the landscape from ca. 7000 B.C. (Willis 1997). The palaeoecological record from three different sites in Hungary, Kos-Mohos Tó, Bátorliget, and Sárret provided evidence of anthropogenic disturbance caused by fire and resulting in forest clearance and different composition of the forest. In all three localities, there is no clear evidence of early cultivation, but the increase in herbaceous plants suggests that animal husbandry and wild game drive were the main causes for the clearance by burning (Willis et al. 1998).

The palaeoecological evidence from the three localities in Hungary (Kos-Mohos Tó, Bátorliget Sárret) revealed that the vegetation during the Full Glacial was forest-steppe, succeed by coniferous forest in the Late glacial. With the onset of climate warming in the early postglacial, the deciduous forests increased and the coniferous species decreased. All three localities are in the Hungarian plain, north of the Opovo site. Composition of the woodland species varied among the regions, with *Tilia* woodland in the east, *Corylus/Quercus* in the north, and *Ulmus/Quercus* parkland in central Hungary (Willis 1997).

The closest recent well-dated palaeoecological investigations in the Pannonian region are from the locality Semenec in the Banat Mountains, the westernmost part of the Southern Carpathians in Romania (ca. 100 km northeast of Opovo), and from the locality Noričke Graba in the Subpannonian phytogeographic region in Slovenia (ca. 500 km northwest of Opovo). At Semenec, (1445 m asl) in Romania, the pollen samples dated to the Middle Atlantic Period (5715–5525 B.C.) are characterized by the Mixed Oak-Corylus zone. The first human impact can be recognized in the Bronze Age (Rösch and Fischer 2000). At Noričke graba (240 m asl) in Slovenia, pollen analysis has shown the decline in lime pollen after 5000 B.C., the appearance of cereal-type pollen grains and soil erosion, and the evidence for human activity after 4000 B.C. (Andrič 2001).

In Slovenia, palaeoecological studies, including pollen analyses and radiocarbon dating, of several localities, indicate changes in landscapes that reflect small-scale clearance, burning, and coppicing, but that Holocene vegetation has no present-day analogues, However, the middle Holocene vegetation (after 6900 B.C.) occasionally "swung" towards the present one and modern-day landscape was a sudden event, according to the investigations by Andrič (2001).

In general, there is insufficient pollen data available from regions relevant for the introduction and development of agriculture into the northern Balkans, including Voyvodina. For the Atlantic period that is characterized by warmer and wetter climate, older analyses of the peat land from central Serbia show that pollen of mixed-oak forest is the dominant type for the elevations below 500 m asl, pollen of beech is for the elevations of 600–800 m, pollen of fir is dominant for elevations of 850–1200 m, and pollen of spruce for elevations above 1400 m (Gigov 1959, 1964; Pantić 1960). Palynological research performed on the Great Hungarian Plain (Szujkó-Lacza 1991, cited in Gyulai 1993) suggested presence of various forest associations during the Atlantic Period (5500–2800 B.C.). Juniper and poplar forest (Junipero-Populetum association) can be regarded as characteristic of the sandy areas, forest of ash and elm (Fraxino-pannonicae-Ulmetum association) is probably an extension of the Balkan forests, and forest of oak and hornbeam (Querco robori-Carpinetum association) is considered the continuation of the Ukrainian steppe flora (Gyulai 1993).

Because no pollen analysis has been performed in the wider Opovo region, Tringham et al. (1992:355–356) draw cautious inferences about the potential vegetation on the basis of research carried out at the Plitvice locality near Zagreb (Culiberg and Šercelj 1981) and from localities in Greece (Bottema 1974). This work suggests that during the eleventh and tenth millennia B.C., steppe vegetation covered lowland Greece and the valley of the Danube. In the ninth through seventh millennia, drier and warmer conditions encouraged the spread of open oak forests in the southern Balkans, while open vegetation of birch and fir, with some oak and hazel predominated in the northern parts of the former Yugoslavia. In the seventh through fifth millennia, the temperate climate was variable, being somewhat drier inland. An increase in precipitation north of the Danube-Sava line likely begun in the fourth millennium B.C.

Tringham et al. (1992:356) conclude that in the late fifth and fourth millennia B.C., during the Vinča period, Opovo's climate was somewhat drier than at present and that it "supported a steppe-like environment with some isolated trees". Scattered ponds and shallow lakes were to be found in the vicinity of the site, some created seasonally by spring floods, while others were permanent. The authors find it curious that the evidence should point to a relatively treeless landscape because rather large numbers of roe and red deer remains were found at Opovo, and their presence in a steppe-like environment is difficult to explain. So, possibly there were dense gallery forests lining the major river courses of the region.

Many archaeobotanical reports are available from central Europe where chernozem or *Schwarzerden* soils occur on the loess substratum, which is similar to the Opovo situation, although the climate was probably more wet than in the southern part of the Pannonian Plain. These analyses include extensive on- and off-site pollen and macrobotanical data. Despite the fact that these regions are situated further north and west of Opovo compared with the sites near Zagreb and in Greece, it seems that the analogy with LBK sites is more relevant. By now, consensus has been reached among archaeobotanists and archaeologists that early LBK settlements were usually established in areas covered by woodland and not by steppe vegetation, as previously argued. The evidence for such a conclusion comes from the various LBK settlements in loess areas. Bakels (1988), and Kalis (1988) argue that the vegetation consisted of a closed-canopy lime forest during the Middle Atlantic Period in northwest Europe. Furthermore, Beug (1986) suggests that deciduous forests of *Quercus* (oak) and elm (*Ulmus*) with a high proportion of lime (*Tilia*) are dominant in pollen diagrams from the samples taken on a loess plateau in southern Saxony. Van Zeist and Van der Spoel-Walvius (1980) also suggest for the loess area of the Paris basin that a fully developed deciduous forest was present in the middle of the Atlantic Period. Although climate and soil substrate are the determining factors for vegetation composition, the authors agree that various types of *Quercetum mixtum* forests (mixed oak forests) were present in the middle of the Atlantic Period, regardless of regional variations. Kreuz (1995) concludes that *Quercetum mixtum* forests existed during the middle Atlantic period, and that frequency of the pollen of four tree genera *Quercus* sp. (oak), *Ulmus* sp. (elm), *Tilia* sp. (lime), and *Fraxinus excelsior* (ash) varied. Hence, the available pollen data justified the rejection of the *Steppenheidetheorie* proposed by Gradmann (1901, 1933), which assumed that steppes were the dominant vegetation type in the areas settled by the first farmers in southern Germany. It has been documented that steppes did not cover loess regions during the Neolithic period in central and west Germany.

Among the Opovo plant macro samples, identified wood charcoal included fragments of *Quercus* sp. (oak), *Ulmus* sp.(elm), Salicaeae (willow), and *Cornus* sp. (Cornelian cherry) trees (Table 4.2). Ubiquitous remains of wood charcoal from the flotation samples, in addition to impressions of wooden posts of considerable sizes in house daub, argue for the existence of trees in the vicinity of the Neolithic settlement; however, they cannot provide evidence of the density of forest cover.

Present-day Vegetation Communities

Most of Voyvodina today is open agricultural land, including the Opovo region. Only individual trees can be found sporadically along the roadsides in the Opovo fields. Crops such as wheat (*Triticum aestivum*), corn (*Zea mays*), and watermelon (*Citrullus vulgaris*) are commonly grown in the fields around the site. The excavation block, in fact, was located in a corn field.

The following account of the present-day vegetation (Figure 4.4) is based on the results of the Project "Mapping of Vegetation of Voyvodina", part of a larger project: Vegetation Maps of the former Yugoslavia, coordinated by the Scientific Council of Vegetation Maps of the former Yugoslavia. The Vegetation of Voyvodina carried out by Professor Stanija Parabućski and her colleagues from the Faculty of Biology, University of Novi Sad (1983, unpublished). In their research on present-day vegetation, they used the Swiss-French school method developed by Braun-Blanquet (*Pflanzensociologie*, Wien 1951). Phytosociology defines plant communities on the basis of their floristic composition and through the ecological conditions that give rise to the association of species. Using this method as briefly described by Küster (1991), the phytosociologists list all plant species occurring in the precisely defined habitat with estimated frequencies of their occurrences. The highest plant community category is 'formation' followed by 'class', 'order', 'alliance', 'association', and the lowest category is 'subassociation'.[4] I use this in the description of the present flora and potential past vegetation of the Opovo site and its environs.

Presently, the only uncultivated land surrounding the site (Figure 4.4) is the lowest part of the dried up oxbow lakes created by rapid change in the course of the river Tamis (Vegetational Map 1983: Section Zrenjanin 3). These lower areas have a solonetz type of soil and often halophytic vegetation. One such depression in the shape of a horseshoe surrounds the elevation on which the Opovo site is located. Similar depressions are found north and south of the site. In these depressions there were two types of wetland associations: *Bolboschoenetum maritimi continentale* Soo (1945) along the edges, and *Scirpo-Phragmitetum* W. Koch (1926) in the lower parts of the depressions.

In the floristic composition of the association *Bolboschoenetum maritimi*, the dominant species is *Bolboschoenus maritimus* (shore rush). In some instances, a common reed

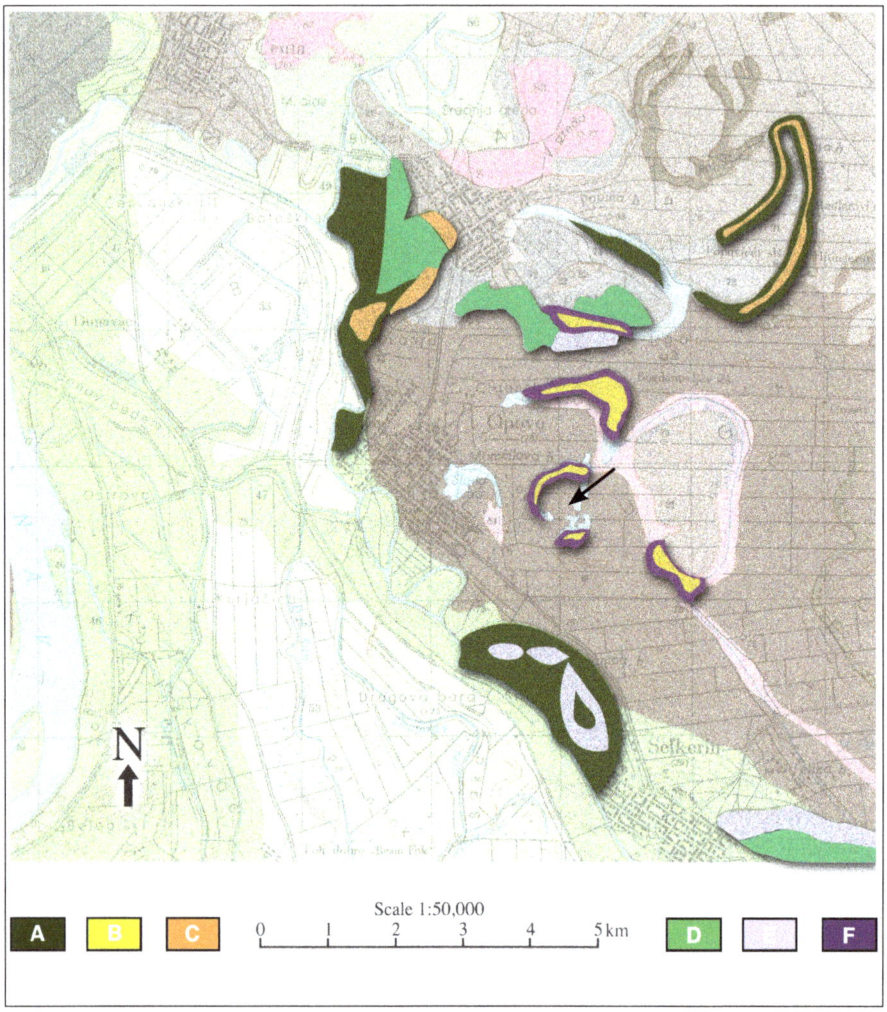

Figure 4.4. Map of present day vegetation around the Opovo site (adapted from Parabućski 1983, unpublished) over the Soil Map of Voyvodina (see Figure 1.8). The location of the Opovo site indicated with an arrow.

Legend:
A – Intentionally forested area of the Europo–American cultivar species of poplar (usually cultivars of *Populus nigra*);
B – *Scripo–Phragmitetum* W Koch (1926)–wetland community of sedges and reeds;
C – *Salicetum albae pannonicum* – riverbank community of white willows;
D – *Hordetum histrix* – halophytic community of wild barley; Unnumbered–arable land under crops;
E – *Glycerietum maxime* Hueck. (1931)–wetland community of sweet grasses;
F – *Bolboschoenetum maritimi continentale* Soo (1945)–wetland community of shore rushes.

Undesignated areas indicate modern farmland.

(*Phragmites communis*) can have a dominant role. Hence, there are two types of this vegetational community, one with a predominance of sedges and the other with a predominance of reeds. Among the accompanying species are *Agrostis alba* (bent grass), *Carex vulpina* (sedge), *Alopecurus pratensis* (meadow foxtail), and *Potentilla reptans* (five-finger rose). Most of these plants are salt tolerant, especially the halophytes such as *Roripa kerneri* (marsh cress), *Atriplex hastata* (orach), and *Lotus tenius* (trefoil), which are also represented in these plant communities.

The other plant association, *Scirpo-Phragmitetum*, is present in the lower, central areas of the depressions where water is present for most of the year, although it can dry during summers. For this plant community the following species are characteristic: *Phragmites communis* (common reed), *Schoenoplectus lacustris* (reed bed plant), and *Typha angustifolia* (cat-tail). Among them, the most common and most abundant is *Phragmites communis*. Along with these plants, the following plants are present: *Bolboschoenus maritimus*, *Carex vulpina*, *Agrostis alba*, and others.

In addition to the plant communities present in the depressions, a similar plant association is present a few kilometers south of the site: *Glycerietum maxime* Hueck. (1931). This plant community usually develops close to the reedy communities, but on slightly higher depressions on the alluvial soil rather than on the solonetz soil. The characteristic species of this association is *Glyceria maxima* (sweet grass), which is also the dominant species. Also, the following species are present: *Schoenoplectus lacustris*, *Heleocharis palustris*, *Butomus umbellatus*, *Galium palustre*, *Veronica anagallis-aquatica*, and *Alisma plantago lanceolata*. Among the accompanying species are *Agrostis alba*, *Roripa silvestris*, *Marsilea quadrifolia*, and others.

Along the Tamis River, Europo-American hybrid poplars (usually cultivars of *Populus nigra*) were intentionally planted. Sometimes one can find *Quercus robur* (pedunculate oak) and *Populus nigra* and *Populus alba* (black and white poplar) present as relics from the previous forests. In rather isolated islands within intentionally forested areas, there are small patches of *Salicetum albae pannonicum* (willow) forests. These forests are indigenous and specific to areas often flooded, where water stands during the growing period up to 2 m deep. The characteristic species of these forests is *Salix alba*, although *Populus nigra* is also present but in small numbers. The bushes, the second canopy, are nearly missing from this plant community. Among the lower plants the dominant species are: *Agostris alba*,

Polygonum mite, and sporadically *Carex elata* and *Myosotis palustris*, etc.

North of the Opovo site, there is a rather small area of saline meadow vegetation of *Hordeum histrix* present. The characteristic species, *Hordeum histrix* (a type of wild barley) usually spreads secondarily, after the regions are used for grazing. In addition, bushes such as *Prunus spinosa* (sloe) and *Crataegus monogyna* (hawthorn) represent degraded elements of the previously existing forests.

Potential Vegetation of Voyvodina

Reconstruction of the potential natural vegetation of Voyvodina, without human impact, presents a problem, since 75% of the land is under agricultural crops. Anthropogenic factors are still very influential in changing the landscape by constructing levees, canals, pastures, etc., and hence constantly altering vegetation. In many area of Voyvodina, a chernozem soil is present, and conditions of its formation have not yet been resolved. Therefore, the tentative reconstructions of potential vegetation are generally done on the basis of pedological, climatological, and hydrological information, and also by using analogies from nearby areas where the "original" vegetation is still present.

In many European countries, phytosociology is used to describe the present vegetation. Particular types of vegetation are described for various habitats using the already mentioned Braun-Blanquet method (1951). Using this method, Horvat et al. (1972, 1974) included Voyvodina (together with the Opovo site) in a forest-steppe region of the association *Aceri tatarici- Quercion* Zolyomi et Jakusc 57. According to their vegetational map (which represents the development of vegetation on mature soils on the ground level under climatic conditions of each region), the natural vegetation of the Danubian plains would be open oak forest, not steppe. For the dry areas of the plains, the subassociation of *Quercion frainetto* is characteristic. Similarly, the potential vegetation of the Pannonian area was assessed as a part of mapping the natural vegetation of Europe by Bohn et al. (2000).

Horvat et al. (1972, 1974) stress that despite the present treeless landscape, the so-called *Steppenwälder* (steppe forests) consisting of maple and oaks would have thrived in these areas. However, the forests would have included many species from the eastern regions, a combination that distinguishes them from submediterranean forests on the one hand, and central European forests on the other. Horvat et al. (1972) suggest that steppe turf covered this region before the beginning of the first millennium A.D. It is possible that this area was deforested at the beginning of the Iron Age and never fully reforested afterwards. The climate of this region is fully continental and dry, similar to the natural steppe regions north of the Black Sea.

Botanists Parabućski and Janković (1978:5–20) have attempted to reconstruct a detailed potential vegetation of Voyvodina only, assuming that there were no anthropogenic factors. According to their vegetational map (Parabućski and Janković 1978:Fig.1), the Opovo site would be located in the marginal area of the vegetational association of *Saliceto-Populetum* sensu lato and *Genisto-Quercetum roboris* Horv. and close to *Tilieto-Quercetum crassiusculae typicum* Slav., which extends to the east and north (Figure 4.5). Furthermore to the east, the plant vegetational association of *Fraxineto pannonicae-Quercetum roboris* [pedunculiflorae] is suggested. Towards the big rivers of the Tisa and the Danube, the association of the *Saliceto-Populetum* s.l. extends to the west. North of the Opovo area, there is a rather small patch of saline vegetation (Figure 4.5).

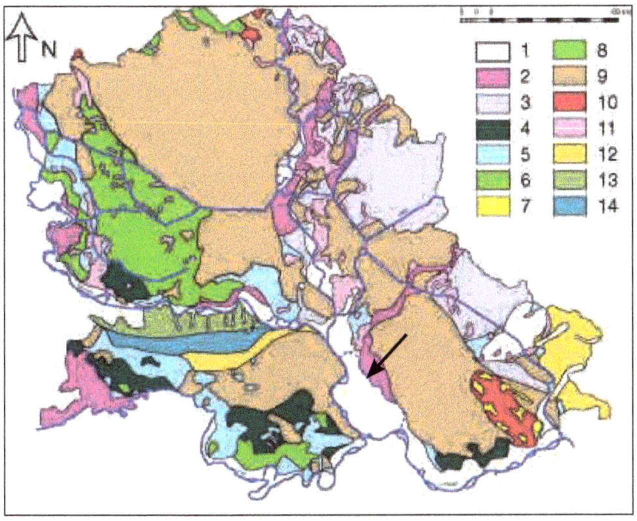

Figure 4.5. Map of potential vegetation of Voyvodina (adapted from Parabućski and Janković 1978). Arrow indicates location of the Opovo site.

Legend:

1. Saliceto-Populetum sensu lato
2. Genisto-Quercetum roboris Horv. 38
3. Fraxineto pannonicae-Quercetum roboris-[pedunculiflorae?] sensu lato
4. Tilieto-Quercetum crassiusculae carpinetosum Slav. 52
5. Tilieto-Quercetum crassiusculae typicum Slav.52
6. Tilieto-Quercetum crassiusculae farnettetosum Slav. 52
7. Querceto-Tilietum tomentosae Stjep.-Vesel. 53
8. Convalario-Quercetum roboris Soó 57
9. Tilieto-Quercetum crassiusculae typicum Slav. 52. Tilieto-Quercetum crassiusculae farnettetosum Slav. 52, Chrysopogonetum pannonicum sensu lato and other associations, Festucion rupicolae Soó 40
10. Chrysopogonetum pannonicum Stjep-Vesel. 53
11. Vegetation on saline soils
12. Quercetum farnetto-cerris sensu lato
13. Querco-Carpinetum servicum Rudski 49
14. Fagetum submontanum Jank. et Mišić 60

The determining species of the vegetational association of *Genisto-Quercetum roboris* Horv. are *Quercus robur* (pedunculate oak) and bushy plants *Genista elata* (greenweed) in the family Papilionaceae. This type of vegetation is characteristic of older alluvial soils, but also of more developed alluvial soils such as pseudogley, hydromorphic black soil, etc. These forests can occupy both areas that are not constantly flooded and areas that are flooded for a short period of time. Underground water during the vegetational period in these forests is on average 1.5 m to 3 m deep. The best examples of these forests are still existing forests in the southwest part of Srem, Bačka, and Slavonia, regions not more than 100 km from the Opovo site. These forests still thrive today in the floodplains of the rivers Danube (Figure 4.6) and Sava, and their tributaries. The existence of these famous "Slavonian oak" (*Quercus robur* ssp. *slavonica*) forests in the floodplain of the Sava River is threatened by drainage programs and the lowering of the water table (Ern 1990). Because the present-day water table in the vicinity of the excavation block at Opovo is at the depth of 4 m, oak forests could have thrived in the Opovo floodplains during the Neolithic period, even if the ground water table has been higher.

In areas where there is less flooding and where the water table is not so high, the association of *Tilieto-Quercetum crassiusculae* would probably be developed as suggested by Parabućski and Janković (1978). Present-day vegetation of this type of forest mainly consists of lime and oak, as described by Slavnić (1952:17–38) for the locality near Belgrade. He divided the association of *Tilieto-Quercetum* into three subassociations. A typical forest would be growing on the degraded chernozem, brown forest soil, or on alluvial and loess terraces. Parabućski and Janković (1978: Fig. 1) indicate these forests thrived in southern Banat and encompassed the wider Opovo region. Because the Opovo site is located on a loess terrace, and degraded chernozem soils are present, we can suppose that such forests consisting primarily of lime and oak trees could have thrived in the Opovo environs where the water table was not very high. If the area around the site had a higher water table, possibly another subassociation of this type, *Tilieto-Quercetum carpinetosum*, could have grown.This type of forest develops in regions with an average annual rainfall of 500–700 mm, tolerating underground water up to 1.5 m. Such types of forests exist in some limited areas of Voyvodina, similar forests grow in Hungary where there is less rainfall and annual temperatures are even lower (Zolyomi 1957, cited in Parabućski and Janković 1978:9).

Parabućski and Janković (1978) believe that because these forests can grow today in a less favorable climate, they could also have grown in the past in Voyvodina. The forests in present-day Hungary belong to the alliance of *Alno-Ulmion*, more precisely, the association *Querceto-Carpinetum*, according to Horvat et al. (1974). This forest of oak and hornbeam develops on soil such as hydromorphic black soil, a non-calcareous meadow soil that has altered loess for the substrate. Opovo has a substrate of altered loess, but the overlying soil is, in most instances, chernozem.

Figure 4.6. *Quercus robur* (pedunculate oak) forest Štrbac in the Danube floodplain, near Sombor, northwest Voyvodina, March 1998 (Photo courtesy of Dragana Vukov).

Carbonate chernozem soil and similar soils would be the best substrate for the development of steppe vegetation. Presently, steppe vegetation in Voyvodina can be found only on the sides of loess plateaus, or sometimes on the plateaus themselves, also beside roads, and rarely between crop fields. These steppes are similar in floristic composition to the association that develops in sandy regions, in particular to the association of *Chrysopogonetum pannonicum* Stjep.-Vesel. 53 and *Festuco-Potentilletum arenaries* Stjep-Vesel. 53. The second association is developed by degradation of the first, that is of *Chrysopogonetum pannonicum*. The structure and development of steppe vegetation is radically changed by intensive grazing and other anthropogenic factors that create a type of meadow steppe. Parabućski and Janković (1978) believe that without anthropogenic factors, then these meadow steppes would develop into steppes and would successively turn into forests of the type *Aceri tatarici-Quercion* Zolyomi et Jakusc 57. The authors stress that climate in Voyvodina is not steppic and that favorable conditions for the development of xerophytic communities are no longer present (Parabućski and Janković 1978).

Along river floodplains, such as that of the Tamis, which is in near the Opovo site, and further west towards the rivers Tisa and Danube, vegetation during the Neolithic period could have been an association of willows and poplar described as *Saliceto-Populetum* s.l. (Figure 4.1). Stands of willow and poplar develop on the recently formed soil, subjected to frequent flooding and where underground water in the vegetational period is only 1.5 m. below the surface (Figure 4.7). In the stagnant waters and canals, *Trapa natans* (water chestnut) still thrives (Figure 4.8) and was certainly present in the Neolithic times, as indicated by the remains of water chestnut fruit from the Opovo archaeobotanical samples (see Figure 2.19).

During pedological research in the Opovo region, chernozem soil was identified in the higher topography, while solonetz soil is characteristic for the more recently developed soils in the lower areas. In the areas of solonetz soil, present vegetation is halophytic. Halophytic vegetation represents secondary vegetational development in places where marshes or swampy forests originally grew and where the water was later drained (Parabućski and Janković 1978). Halophytic vegetation is discussed in descriptions of present-day vegetation. If these areas were forested in the past, they probably consisted of poplar and willow, e.g., *Saliceto-Populetum* s.l. On the other hand, if marshes were present in the past, then the representative communities could have been *Scripo-Phragmitetum* and *Glycerietum maxime* vegetation, as noted.

Archaeobotanical, Archaeozoological, and Archaeological Data for the Opovo Environs

It is rarely possible to translate modern phytosociological associations into archaeological ones because archaeobotanical specimens almost never equal one plant community but rather represent the result of human activities across the landscape (Küster 1991:17–26; Hillman 1991:27–41). The Braun-Blanquet method (1951) of determining plant communities in the landscape is very useful for describing present vegetation and for making vegetational maps of the regions; however, the phytosociological system cannot be of immediate use for reconstructing past vegetation. Nevertheless, the presence of some plant species in archaeobotanical samples points to edaphic and hydrological conditions in the region, as has been suggested by van Zeist (e.g., van Zeist et al. 1986, cited in Küster 1991:19).

Wood charcoal is the most ubiquitous and often the most abundant taxon in archaeobotanical samples collected across the excavation area at Opovo. In some samples, large quantities of wood charcoal were found. Larger wood charcoal specimens from the Opovo site were identified by Neal Lopinot (personal communication, 1997). He analyzed 20 wood fragments from six samples of various proveniences (Table 4.2). *Quercus* (oak) fragments were the most ubiquitous and most abundant. The Salicaceae (willow/poplar) and *Cornus* (dogwood) are the second and third most ubiquitous wood taxa. One sample contained three fragments of wood identified as the *Ulmus*/Ulmaceae. Such a combination of wood taxa and the percentage presence (ubiquity) of wood charcoal confirm the hypothesis that mixed-oak forests grew somewhere around the site. Willow forests probably thrived along the rivers and depressions of palaeomeanders. The evidence can point to

Figure 4.7. Saliceto-Populetum s.l. association (willows and poplar forests) along the riverbanks of the Tamis river near modern village of Opovo, 1998 (Photo courtesy of Slobodan Marković).

Figure 4.8. *Trapa natans* (water chestnut) plant in front of the dense cover of the plants at Agli Bara, an ox-bow lake of the Danube, Voyvodina, Serbia, August 2001.

Table 4.2. Wood charcoal taxa (count) identified from the selective Opovo flotation samples.

Locus	1418	1574	1782	1887	1898	Loam pit
Quadrant	58/109	65/115	63/115	60/117	65/117	
Layer	7	7	11	10	10	
Feature	41/42	House 6	30	61	59	
Quercus sp		2	19	12	7	20
Cornus sp.		18			8	
Salicaceae			1	4	3	
Ulmaceae				3	1	
Type A	20					
Type B				1		
Type C					1	
Total	20	20	20	20	20	20

Note: Charcoal fragments were identified from six samples by Dr. Neal Lopinot. Type A is extremely knotty or gnarly wood, probably diffuse or semi-diffuse porous with numerous and variable rays (1-5 cells wide); Type B is a finely diffuse porous type of wood, but vessels are much more scattered than in Salix; Type C is a finely porous, semi-diffuse porous, or ring porous wood.

exploitation of forests in the proximity of the site (Figure 4.1).

The argument for the existence of drier areas can also be supported by the frequent presence of both stones of *Cornus mas* fruit and wood charcoal in many samples (Figure 2.13). *Cornus mas* is a common species of the deciduous oak forests (Figure 4.1). Moreover, this species is thermoxerophitic, indicating a preference for warm and dry locations (Šilić 1988b:140). However, presently not a single *Cornus* plant has been observed near the site. Other evidence for the existence of such drier areas in the vicinity of the site is the presence of the cereals, legume and flax seeds. Cereals, legumes, or flax cannot thrive if their root systems reach underground water (Figure 4.1).

Additionally, the evidence from wild fauna indicates that forests existed in the Opovo region. The identified bones of red deer, roe deer, and wild boar, which are primarily forest species, allowed Russell (1993:39) to accept that a mosaic of light forest with relatively small patches of steppe was the dominant vegetation type of the region, as suggested by botanists. The brief survey of red deer, wild boar, and roe deer feeding habits supports the hypothesis about the Opovo forested environs.

Red deer (*Cervus elaphus*) is the most abundant taxon in the Opovo samples (Russell 1993). Their natural habitat is forested land. In lowlands, red deer live in oak forests where they can find an abundant supply of acorns to feed upon. In addition to acorns, red deer feed on young saplings, leaves, or fruits of bushes such as of dogwood, elderberries, and blackberries. Besides the oak forests, red deer also inhabit natural stands of willow and poplar whose young saplings they readily eat. From the bottom forest layer they feed upon *Brassica*, *Anemone*, *Plantago* species, and grasses (Poaceae). They usually feed during the night and ruminate during the day, hidden in the forests (Novaković 1996:15–17).

Bones of the pig/wild boar (*Sus scrofa*) are the second most abundant taxon identified among the Opovo faunal remains. Most of the pig bones belong to wild boar and there was little if any hybridization between domestic and wild pigs occurring at the Opovo site, according to Russell (1993). Wild boars are commonly found in the wider region and their natural habitat is humid forest with dense bush underlayer. They usually eat acorns, mushrooms, tubers, small vertebrates, and they are also scavengers. Close to the ponds or canals, wild boars also readily eat water chestnuts. Nowadays, wild boars are known as notorious maize and potato raiders, but in the past instead of maize fields, they could have invaded cereal fields. They are most active during twilight and night (Novaković 1996).

At Opovo, many bones of roe deer (*Capreolus capreolus*) have been identified (Russell 1993). Their natural habitat in the past were forests, but among ungulates, roe deer adapt best to environmental changes. Locally, roe deer inhabit low-lying, wetland-forests up to the mountains, and often live on the edge of natural vegetation and fields approaching human settlements. During the summer, roe deer often raid crops, and after harvest time, they return to the forests. They graze on many more grasses, including cereals, than red deer, and they eat both during the day and night, especially at twilight (Novaković 1996:15–17).

Figure 4.9. Large wood impressions in house daub from Opovo. Mirjana Stevanović cleaning daub (Tringham ©Creative Commons License).

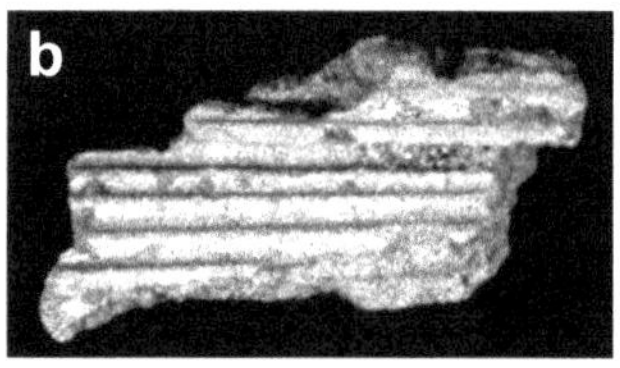

Figure 4.10. *Phragmites communis* (reed) impressions from Opovo:
a) in house daub (lower left) (Tringham ©Creative Commons License)
b) detail of the stem structure in daub (length ca. 2 cm).

Additional information supporting the hypothesis of the existence of forests is provided by lithic analysis of stone tools, many of which were used for wood processing at the Opovo site (Tringham et al. 1985, 1992).

Moreover, the organic building materials used in the Opovo structures also reflect mixed vegetation and the presence of forests, according to Stevanović (1996, 1997a). The wood impressions in house daub indicate that wood used for house construction consisted of branches, as well as timbers that were split from thick tree trunks (Figure 4.9). Although it is impossible to identify the tree species from impressions in daub, Stevanović (1996:188) suggests oak was used for primary and secondary house posts at Opovo. Also, a preserved fragment of charcoal that belonged to pine tree and a fragment of a tree bark were identified from an oven floor (Stevanović 1997a:358).

Marshes could have been present in the low-lying areas with stagnant water (Figure 4.1). Impressions of *Phragmites communis* (reeds) found in the Opovo house rubble indicate there were marshes around the site (Figure 4.10). Furthermore, harophytic algae that were observed during petrological clay analysis confirm that clay from marshland sediments was often used for house construction (Stevanović 1996:168; 1997a:358). The existence of wet areas in the Opovo environs is also indicated by the presence of fruits from an aquatic plant *Trapa natans* (Figures 2.19 and 4.1). Nevertheless, there could have been some natural openings in vegetation on drier land, possibly covered with steppe vegetation or meadow, as has been suggested by Kreuz (1995) for central European Neolithic sites.

Presumably, not much saline vegetation was present in the area in the past, because saline vegetation usually occurs with a systematic lowering of the water table. Salinity of the soils is usually associated with irrigation and more recent drainage. It is not clear if the soil salinity increased in the prehistory as a consequence of removal of woodlands on flatter sites, but if flooding occurred frequently then the salt would have been leached out (Shiel 1997).

***

All evidence cited above (Bakels 1988, Kalis 1988, Kreuz 1995) from central and western Europe seems to imply the presence of forests, and that some were cleared for crop cultivation, construction of houses, and for fuel during Neolithic times. The more recent palaeoecological record from three different sites in Hungary indicates anthropogenic disturbance and evidence of fire and forest clearance from 7000 B.C. The clearance by burning and increase in herbaceous plants is, however, attributed to animal husbandry and wild game drive rather than to crop cultivation (Willis 1997). On the other hand, pollen data from Southern Balkans indicates there is no evidence for major forest clearing for crop cultivation in these regions until the end of the Neolithic and beginning of the Bronze Age and later periods (Willis and Bennett 1994).

Only a very few of the identified plant species from the Neolithic period belong to plants characteristic of present-day treeless vegetation, and hence point to openings of arable land once surrounded by shrubs and trees that shaded some of the crops. Other authors (Jager and Neuhausl 1992a in press, cited in Jager and Neuhausl 1992b), however, suggest there were fields of various sizes interspersed with small groups of shrubs and single trees. These authors believe this argument better explains the evidence of erosion since Neolithic times. Nevertheless, it seems certain that the first farmers had the technology for clearing forests and cultivating crops on those lands with or without use of the ard, for which there is no evidence in the Neolithic period.

The Opovo site is a a relatively small Late Vinča hamlet, ca. 5 ha (Tringham 1994) compared with other large Vinča sites, e.g., Selevac ca. 53 ha (Tringham and Krstić 1990). Tringham (1994) estimated that the Opovo settlement was occupied by not more than 10 households at any one time for 200 years (five-seven generations), in contrast to the larger Vinča sites which were occupied by hundreds of households. Assuming that 50–250 people could have lived in the settlement at one time, there were probably enough uninundated areas suitable for crop cultivation and animal grazing, especially if population was low.

In conclusion, multiple lines of evidence including present-day botanical, faunal, archaeobotanical, archaeozoological, and archaeological data from the Opovo excavations indicate that there is no reason to suggest rather treeless vegetation to have dominated the Opovo region in the Neolithic period during the occupation of the Late Vinča settlers.

Although pedological investigations have shown that the fluvial landscape near Opovo was characterized by either stagnant or running water, one may assume that, in addition to marshy areas, there were riparian mixed oak forests and potentionally dry areas (steppe) suitable for farming and animal grazing. Indeed, archaeobotanical samples show the presence of cereals and legumes along with wild fruits and aquatic plants (Figure 4.1).

CHAPTER 5. HUSBANDRY

Archaeobotanical samples can provide information on the way crops were managed in fields and processed in a settlement. Such an approach was pioneered by Dennell (1974), and elaborated and substantiated with detailed ethnographic studies of crop processing by Hillman (1981, 1984), Jones (1984), and others. Composition of archaeobotanical samples from various archaeological contexts is very similar to the samples observed in ethnographic studies. Thus, by reference to modern analogies, archaeobotanical samples can provide valuable clues to crop husbandry (Hillman 1984). In addition to the ethnographic and historical data available from the literature, ethnographic observations that I mention without references are from personal experience and from information provided by agronomists and peasants. Most of the information is drawn from the Voyvodina region where the Opovo site is located, and from the hilly region of Banija in Croatia where traditional farming was practiced until recently.

After the identification of plant taxa and the composition of plant remains from various features presented in the preceding chapter, in this chapter, I attempt to infer the agricultural practices of the Opovo inhabitants during the Late Neolithic period. I concentrate mostly on the husbandry of the main crops whose remains have been found in the Opovo samples: emmer, einkorn, lentils, vetches, and flax. According to Hillman (1984), thirty distinct operations are involved in growing a crop and converting it to food. Although the comparative data that I assembled are from temporally and spatially diverse sources, I attempt to determine the possible location of the Opovo fields and the probable cereal yields, as well as evidence for the use of straw and chaff in the Opovo settlement. In addition, I discuss several operations including plowing, sowing, crop rotation, harvesting, dehusking, pounding, and storing during the prehistoric occupation of the site.

Cereal Cultivation

To predict the area of land needed for cereal cultivation and the possible yields, one must estimate the potential population of the settlement on the basis of the houses and densities.

As mentioned earlier, the Opovo site is a relatively small Late Vinča site (5 ha) in comparison with other Vinča sites (e.g., Selevac, Poptoranj) which are 50–100 ha large (Chapman 1981; Tringham 1994). It is difficult to extrapolate from the magnetic survey (area of 45×200 m) and the excavation data (block 16×20 m) the number of houses built at the Opovo site during one building horizon. The excavated houses were not large and densely placed (Figure 1.9). Their sizes ranged in surface area from ca. 24–40 m² (Stevanović 1996:185). The few contemporary houses discovered at Opovo were built close together, only a few meters apart. Tringham (1994) estimated that up to ten households occupied the settlement at any one time.

Population estimates are 100–300 people per village (2–5 ha large) for the Neolithic period in Thessaly (Demoule and Perlès 1993). For the late Vinča period, estimated population range between 150–400 people for settlements 3–3.9 ha large (Chapman 1981). Considering the size of the excavated houses at Opovo, we can assume that 5–6 people could have lived in one house. If at least few houses were inhabited, we can suggest that at least 50 and up to 250 individuals lived in the Opovo settlement at one time.

The prehistoric farmers' dependence on cultivated grain and yields and the available land remain unknown. But for reconstructing prehistoric agricultural activities, researchers have used ethnographic, historic information, and data from experimental cultivation of emmer and einkorn to estimate prehistoric crop yields. Emmer and einkorn were the principal cereals at the Opovo site.

Yields

Yields of grain are traditionally measured in various small units of weight (not in tons), and commonly as a ratio between the amount of sown grain to the amount of harvested grain (Sigaut 1992:395–403). The ratio in traditional agriculture varies between 1:2 and 1:3 in bad years and between 1:8 and 1:10 in good years, regardless of the amount of seeds sown per land unit. Measurements of yields relative to seed grain (fold) are standard for most villages where traditional agriculture is practiced (Ertug-Yaras 1997:Table 27; Watson 1979:67; personal observations). However, I employ tons or kg per hectare in order to standardize the results and compare them with other ethnographic and historic data about wheat crop yields.

Bakels (1992) estimated for the minimum area of land needed by 50 people for agriculture and cattle herding in the early Neolithic LBK settlements of Central Europe (ca. 5400–4900 B.C.). She based her calculations on historic evidence from Russia and Canada, on the results of experiments with einkorn yields at Butser farm in England, and on the FAO data on basic requirements of "average people". She produced a table (Bakels 1992:Table 1) in which she calculated that not more than 11 ha were needed for 50 people to grow wheat if the yield was 800 kg per hectare and if the diet were 65% wheat. An additional 150 hectares would have been necessary for cattle to graze. Bakels (1992) concludes that only a dozen hectares of cleared land were required for an LBK settlement of average size, and hence the first farmers had limited impact on the.

Gregg (1988) used historic data from 1850–1900 from the Württemberg region in Germany in estimating crop yields. She computed that an average yield was 845 kg/ha for emmer-spelt winter yield, while Gregg's adjusted yield of 556 kg/ha for spring emmer-spelt wheat is lower than the Bakels's estimates (1992).

Gross et al. (1990) simulated supplies necessary for 35 settlers in a settlement of seven houses over a 25-year period in the Late Neolithic of the Zurich Lake area (Pfyner Dorf AKAD J dated to 3700 B.C.). They based the calculations on the actual MNI of animals and the calories provided by meat. By changing the values of calories provided by cereals and gathered plants, they suggested a model in which more than 52% of calories would be provided by cereals, ca. 31% from meat and milk, and ca. 16% from nuts. In order to provide a daily intake of 1244 kcal from cereals per person per day, Gross et al. (1990) estimated that 7 ha of land were under cultivation if the yield was 650kg/ha, based on ethnohistoric yields cited in Gregg (1988).

Hillman and Davies (1992:Table 1) have compiled data on the amount of grain for a family of five. They based their calculations on the caloric contents of modern wheat, and on the assumption that 25% of one's total caloric requirements are provided by grain. According to these computations, a family of five would need 330 kg/per year of wheat. If the yield was 500 kg/ha, then 0.75 hectares had to be sown; if yields were higher, 1,000 kg/ha, then 0.4 ha would be sufficient for five people for one year.

Additionally, there are now data available on emmer and einkorn yields on archaeological experimental farms in England and France that can serve as analogies for prehistoric yields. At the Butser farm in England, the average yield of autumn-sown emmer grown in fields of different microclimatic zones during fifteen seasons was 1.65 t/ha (Reynolds 1992:Table 5). The ratio between seed sown to seed yield varied from 1:4 up to 1:60. However, the average ratio in three climatic zones is above 1:20, a higher ratio than ethnographic evidence for the Balkans and the Near East. As we have seen from previous ethnographic examples, a ten fold yield constitutes exceptionally good harvest. Similarly, the historic data indicate much lower yields than those obtained in the Butser farm. For example, the Roman agriculturalist Columella (*On Agriculture*, Book II, III) writes that a wheat yield from an average farm in Italy was about 530 kg/ha. Likewise, data for Medieval Britain from the thirteenth and fourteenth centuries show that the average wheat yield was ca. 530 kg/ha (Titow 1972, cited in Russell 1988:111). Good yields that were obtained at the Butser farm were empirical and were not extrapolations, although Reynolds (1992:392) cautioned that they should be regarded as probabilities before their conversion into the prehistoric data. One reason for such high ratios at the Butser farm is that the climate in England allows much better tilling of wheat than in drier areas. In addition, Reynolds was an exceptionally careful and diligent farmer. Reynolds (1979:62–63) sowed emmer "in seed furrows in the tilth" in order to hoe out the weeds. He justifies such a practice with a wealth of ethnographic evidence from the temperate zones of Europe to support wheat sowing in furrows. He was able to control weed infestation and seed density. Nevertheless, such a type of wheat sowing is uncommon in the Balkans and in many other parts of Europe and the Near East where wheat is sown by broadcasting and weeds are removed by pulling and never by hoeing, although we cannot exclude the possibility of sowing seed grain in furrows in prehistory of the Balkans.

Similarly, high yields of ca. 2500 kg/ha of winter bread, more than 20 fold, were obtained at the experimental plots near Stuttgart where forest was newly cleared by the slash-and-burn method (Rösch et al. 2002). The continuous cultivation of cereal crops on the same plot produced zero yields due to the heavy infestation by weeds. The yields obtained from the three-field system produced less than 1000 kg/ha. Summer cereals produced lower yields than the winter cereals.

The oldest historic documents (that I could find) for the region of Banat where the Opovo site is located date to 1858 (Hegediš and Čobanović 1991:310). During that period, already more than a million hectares were under crop cultivation. The main crops were maize and wheat, although a mixture of wheat and rye (*napolica*) was still sown on hundreds of thousands of hectares. The average wheat yield was 1.41 t/ha, and the average yield for a mixture of wheat and rye was 1.51 t/ha (Table 5.1). Data from the twentieth century show that only wheat and maize yields significantly improved in comparison to the yields in the nineteenth century; the yields for other cereal crops did not change significantly. Also, there are data for a small amount of land (14 ha) that was still under emmer cultivation from 1966-1975 (Table 5.1). The yield of emmer was 1.71 t/ha (Savezni Zavod, *Statistički Bilten* 95, 1957). Emmer was grown in small quanties in Montenegro until the 1970s (Figure 5.1).

Figure 5.1. Emmer (*Triticum dicoccum*) field near Nikšić, Montenegro in the 1970s (Photo courtesy of Ljubo Pavičević).

Table 5.1. Crop yields in Voyvodina from historic documents.

	Banat 1858[1]		Voyvodina 1930–1939[2]		Voyvodina 1966–1975[3]	
Crop	Area (ha)	t/ha	Area (ha)	t/ha	Area (ha)	t/ha
Wheat	291,291	1.41	660.000	1.37	427.330	3.56
Wheat+Rye	205,844	1.51				
Rye	24,710	1.57	6.000	1.19	2.201	1.86
Barley	72,610	1.70	35.000	1.52	23.458	2.58
Oats	280,034	1.80	24.000	1.23	262	1.06
Millet	325	1.28			143	1.70
Millet (shown after wheat)					27	1.10
Emmer					14	1.71

[1] Banat is a region within Voyvodina where Opovo is located. Data for 1858 after Hegediš and Čobanović (1991:310).
[2] Data for 1930-1939 from Statistički bilten 95 (1957).
[3] Data for 1966-1975 from Statistički bilten 1049 (1977).

In addition to historic data, precise ethnographic data from the 1950s from Banija, a region in Croatia where traditional agriculture was practiced until the 1970s, can serve as an analogy for approximate yields when no pesticides or mineral fertilizers were used (Borojević 1995). In the Neolithic period because of the limited technology, grain yields may have been low in Opovo; however, they were possibly higher than in the hilly regions of Banija. As mentioned earlier, the main crops in the Opovo region were *Triticum monococcum* and *Triticum dicoccum*. Unfortunately, there is no precise information on the yields of *Triticum dicoccum* from Banija, although emmer was grown in this region in limited quantity until World War II. Therefore, I used data on yields of old varieties of bread wheat grown locally (S. Borojević 1995:Table 2). These yields were calculated from counted spikes and weighed grain from 1 m^2 taken from various plots. Although the yields extrapolated to hectares on the basis of 1 m^2 might not have been completely realistic, they were more reliable than the villagers' estimates. Not a single peasant was able to tell what the exact wheat yield per land unit was because the fields were smaller than standard land units (*jutro*=5,500 m^2), and because threshing was done after sufficient amount of wheat from various plots was harvested. In 1956, which was considered a bad year (vegetation period during 1955/1956), the minimum number of wheat spikes per 1 m^2 on one peasant's field was 141, the average number from another farm was 229 spikes, and the maximum number was 291 spikes per 1 m^2 on the third peasant's farm. In the following 1956/1957 year, which was a rather good year in Banija, the grain yields were higher even though the number of spikes per 1 m^2 was about the same, because the grain number and weight was higher per spike (Borojević, 1995).

In order to calculate the possible yield of the Opovo crops instead of old bread wheat varieties, I took whole spikes of *Triticum monococcum*, *Triticum dicoccum*, and *Hordeum* from the Anatolian plateau in Turkey, obtained from Mark Nesbitt's reference collection. I counted the number of grains per spike and weighed the grains per spike in grams. In addition, I took spikes of Triticum *monococcum* from the Agricultural Experimental Station in Novi Sad in Voyvodina (the former Yugoslavia), where the soil is highly fertile. Also, I included *Triticum aestivum* ssp. *vulgare* spikes from Anatolia (T) and from Novi Sad (NS), and *Triticum spelta* from Novi Sad (Table 5.2). I multiplied the actual spike grain weight of *Triticum monococcum*, *Triticum dicoccum*, *Triticum aestivum*, *Triticum spelta*, and *Hordeum vulgare* specimens by the number of spikes per 1 m^2: 141, 229, and 291, for three fields respectively from the ethnographic crop data from Banija (Table 5.3). As a result, I was able to extrapolate yields per hectare for *Triticum monococcum*, *Triticum dicoccum*, and other crops. The einkorn yield calculated on the basis of the Anatolian spikes is 0.47 t/ha (141 spikes per 1 m^2), 0.77 t/ha (229 spikes per 1 m^2), and 0.98 t/ha (291 spikes per 1 m^2), while the einkorn yield calculated from the Novi Sad spikes was almost double. Such a discrepancy in einkorn yields can be explained as a consequence of better soil and tillage in the Agricultural Experimental Station in Novi Sad compared with rural Anatolia. Furthermore, the yield of emmer is almost double than of einkorn, and it varied from 1.22 t/ha, to 1.99 t/ha, up to 2.53 t/ha. *Hordeum vulgare* spp. *hexastichum* yields range from 1.42 t/ha, to 2.31 t/ha up to 2.93 t/ha (Table 5.3).

If one imagines that there was a similar low density of cereal crops per 1 m^2 at Opovo as ethnographically documented in Banija, then it is understandable that various weeds were able to grow in crops. Seeds of several weed species such as *Bromus*, *Chenopodium*, and *Galium* were indeed found in the Opovo archaeobotanical samples (Table 2.1).

Wheat yields similar to the ones I calculated were obtained by peasants in the village of Kizilikaya in central Anatolia in the 1990s (Ertug-Yaras 1997:Table 27). These are realistic figures from a questionnaire circulated among 30 households of different income levels. The mean wheat yield among 22 households was 0.7 t/ha. The average ratio

Table 5.2. Cereal grain and chaff measurements per spike.

Cereal Species	Actual Measurements per Strike			
Specimens:	Grains Ct.	Chaff Wt. (g)	Grains Wt. (g)	Chaff/ Grain Ratio (g)
Triticum monococcum (T)[1]	13	0.13	0.34	0.38
Tr. monococcum (NS)[2]	19	0.14	0.62	0.23
Tr. dicoccum (T)	35	0.30	0.87	0.34
Tr. dicoccum (T)	25	0.16	0.69	0.23
Tr. aestivum (T)	27	0.27	0.88	0.31
Tr. aestivum (NS)	42	0.75	2.12	0.35
Tr. spelta (NS)	26	0.42	1.07	0.39
Hordeum vulgare hexastichum (T)	31	0.21	1.01	0.21

[1] T=Turkey

[2] NS=Novi Sad, Voyvodina

[3] Number of spikes per 1 m^2 is from ethnographic data from Yugoslavia for 1955/1956 (after S. Borojevic 1995:71-72).

between sown and harvested wheat was 1:5, the minimum was 1:2, and the maximum was 1:13. Since 1994 was a rather dry year, in good years the average ratio was somewhat higher. Of 115 tons of wheat yield in 1994, 41 tons were saved for household consumption and for the next year's seed (Ertug-Yaras 1997).

There is available data for the quantity of wheat grain necessary for making flour in Anatolia. An average for 163 individuals in 30 households was 114 kg of wheat grain per person, per year (Ertug-Yaras 1997). One-quarter of the wheat grain weight was bran that must be separated during the milling process. That amount of grain produced 88 kg of flour per person per year. The maximum amount of flour per person was 106 kg, and the minimum was 28 kg per person in a household that did not have its own wheat fields. In contrast, there were households that did not sow their own wheat but consumed 60–90 kg of flour. According to the responses elicited by the questionnaire, there was no strong correlation between the amount of bread consumed by each family and their wealth and the amount of land owned. There is slightly lower consumption of flour by the wealthier group, and the middle-class consumed bread slightly over the average, and the poor consumed less than the average amount. The amount of bread consumed per household increased during special occasions when additional bread was baked for workers or guests.

Estimates of the minimum amount of wheat consumed at the Opovo site per person are very approximate because we do not know the size of the prehistoric Opovo population. Extrapolations can be made on the basis of Hillman and Davies (1992) information, that 330 kg of wheat per year is sufficient for a family of five if the diet consists 25% cereals intake. Ethnographic data from Anatolian village show the average consumption of wheat grain was 114 kg per person per year (Ertug-Yaras 1997), and simulations for the Neolithic settlement in Switzerland estimated the consumption at 130 kg of cereals per person per year (Gross et al. 1990). If the Opovo inhabitants consumed coarsely ground wheat grain as porridge or bulgar rather than leavened bread, then less wheat would have been necessary. Also, making pita bread requires less flour because approximately one-third of the mixture is liquid. More flour is needed to make leavened bread, because only one-fourth of the mixture is liquid. Several grinding stones have been recovered from the Opovo site, but it seems clear that grinding implements were not exclusively used for grinding grain on the basis of wear patterns and the choice of rock types, according to Voytek, co-author in Tringham et al. (1992:379). The absence of grinding stones intended for exclusively grinding of grains might indicate that less grain was ground into flour and that most was coarsely ground, perhaps in wooden mortars in the Opovo settlement. Such a practice of grinding grain may imply that wheat was consumed more in the type of bulgar and/or coarse pita bread than in a form of a fine-leavened bread. Bulgar is cracked wheat, a staple throughout the Near East. There is a wealth of both historic (White 1976) and ethnographic evidence (Ertug-Yaras 1997; Hillman 1984) documenting preparation and use of bulgar. Some types of

Table 5.3. Extrapolated cereal yields on the basis of grain and chaff measurements.

Cereal Species	Extrapolated Yields (g) per 1m^2				Extrapolated Yields (tons) per 1 ha			
Specimens:	Grain Wt. (g) 141 spikes[3]	Chaff Wt. (g) 141 spikes	Grain (g) 229 spikes	Grain (g) 291 spikes	Chaff (t) 141 spikes/1 m^2	Chaff (t) 141 spikes/1 m^2	Grain (t) 229 spikes/1 m^2	Grain (t) 291 spikes/1 m^2
Triticum monococcum (T)	47.94	18.33	77.86	98.94	0.479	0.183	0.779	0.989
Tr. monococcum (NS)	87.42	19.74	141.98	180.42	0.874	0.197	1.420	1.804
Tr. dicoccum (T)	122.67	42.30	199.23	253.17	1.227	0.423	1.992	2.532
Tr. dicoccum (T)	97.29	22.56	158.01	200.79	0.973	0.226	1.580	2.008
Tr. aestivum (T)	124.08	38.07	201.52	256.79	1.241	0.381	2.015	2.561
Tr. aestivum (NS)	298.92	105.75	485.48	616.92	2.989	1.058	4.855	6.169
Tr. spelta (NS)	150.87	59.22	245.03	311.37	1.509	0.592	2.450	3.114
Hordeum vulgare hexastichum (T)	142.41	29.61	231.29	293.91	1.424	0.296	2.313	2.939

bulgar are made from unboiled wet grains, but, for other types wheat grain is first boiled, dried, dehusked in mortars and mills, and then cracked in mills. There is now available archaeological evidence of such practice. Broken grains of barley and wheat, found at the Late Neolithic site of Kapitan Dimitrievo in Bulgaria, were probably prepared as bulgar (Marinova 2001). At the Bronze Age site Mesimeriani in Greece, glume wheats were broken prior to charring and perhaps treated with water (Valamoti 2002). Bulgar is no longer locally made or consumed in the Opovo region. However, a type of cooked wheat that is usually ground (*žito, koljivo*) is consumed by Orthodox Christians including the present-day Opovo region to symbolize death and resurrection on various religious occasions.

It is likely that the Opovo inhabitants consumed their cereal grains primarily in the form of groats, porridges, and whole-grain flat breads. As to why the Opovo settlers might have preferred coarsely ground grains to more finely ground flour, I might suggest several answers. One of the possibilities is simply the taste preference. Another explanation could be that since the grinding stones are heavy, and the nearest quarries are 40 km or more away, it was difficult to transport many grinding stones from their sources, and the Opovo inhabitants had to rely more on wooden mortars and use grinding stones for multiple purposes.

Straw and Chaff

Mineralized (silicified) chaff from domesticated glume wheat, probably of emmer and einkorn, and straw remains and their omnipresent impressions in house daub from the Opovo site, prompted me to investigate further the amount, function, and importance of chaff and straw used in tempering clay used for building Opovo houses.

Opovo houses were built using the wattle-and-daub technique (Figure 1.10), which required use of wooden posts, twigs, and house daub for plastering the wooden construction of walls. Stevanović (1996, 1997a) examined the Opovo house structures in detail and found that plant matter was used for tempering the clay including large quantities of chaff. The Opovo houses were destroyed by fire and in the burnt rubble plant material was preserved in silicified form or as hollows known as pseudomorphic vegetal voids (Figures 4.9, 5.2–5.4).

Although there are no techniques to identify the actual quantity of straw and chaff used, visual inspection of large quantities of daub verified that great quantities of wheat chaff were used for tempering Opovo house daub (Figures 5.2 and 5.3).

The addition of straw and chaff to clay is documented archaeologically, ethnographically, and in historic documents. In the Bible (*Exodus* 5:6–9), Pharaoh is mentioned punishing the Hebrew captives by requiring them to make the usual quota of mud bricks without providing straw. Similarly, the use of straw, clay, and water as the basic ingredients in brick making is depicted on the wall of the tomb of the of grand Vizier Rekhmire in Thebes, Ancient Egypt, dated to ca 1430 B.C. (Strouhal 1992).

In experimental archaeology, straw was used to construct an experimental wattle-and-daub circular house, 6 m in diameter, at Butser farm in England (Reynolds 1979:30–35). The house was built according to the actual plans and excavation information of the Iron Age house of Maiden Castle in Dorset. For that purpose, daub was made following the recipe: 3.5 tons of clay, 3.5 tons of earth, 40 bales of straw, the hair from 40 pigs, sundry brambles, grass, hay, roots, and other vegetable matter (without stating more precise volume or weight measurements).

Figure 5.2. House daub with twig and cereal chaff impressions from Opovo.

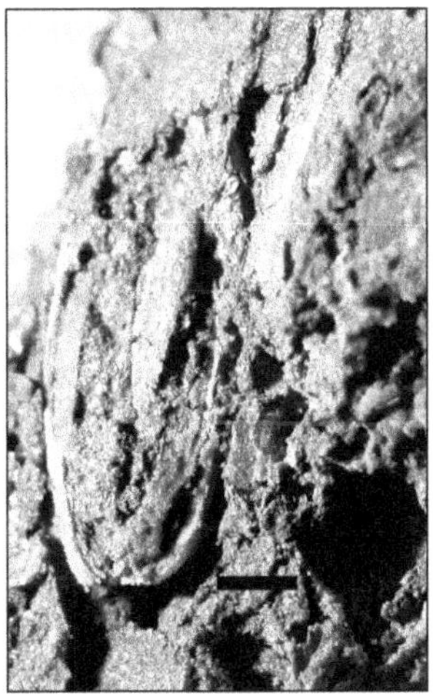

Figure 5.3. Mineralized *Triticum* sp. chaff in house daub from Opovo (scale 1 mm).

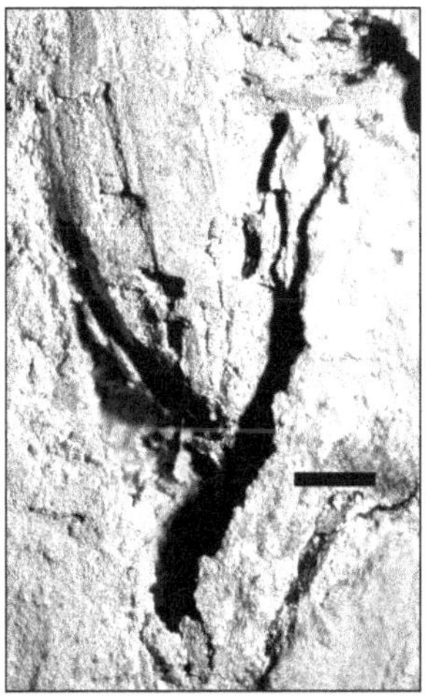

Figure 5.4. *Triticum monococcum/dicoccum* spikelet impressions on a fragment of a large pot from Opovo (scale 1 mm).

Table 5.4. Estimates for the amount of straw needed to temper 1 m^3 and 12 m^3 wall daub.

Straw	Quantity of straw for 1 m^3 daub wall			Quantity of straw for 12 m^3 daub wall		
Percentage (%)	Volume straw (Liters)	Weight (kg) if 3 lit. straw=100 g	Weight (kg) if 3 lit. straw=150 g	Volume straw (Liters)	Weight (kg) if 3 lit. straw=100 g	Weight (kg) if 3 lit. straw=150 g
13	130	4.3	6.5	1560	51.6	78
17	170	5.6	8.5	2040	67.2	102
20	200	6.6	10	2400	79.2	120

Ethnographic studies using local clay can serve as analogies for estimating the clay-straw ratio used in the construction of the Opovo houses. The exact clay-straw proportion depends on the availability of straw and the properties of the clay, and probably on the local house construction traditions. For this study, the detailed information provided by the architect Ginder (1996) was used to calculate the amount of straw and chaff used for tempering the daub. Ginder studied and documented the traditional construction of mud (*pisé*) houses locally built in Voyvodina in the 1960s and the 1970s, in the same general region of the Opovo site. According to Ginder (1996:61, 81), for the construction of 1 m^2 of wall daub plaster 2 cm thick, the following quantities are used:

20 liters of clay + 3–4 liters of cut straw or chaff,

or 1 m^3 of clay + 0.15–0.20 m^3 of cut straw or chaff.

The following question arises: How many loads of straw and chaff were required to build an Opovo house? The exact quantity of additional straw used for tempering depends on the clay properties, but according to data, the proportion was 13–17% of straw to 83–87% of clay (Ginder 1996:61, 81). The mud walls in Ginder's investigated required much thinner layers of plaster than did the wattle-and-daub construction used at the Opovo site. Stevanović (1996:180) has estimated that approximately 12 m^3 of clay were used for an idealized Opovo house 8×4×2 m large, and suggested 20–30% of organic matter was added to clay.

I used bread wheat (*Triticum aestivum* ssp. *vulgare*) straw instead of emmer and einkorn straw to calculate the weight from volume data provide by Ginder (1996). Although bread wheat straw has bigger and heavier internodes, the stalk is hollow. On the other hand, emmer and einkorn straw are considerably thinner than bread wheat straw, but are not hollow.

If we extrapolate from the liters to the actual weight of bread wheat straw, then

3 liters of straw = ca. 100–150 g of straw

1 liter of straw = ca. 33–50 g of straw.

Providing a range of 33–50 g (instead of exact weight) allows for various weights of wheat straw and chaff.

In Table 5.4 are presented calculations for the amount of straw needed to temper 1 m^3 and 12 m^3 wall daub, if 13–17% straw was added (according to Ginder's data 1996) or if 20% straw was added (according to Stevanović's estimates 1996). In total, for 1 m^3 of daub wall, 4.3–10 kg of straw was added. For an idealized Opovo house for which 12 m^3 of wall daub was used, 51.6–120 kg of straw was needed for tempering, depending on the straw: clay ratio and the weight of straw per volume. I rounded to 75 kg of straw as the minimal estimate for the weight of straw need to temper an Opovo house. If we suppose that similarly to the Butser circular house (6 m diameter), 40 bales of straw were added to 7 tons of clay and earth, then much more straw would have been required to construct one Opovo house, probably exceeding the highest estimate of 120 kg of straw.

In order to estimate the area of land under cereals needed to produce 75 kg of straw, we must estimate the grain: straw ratio of prehistorically grown wheats. These stalks were likely to taller than today's wheat varieties, but the data from the historic and ethnographic data of grain: straw ratios provide the estimates for emmer and einkorn straw: wheat ratios in the past. Gregg (1988) cites historical data (the second half of the nineteenth century) from the Württemberg region in Germany that show kernel and straw yields for einkorn and emmer/spelt. The kernel: straw ratio is higher than 1:2 kg/ha. Known in agronomy as the harvest index, it represents a ratio between the yield of straw (t/ha) and yield of grain (t/ha), and is usually expressed in percentages. The harvest index varies among different species and varieties of wheats, but it is generally known that more straw than grain is produced in world cereals. According to present-day measurements (Petersen and Munck 1993:442), the grain:straw ratio (dry matter) is on the average 1:1.51 (1.04–2.22) in spring wheats, while in winter wheats it averages 1:0.92: (0.68: 1.63). If we round off these precise figures and assume that the grain: straw ratio was about 1:1, then 5 kg of wheat grains would produce 5 kg of straw. These are extrapolations for the amount of straw, but there was more straw in prehistoric cereals than in modern cereals, and the calculations include estimates of a grain: straw ratio of 1:2.

In order to predict the area of land needed to produce 75 kg of straw, besides grain: straw ratio, we must also estimate the grain yields of prehistoric cereals (Table 5.3–5.5). A quantity of straw and chaff (75 kg) needed to temper one Opovo house, plus estimates for the land surface under cereal cultivation allowing for different straw: grain ratios and different yields per ha are presented in Table 5.5 (The model assumes that straw and chaff were not imported to

Table 5.5. Estimates for the area of land needed to produce 75 kg of cereal straw.

Straw (kg)	Grain:Straw	Yield kg/ha	Land surface m²
75	1:1	500	1500
	1:2	500	750
	1:1	800	938
	1:2	800	469

the Opovo settlement). A quantity of straw and chaff sufficient for tempering one Opovo house could have been obtained from sowing cereals on approximately 1,500 m² field. If the grain: straw ratio was 1:2, the area would be half in size, i.e., ca. 700 m² large. Even if the yields were 800 kg/ha and the grain: straw ratio remained 1:2, an area of 469 m² had to be under cereal cultivation to produce 75 kg of straw (Table 5.5).

If the Opovo wattle-and-daub houses were constructed in the late spring when the clay was still naturally wet, then the straw and chaff had to be saved from the previous harvest. Straw could have been stored in stacks in the open air or in sheds. However, if the houses were built after June, then straw from that year's harvest could have been used. If we assume that one or two houses from one building horizon were constructed in one season, then at least several thousands of square meters must have been under cereal cultivation at one time. Besides being used in house wattle-and-daub constructions (Figures 5.2 and 5.3), chaff was also used as temper for large jars that were made of local clay (Figures 5.4 and 5.10). However, chaff was not the most commonly used temper for pottery (Tringham et al. 1985:436–437; Tringham et al. 1992:Table 5). Under closer examination of pottery fragments, it is evident that wheat chaff, possibly of *Triticum monococcum* or T. *dicoccum* was used for tempering (Figure 5.4). Straw and chaff can be crucial winter fodder for many domestic animals when there is not enough hay or other fodder available (P.J. Watson 1979; personal observation).

Straw can also be used for thatching roofs, but straw roofs last only from 8–10 years while reed roofs can hold up for 40 years (Ginder 1996). Since no remains of roofs dating to the Neolithic period have been found in southeast Europe, it is usually assumed that they were made from perishable organic materials. Stevanović (1996:192) believes that reeds were used in house-wall construction as well as in roofing in Opovo. Reeds are nowadays readily available locally and that was probably the case in prehistory, because reed impressions (Figure 4.10) and silicified fragments were found in daub from the Opovo houses (Stevanović 1996, 1997a; personal observation). However, we do not know the density of reeds in the waters surrounding the prehistoric settlement.

Stevanović (1996:190, 1997a:359), who studied the Opovo house in detail, offers different explanation for the use of straw and chaff in house construction during the Neolithic period and states that it "might be of a nonaccidental and nonutilitarian nature" and might have a symbolic nature that is "expressive of the close connections and interdependency between agriculture and a domestic way of life". According to Stevanović (1996, 1997a), although straw and chaff were used in considerable amounts in house construction at Opovo, the use of straw and chaff might not have been necessary for tempering clay. However, in the report from her later experimental work at Çatalhöyük, Turkey, Stevanović (1997b:4–5) writes that bricks made with the addition of straw and chaff once dried "were very hard and compact" and that various types of soils or soil mixtures "if used for brick manufacturing without vegetal material produce unsuccessful bricks".

In house construction, straw serves to ensure strength and durability of structures, similar to the use of steel reinforcement in concrete as shown by numerous historic, ethnographic, and experimental evidence. Chaff of em mer and einkorn was available byproduct of processing glume wheats at Opovo. The use of chaff and straw and its utilitarian function as a construction material has been attested in various parts of the world. If brick makers in widely different contexts, including the Opovo daub makers, all recognized the functional/technical importance of using straw and chaff, then the differences among historic and ethnographic contexts give the functional/technical argument credibility, regardless of differences in culture, period, religion, or region. If they made mud bricks, they recognized the importance of adding straw and chaff, as attested by later Stevanović's (1997b:4–5) own experimental work at Çatalhöyük, Turkey. Making daub and adding straw and chaff, however, could have been accompanied by rituals performed during house constructions.

Plowing

There is no direct evidence for the use of ards in the Opovo settlement, nor there is any in the Balkan prehistory until the Bronze Age. (Ards or scratch plows are simple wooden devices that make furrows in the soil between 10–30 cm deep.) The Opovo site is surrounded by chernozem soils, and probably the chernozems (on a loess substrate) were present in the Neolithic period. It is commonly thought that chernozems are too heavy to be cultivated manually. According to Chapman (1981), heavy soils like chernozems cannot be cultivated without the use of ards pulled by draft animals because of the existence of a turf level several centimeters thick. Thus, a mastery of chernozem cultivation, implying the use of the ard, is a feature of the Later Neolithic period in the Balkans (Chapman 1981).

For the Opovo site, the only suggestion that cattle may have been used for traction is the pathological evidence of a single instance of osteoarthritis in the hip of cow, and perhaps the evidence of leg trauma that may have resulted from traction in one female and in one male bovid specimen. Otherwise it is difficult to support the evidence of traction, which is not indicated by age/sex structure of mortality profiles (Russell 1993:397).

Sherratt (1981:261–305) reviews pictorial, iconographic, and some archaeological evidence for plow use in the Old World. He argues that the use of the plow and animals for traction (and milking) were inventions of the 'Secondary Products Revolution of the Old World' dated to the Bronze Age. Chapman (1983:107–122) strongly criticizes Sherratt's article (1981) for his inductive approach and for ignoring the archaeological evidence of plow use both in Europe and the Near East, dating back one or two millennia earlier than cited by Sherratt. Later, Chapman (1990:25) suggested tripartite development of early agriculture, including the use of the ard, for the Near East and the Balkans. The first stage of development would be represented by use of wooden digging sticks or hoes in Anatolia ca. 6500–5500 b.c. or later; the second, by the use of stone hoes and antler mattocks in temperate southeast Europe ca. 5300 b.c. onwards; the third, by the use of wooden or antler ards in southeast Europe from ca. 4200 b.c. (4920 B.C.) onwards. The occupation of the Opovo site (4700–4500 B.C.) would fall in Chapman's (1990) suggested third stage represented by use of ards in southeast Europe. Halstead (1995a:8–22) discussed the economic and social implications of ox-drawn plows in the Mediterranean. He references the plowing scene from Vounoua in Cyprus, and one of yoked cattle from Nemea in Greece dating to the third millennium B.C. (Pullen 1992). The osteological evidence of improved survivorship of male cattle in the third millennium B.C. in Greece may indicate the raising of oxen (Halstead 1987). He suggests, on the basis of his ethnographic research in Assiros in central Macedonia, and the gathered historic data, the maximum area cultivable would have been 2–4 ha by hand and around 10 ha with oxen. The minimum area to support a family of a subsistence farmer is around 2–3 ha, and that area is correspondingly cultivable by hand (Halstead 1995a:12).

Simulation models were used to estimate the area under cereal cultivation, for example, for a small late Neolithic village in the Zurich area, where there is no evidence of plowing. Gross et al. (1990) estimated that 6 ha were under cereal cultivation, assuming that 35 people lived there. Sixteen adults would have to invest 125 working days (8 hours per day), to obtain 900 kcal from cereals per person/per day (Gross et al. 1990).

Additional ethnographic evidence of using cows instead of oxen for pulling plows can point to various uses of animals for traction. I note the more prevalent use of cows as traction animals in the ethnographic past and present in Serbia. The practice has been briefly mentioned by Russell (1993:397) for Serbia and by Halstead (1995a:17) for Greece. The use of cows for several purposes, although limiting milk production and effectiveness of plowing, has been practiced for centuries in Serbian villages with varying degrees of arable land potential. Moreover, cows are used for pulling heavy loads of wood from the mountains and then instead of two cows, four cows are yoked. Only in the high mountains, where the terrain is very steep, were peasants forced to have oxen that would pull heavy loads uphill. In general, in central Serbia, very few households had oxen, even the ones who were better-off and who could have afforded to buy a pair did not have them (personal observation and personal communication with peasants of the Kraljevo region). Also in the same regions where cows are used for traction, milk products are produced and consumed on a regular basis in great quantity by all generations. Likewise, cows are yoked in Slovenia, an Alpine region further west.

In contrast, several hundred kilometers to the west, in the hilly region of Banija (Croatia), a completely different folk tradition exists. There, yoking cows is considered a sin. Banija is considerably smaller than central Serbia, and hence is less diverse, but major farming land is available in the river valleys as well as on hilly slopes. In the Banija region and surrounding countryside, cows are never used for traction as they are in Serbia. The Banija region was populated with small landholding Christians, and in many areas, the population was of the same ethnic origin as in central Serbia. Milk was readily consumed in this area, although milk products may have been less diverse than in central Serbia. All of the households had limited amounts of arable land. For example, during the 1920–1960s, one six-member household owned ca. 4 ha of arable land (plus ca. 2 ha of woodland) and was considered rather well off. In addition to cows, which were used exclusively for milking, this family owned a pair of oxen and a pair of horses. Middle-income farmers who had even less arable land kept only horses. Only poor households had neither oxen nor horses. Although they usually had two cows, poor farmers worked five to seven days in return for plowing service of a half of hectare (Slavko Borojević and Milan Borojević, personal communication 1997).

It is commonly assumed that animals were used to pull of ards in Balkan prehistory since the beginning of plow-use, but there is ethnographic evidence for human-pulled ards in some parts of China and in present day Ukraine, where an ard is still pulled by a man while another man or woman turns furrows (Patty Jo Watson, personal communication 1997). The human-pulled ard may or may not be more efficient than using other tools such as digging sticks, hoes, spades, or so-called foot plows. Human traction of ards would not be reflected in animal herding strategies and thus might not be visible in archaeozoological records.

The area of ca. 1.5 ha, that I suggested for the Opovo site, could have been worked by the members of one household of six persons either manually or with the help of animal-drawn plows, based on ethnographic and historical data (Halstead 1995a). According to Russell (1988:113) who gathered data on various hand tillage methods, the number of hours necessary to till 1,000 m^2 varies between 16 and 50 hours in different parts of the world. If we assume that for tilling heavy soils (e.g. chernozem around Opovo), approximately 40 hours were necessary per 1,000 m^2, then for 1.5 ha, 15 times more hours were needed to till this area. One household would have needed to spend between 10–15 days working 8 hours per day in fields totalling 1.5 ha. This would represent a major labor investment in September before winter-sowing of cereals, and might necessitate additional help if a household could not provide sufficient labor. With an ox-drawn plow, on the other hand, one person could till the area of 1.5 ha in just a few days, given that a pair of oxen can plow between 0.1 and 0.3 ha per day (Halstead 1995a:14). If cows and not oxen were used for plowing, the time would have been longer considering the cows' lower work potential. Given the present evidence, we cannot say whether the animal-pulled ard was necessary to till chernozem soils at Opovo, or whether the soil was worked manually at the expense of more time and energy.

Sowing

The principal crops identified in the Opovo samples are *Triticum dicoccum* and *Triticum monococcum*. In many samples and in almost all features, emmer and einkorn were found together in various proportions. Both wheats can be either autumn or spring-sown in this region. After the introduction of agriculture in the Balkans occurred in the sixth millennium B. C. (Dennell 1992; Kroll 1991a), probably both winter and spring habit wheats were developed (by the fourth millennium B.C.). Barker (1985:254–255) suggests that traditional Mediterranean agriculture in the early stages included cultivation of autumn-sown crops and raising sheep and goats; a thousand years later, it included a different form based on spring-sown crops and keeping cattle.

In many Balkan Neolithic sites, emmer and einkorn were found together; therefore, they were likely grown together as mixed crops (Hopf 1974). These deliberate mixtures, or so called "maslins" (Jones and Halstead 1995), were grown to reduce the risk of failure of one of the major crop components under different growing conditions. Monocrops were not as common in traditional farming as they are presently, and even monocrops included minor contaminants that were either difficult to remove or were deliberately tolerated (Jones and Halstead 1995). P.J. Watson (1979:76) mentions that in some growing and harvested crops in the Iranian village she studied, wheat seed sometimes contained so much barley that it was difficult to tell whether a field was supposed to be a stand of wheat or one of barley. Also, in the wider area around Opovo in the Banat region, a mixture of wheat and rye was grown until recently (Hegediš and Čobanović 1991).

Presence of characteristic weeds can indicate whether crops were sown in winter or spring. Comparisons are made on the basis of modern analogies. In winter-sown cereals, winter annual weeds germinate in autumn and grow together with cereals in spring. Maturing cereals and the accompanying winter weeds inhibit many of the spring-germinating weeds from developing. In contrast, in the spring-sown cereals, the weeds that had germinated the preceding autumn are destroyed by spring plowing. Later, the spring weeds and some of the unplowed winter weeds can grow together in the crop until harvest. Thus, some weeds can accompany both winter and spring crops. In many instances, indicator weed species have been found in archaeobotanical samples and were used to reconstruct sowing practices (e.g., Reynolds 1979; Wasylikowa 1981:11–23). The seeds of *Bromus arvensis*, *Galium aparine*, and *Lolium temulente* that have been identified in the Opovo samples would point to winter-sown cereals. (In all features where these seeds were identified, cereal grains or spikelet bases/forks were found.) These weed seeds were found in the same samples together with cereals, except for one sample from Feature 41 and one from Feature 51 in which no wheat remains were found. The one from Feature 41 consisted almost entirely of weed seeds except for seven barley seeds. Feature 41 yielded a total of 26 barley seeds. However, the one sample containing barley seeds is from the same locus (1679) where many wheat seeds were recovered, implying that grains from this one sample might not represent a significant difference in crop growing, but rather the way the cereal byproducts were discarded in the pit.

Barley is usually a spring-sown crop in the northern Balkans and the Pannonian Plain due to the region's severe winters and barley's low resistance to cold. Barley can be autumn-sown, but it is sown only late in autumn to prevent the full development of seedlings. In this way, barley can overwinter even if the winters are harsh. When barley is autumn-sown, some leaves are damaged by frost, which may affect the yield; nevertheless, winter-sown cereals tend to have higher yields than spring-sown cereals and are less affected by droughts. With an atypical weed composition, as has been found in the Opovo samples, it is very difficult to conclude whether all the cereals were winter-sown. I think, however, that some cereals were winter-sown, while some were spring-sown.

In southeast Europe and most of the Near East where there is no irrigation, cereals are broadcast sown today and in recent history (see Figure 5.1). According to Sigaut (1975), ard cultivation tends to be associated with broadcast sowing while manual cultivation is paired with dibbling or planting. Ethnographically, there is no evidence of sowing in furrows in the Balkan region. Cereal crops may be hand-weeded only while the seedlings are young. Later, when the cereal plants are developing, it is often impossible to distinguish between weedy grass plants and the main crop; therefore, cereal crops are not weeded before harvest. Because the cereals are broadcast sown and not hoed, they are not manured. The number of undesirable weeds would increase with manuring, which would inhibit the full growth of the crop. Soil fertility is thus enhanced prior to the sowing of cereals either by crop rotation or by leaving the fields fallow. However, Jones and Halstead (1995:109) note that newly fallowed or manured fields were favored for wheat sowing on the Greek island of Amorgos. This practice of manuring fields was probably followed by intense plowing before wheat was sown.

In general, wheat is a winter crop and nowadays is sown in October in the Voyvodina region. Spring cereals are sown in March. Usually, 110–150 kg of wheat were broadcast sown per half hectare, which is equivalent to ca. 300 kg per hectare in the Banija region (Croatia) where traditional framing was practiced until modern times. Broadcast sowing requires a greater quantity of seed because many do not germinate. The losses were great, ranging between 35–70% in 1956/1957 on the plots of various farmers. The greatest losses were on bad and cold soils, while on better prepared and more fertile soils, the losses were smaller (S. Borojević 1995:68). Similar figures for the amount of sown grain in traditional farming are given by Sigaut (1988:6) and Watson (1979:74). In the northern Balkans, sowing is traditionally done by the members of a household, both by men and women. Usually there is no need for outside help to sow cereals for one household. Sowing of cereals on the suggested 1.5 ha for each Opovo household could have been accomplished in a day or two.

Harvest

Wheats are harvested by the end of June or the beginning of July in the southern Pannonian Plain. It is possible that einkorn and emmer were harvested a few weeks earlier in June to prevent the shattering of the fully ripe spikes. Barley ripens a few weeks earlier than wheat and is harvested in June. The weed composition from various Opovo features in which cereal remains were found is not indicative of the way cereals were harvested. Most of the weed seeds found in the Opovo samples came from weedy plants that varied from 20–80 cm or 30–150 cm in height (see Chapter 2), and none of them are exclusively low plants. Harvesting by breaking spikes or by uprooting has been documented ethnographically and experimentally (Ertug-Yaras 1997; Hillman 1984; Reynolds 1979). Typical, low, weedy plants do not occur in the Opovo samples, and their absence may indicate that the plants were not cut or broken near the ground, and also that most were not uprooted. In the Opovo samples, there are no culm nodes that would point to uprooting, and many weed plants can be avoided by uprooting. Even when only cereal spikes are harvested, often the remaining straw is cut and used for fodder or fuel. It is possible that straw was cut after the spikes were harvested, but then many of the weed seeds probably would not have been present in the samples since straw and the accompanying weeds would not have been processed with cereal grains. I was not able to determine whether only wheat spikes were harvested, or at what height the cereal plants were harvested based on the projected height of weed plants identified from the Opovo samples.

Traditionally, cereal harvests are the busiest times the year in the region. The Pannonian Plain has very hot summers, but summer rains are common. If we assume that a similar rain pattern existed during the Neolithic period, then harvests were performed in a short period of time. Before the introduction of machinery, cereals were harvested with metal sickles by 10–15 household members, relatives, and neighbors. Both men and women participated in cutting cereals and binding sheaves and forming stacks. Harvesting labor was shared among households. Harvesting was done in one day on various plots of one household. On the following days, the team continued harvesting the cereals of other households. If in prehistory a type of sickle with flint inserts was used, then an average of 20 hours would have been needed for cereal harvesting of 1,000 m^2 and binding sheaves, based on the experimental data gathered by Russell (1988:Table 20). Similarly, on the Greek island Amargos, between 1–3 man days were needed for harvesting 1,000 m^2 of cereals, but with metal sickles (Halstead 1995a). If we assume that 1.5 ha were under cereal cultivation per household in the Opovo site, then approximately 300 hours were needed for harvesting and binding of cereals. One household of 5–6 persons may have needed outside help to avoid prolonging the harvest and losing grain.

Threshing

In the Balkans generally, harvested wheat was left in the fields for a few days, and then it was threshed on threshing floors. Commonly, each household had its own threshing floor in a field close or adjacent to the house. Harvests from one household various fields were brought to one threshing floor, where horses, donkeys, or cattle would trample cereals. Winnowing of cereals and coarse sieving was performed on the threshing floor as well. In the Opovo prehistoric settlement, threshing floors could not have been located between the houses or inside household compounds because the (excavated) houses were built close to each other.

According to the ethnographic data from an Iranian village, ca. 300 kg of cereals were threshed by trampling of oxen, then winnowed and sieved in 16–21 hours (Watson 1979:82). In the case of the Opovo site, I suggested that 900 kg of wheat might have been needed for a household of six, indicating 60 hours were necessary to thresh that amount of cereal. Thus, one household had to spend several days threshing their harvested wheat. But because the

Figure 5.5. Blade segments with sickle gloss from Opovo (Photo courtesy of Barbara Voytek).

animals could be driven by one person, and sieving and winnowing performed by another person, there was probably no need for outside help, unless all threshing was done by flailing. Historic data about the efficiency of threshing and winnowing wheat using flails versus threshing by animal trampling indicate comparable hours spent with either method (Russell 1988:Tables 30, 31, 32).

In wetter climates, cereals can be threshed by flailing inside barns where cereals are stored in sheaves. However, there is no evidence for the existence of such barns prior to medieval times in the colder parts of Europe (Sigaut 1988). Also, cereals could have been stored in the fields in sheaves, and then gathered, forming stalks (piles) in the fields. However, wheat from the stalks cannot be taken when needed: a whole stalk would have to be shattered. Considering that the cereal remains from the Opovo samples are indicative of daily processing of hulled grains, I doubt that the whole cereal sheaves were stored in the field. However, it is possible that straw was stored in stalks in the open air, similar to storing whole cereal sheaves in stalks, as documented ethnographically by Sigaut (1988).

In historic times, there is no evidence for the existence of threshing sledges in the Pannonian region, although such implements are ethnographically documented in Bulgaria, Greece, and Turkey (Ataman 1992:305–319; Skakun 1992:289–303). At the Opovo site, einkorn and emmer were retrieved and these glumed wheats needed to be dehusked for human consumption. It is common, especially in wet climates, that glumed cereals are not fully processed, and that the whole spikelets are stored (Hillman 1984, Sigaut 1988). Grains in glumes preserve better than naked ones. Also, seed grain is not dehusked before planting, so it would be stored with the glumes intact for the following year's sowing.

Perhaps there were threshing sledges during the Neolithic period in the Opovo region. On the basis of the lithic analysis, it was evident that chipped stone blades were reused as borers, sickles, or scrapers. Even macrocrystalline rock was sometimes used to make chipped stone tools. Such a cautious use of lithics is understandable given the nearest stone outcrops are ca. 30–45 km away from the Opovo site. The majority of the chipped stone assemblage consists of blades. Blades and blade segments that had sickle gloss (Figure 5.5) comprise 14% of a total assemblage (Voytek in Tringham et al. 1992:377–379). More sickles were found in the lower than in the higher layers, perhaps indicating that more cereals, grass, or reeds were harvested in the earlier periods of the settlement.

Dehusking

On the basis of plant remains found at the Opovo site, einkorn spikelets were probably dehusked after threshing, in separate sessions near house compounds. The evidence of such dehusking are the finds of numerous spikelet forks/glume bases in several pits, especially in Features 31, 52, 59, and 61 (Figures 2.6–2.9). It is probable that spikelets were parched to facilitate the dehusking process. However, it is also conceivable that glumed einkorn and emmer seeds were pounded in wooden mortars and later winnowed and fine sieved. Harlan (1967:197–201) suggests that separation of emmer and einkorn wheat grains from their glumes can be achieved by pounding in wooden mortars, resulting in coarsely ground grain suitable for making groats or porridge. Ethnographically and historically, wooden mortars are commonly used for such purposes, and the pounding is usually performed by women (Ertug-Yaras 1997; White 1976). Experimentally, it was documented that dehusking of hulled cereals is much more efficient in wooden mortars than on stone querns (Meurers-Balke and Lüning 1992). Byproducts of pounding and sieving were most abundant in the above-mentioned pits at the Opovo site. Most of the wheat spikelet forks might not have been the result of accidents of parching, because one would also expect to find more charred cereal grains in those samples. In the Opovo samples, pits from the northern part of the excavation block contained many more spikelet forks than wheat grains, while the pits from the southern part had more grains than spikelet forks (Figures 3.11 and 3.13). Probably the spikelet forks found in the pits are byproducts of daily cereal pounding that were thrown into fires and deposited with ashes. Grinding stones from Opovo show little evidence of use in grinding wheat or other grains (Figure 5.6), although the recovered hard compact sandstone and gabbro grinding stones stones would have been suitable for such grinding (Voytek in Tringham et al. 1985:440). Most of the pounding and coarse grinding might have been done in wooden mortars, considering the rarity of stone grinding equipment and the distance of 30–45 km from the nearest stone source. If we assume that there were no additional grinding implements used for cereals elsewhere in the Opovo settlement, then we can postulate that the milling of wheat grains into flour might not have been very common; instead the inhabitants probably ate porridges

Figure 5.6. Grinding stone from Opovo (Tringham ©Creative Commons License).

and cakes produced from whole wheat flour rather than bread made from fine flour. Because generally people eat smaller amounts of cereals if they consume porridges than if they eat bread (White 1976), perhaps a smaller amount of cereals was consumed per person in the Opovo settlement, since bran was not separated in the pounding processes.

Storing

Grains of hulled wheat are usually stored in their glumes. According to Sigaut (1988), underground pits were a standard means of hermetic storage in many parts of Europe, including the Danubian Plain, Balkan Peninsula, and Ukraine until the 1800s, and they continued to be used until the nineteenth and twentieth century. None of the pits excavated in the Opovo settlement were used specifically for grain storage. This differs from the pattern found at other Vinča sites where individual storage pits, pots, and/or rooms were excavated, e.g., Divostin, Selevac, Banjica, and Vinča-Belo Brdo (Garašanin 1979; McPherron and Srejović 1988; Tringham and Krstić 1990). At the site Vinča-Belo Brdo, the group of bell-shaped pits were designated as silos (Figure 5.7). Another silo, from the Žarkovo site was a deep conical pit covered with a conical wooden lid. Monumental pithoi were discovered also at the site Vinča-Belo Brdo (Figure 5.8) indicating that food was stored above ground (Garašanin 1979:159). At Staro Selo/Selevac, two clay-lined pits were interpreted as silos, Silo A and B, (Figure 5.9). The Silo A was semi-circular in shape (1.3×0.7 m), domed with internal division, and it contained a large amount (ca. 75 cm^3) of charred einkorn grain mixed with a small amount of wood charcoal and lentils seeds (Hopf 1974).

Figure 5.7. Group of bell-shaped pits designated as silos at Vinča-Belo Brdo (after Vasić 1936).

In Opovo, pottery jars (Figure 5.10) were considered as the most likely mode of storage (Kaiser in Tringham et al. 1992:374–377). Because not many large jars were recovered at the Opovo site and more serving and cooking vessels were found, Kaiser believes that this trend indicates less long-term planning than at other sites. This might be true, although long-term planning of cereal storage usually implies provisions for one year. It is also possible, however, that some of the bell-shaped pits such as Features 41, 46, and 48 (see Figures 3.4, 3.6, 3.7) were used at one point as storage pits and later as rubbish pits. Secondary use of storage pits has been documented ethnographically and archaeologically. Watson (1979) mentions that storage pits were found in several houses and were intended to hold surplus unmilled grain in the Iranian village, but because of poor harvests, the grain pits were usually stood empty except for trash thrown into them (rags, sticks, leaves). It is difficult to determine whether storage pits did not exist

Figure 5.8. Monumental pithos from Vinča-Belo Brdo (after Vasić 1936).

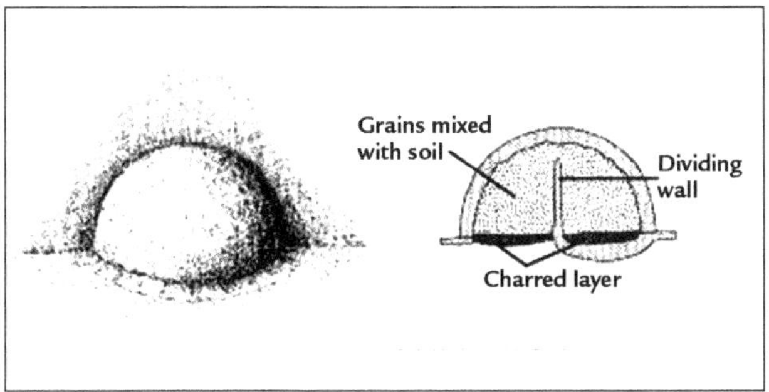

Figure 5.9. Two clay-lined pits interpreted as silos, Silo A and B from Selevac (adapted from Hopf 1974).

at Opovo due to the inferred damp conditions, as argued by Kaiser (Kaiser in Tringham et al. 1992:377), or whether some of the pits were used for storing grain prior to their secondary use. Although there were marshy areas in the Opovo environs, the ground water might have been below the depth of the pits that were dug around the houses because the settlement was located on a well-drained hillock in the Neolithic period.

On the basis of pottery analysis, it was suggested that there was a decline over time in the importance of storage in ceramic containers (Kaiser in Tringham et al 1992). The exception was House 4 from BH 3 in which a large number of whole large pots and big sherds were found. In Feature 2, a large clay container was found in the corner of the internal wall (Stevanović 1996:143). The lack of evidence for storage might indicate that storing was not common at the Opovo site. It can be argued that cereals were stored above ground and/or outside the excavated area, and left no traces. However, as mentioned by Kaiser, grains could have been stored in perishable materials such as basketry, leather, or cloth bags, although there was no direct evidence for such receptacles (Kaiser in Tringham et al. 1992:377). Woven linen textile was found in House 4, demonstrating that cloth was used (see Figure 1.13). Considering that glume wheats are usually stored undehusked, storage bags would have required coarser cloths. Also, large quantities of grain could have easily been stored in wooden chests, a common practice in the region until modern times. If we assume that for a household of six, about 900 kg of wheat was needed for a year's consumption and an additional 150-300 kg for the next year's seed stock (as discussed in Chapter 5), then one household had to store approximately 1,200 kg of cereal grain (about 40 bags weighing 30 kg each). These are approximations for bare grains only, but in the Opovo samples, grains of hulled wheat were found. The weight and volume of hulled grains is approximately one-third greater than naked grain.

The Location of the Fields

The main crops in Opovo were *Triticum dicoccum* and *Triticum monococcum*. The land available for cultivation near Opovo during prehistory was not measured in

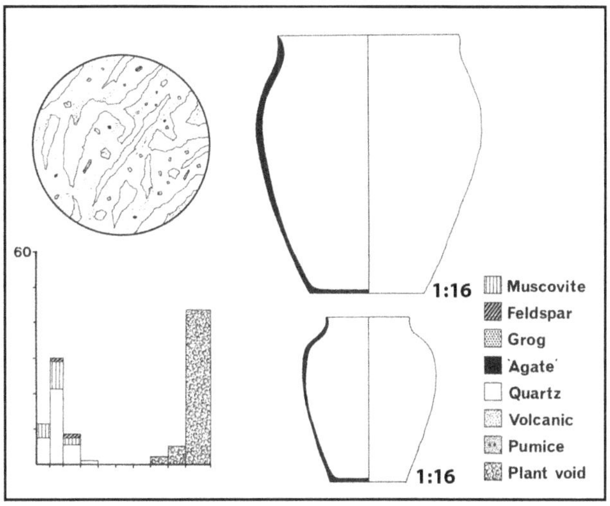

Figure 5.10. Diagram of large pots at Opovo, possibly used for storage (after Tringham et al. 1992).

hectares, and there were probably no single large fields measuring hectares in size (although I continue to use ha for comparative purposes).

As we have seen from the convincing historic, experimental, and ethnographic data, it is likely that the wheat yield at Opovo ranged from 500 kg to 1,000 kg/ha during the Neolithic period. That means that one household had to sow around 1 or 1.5 hectares per year to meet family needs and up to an additional 0.5 hectare to save seed for next year's sowing. Considering that people in the prehistoric settlement of Opovo ate meat, as documented by the finds of animal bones (Russell 1993), it is possible that they sowed wheat in a smaller area than the suggested 1.5 ha. In addition, the Neolithic people ate the seeds of legumes, like *Lens culinaris* and perhaps of some other Fabaceae species, and water chestnuts, which were also found in the Opovo samples.

At the Late Vinča site of Opovo, the yields could have been different from the ones described by Bakels (1992), Hillman and Davies (1992), Gross et al. (1990) and others. Even if the diet of the Opovo inhabitants consisted of 65% cereals and it could have been less, as shown by Hillman and Davies' ethnographic data (1992:123), there were dozens of hectares of highly fertile chernozem suitable for crop cultivation around the small Neolithic settlement of Opovo. As mentioned earlier, the area of the hillock where the site was excavated is about 20 ha. Additionally, there are similar areas with higher elevations (60–80 m) in the close proximity of the site (2-3 km radius), which also could have been suitable for crop cultivation (Figure 1.8). Two kilometers south of the excavation block, there is an area of about 50–60 ha that has an elevation around 70 m. There are also elevated areas to the west and east of the excavated houses. So even if 250 inhabitants lived at one time in the Opovo settlement, there would have been sufficient land for cereal cultivation within one hour's walking distance.

Thus, even though marshy areas surrounded the Opovo site as suggested by Tringham et al. (1985; 1992), there was sufficient potential area for fields in the radius of 2–3 km^2. The area of only 1 km^2 comprises 314 ha of land containing dozens of hectares of highly fertile chernozem suitable for crop cultivation and fallow practices, given sufficient labor.

Contrary to modern Western agricultural practices of using single large plots, small–holders usually have dispersed fields in various locations. One advantage of the dispersed field system is the reduced risk of total crop failure. Also, the fallow system implies the practice of having fields in different locations, although a part of a larger field could be cultivated and the rest left fallow. Furthermore, dispersed fields were often the consequence of different inheritance systems and not necessarily planned risk-reducing strategies and often some fields were located on poor soils (Slavko Borojević, personal communication 1997).

Perhaps we will never determine where the Opovo fields were located in prehistory, nor their crop yields during the Late Neolithic period. But on the basis of local topography and the reconstruction of potential vegetation (see Chapter 5), it is likely that there was sufficient land for cereal cultivation for the inhabitants of a small Neolithic settlement like Opovo.

Time Investment Estimates for Cereal Cultivation

To cultivate 1.5 ha of wheat, which produce around 900–1,200 kg of grain, I estimate 1,000 human hours had to be spent in fields. Hand tilling would necessitate 600 hours, broadcast sowing 20 hours, flint-sickle harvesting 300 hours, and threshing 60 hours, based on ethno-historic data (cf. Gregg 1988; Halstead 1995a; Russell 1988; P.J. Watson 1979). However, if animal-drawn plows were used, considerably less time would have been spent in plowing, probably only ca. 30–40 hours. Neither the time spent going to and from the field nor the time spent in dehusking and grinding grains were included in the 1,000-hour sum. If the working day was around eight hours long, then one person needed to work four months in fields totaling 1.5 ha. If in one household there were more adults capable of laborious field work, then each person probably worked one or two months. This amount of time, needed only for wheat cultivation, could have been provided by one household; however, outside help would have been needed for activities, such as hand tilling before heavy rains and frost in the fall, and quick harvesting and threshing between summer rains.

Legume Cultivation

Among crops, seeds of *Lens culinaris* and small *Vicia*-type seeds were found in the Opovo samples, although in rather small quantities. Legume seeds are known for their high protein content. Furthermore, the green parts of legumes can be used as nutritious animal food. Lentils from the Opovo samples belong to the domesticated form *Lens culinaris* ssp. *macrosperma* and were probably consumed by humans.

Both lentils and vetches are spring-sown crops. In the Voyvodina region, lentils and vetches are sown in March. As already described, lentils are small, upright plants that prefer dry, warm places and cannot tolerate high ground water. Wet conditions cause intensive growth, longer vegetation period, and susceptibility to diseases. Therefore, lentils had to be sown in places in dry field conditions. Lentils tolerate soil drought better than air drought. Weeds around lentil plants are usually weeded out because they would hinder the development of the small, upright lentil crop. Because lentils develop slowly at the beginning of their vegetation period, weeds can seriously affect lentil growth. After germination, lentils are weeded once or twice. Lentil plants use minerals from the soil very efficiently, thus manuring is not recommended. Manuring also encourages growth of the green parts and is used only when lentils are grown for fodder and not for seed production (Jeftić 1986).

Lentils were probably sown by ancient cultivators close to the Opovo settlement in small, dry plots that were easily accessible for the additional tilling that lentils need. Lentil pods ripen irregularly and some dehisce even in modern varieties. They are usually harvested before the pods are fully ripe and the plants are left in the field for three to four days before they are transported to the threshing floors. Harvested lentils cannot be left long in the fields because the pods would dehisce, greatly reducing seed yields. Lentils are usually harvested in the south Pannonian Plain in June, or in August if sown after cereals. Threshing of legumes is usually done on the same floors where cereals are threshed. After being threshed, winnowed, and sieved, lentils do not need to be parched, dehusked, and ground and can be immediately stored or cooked. With fewer steps in processing legumes than glumed cereals, the legumes are often less well represented in archaeological material, unless lentil storage spaces are found. Although crop rotation between pulses and cereals is commonly practiced and ethnographically documented, the number of contaminating seeds of legumes in cereal crops is very limited, and can be easily removed by sieving (Jones and Halstead 1995:111). Lentils found in the Opovo samples were perhaps separate intrusions in pits not mixed with cereals on the threshing floor, unless food preparation required mixture of the cereals and lentils.

If the identified vetches from the Opovo site belong to the domesticated forms, they were also sown in late spring. Vetches are commonly used for fodder, although they can be detoxified and used as human food in times of need. It is possible that some vetches were intrusions in the einkorn crop because many of the vetches, such as *Vicia villosa*, are common weeds in winter cereals. Most of the vetches are vine plants that accompany standing plants such as cereals or weeds of lentils.

Flax Cultivation

Five *Linum* cf. *usitatissimum* seeds were recovered from three pits. In addition, a fragment of linen cloth (see Figure 1.13) and a fragment of linen cord were found in House 4; the first and the only preserved linen fibers from the Vinča culture (Tringham et al. 1992). In various Opovo features and particularly in the northwest corner of Feature 3, a pile of several conical clay weights was found which were possibly used as weights for a vertical loom (Stevanović 1996:148). Weights were found on many other Vinča sites (e.g., Selevac, Banjica) and were interpreted similarly (Tringham and Krstić 1990). These finds indicate that flax was used for making fibers by Vinča people and that was probably cultivated as a crop.

Varieties of flax cultivated for fibers are usually taller than those cultivated for seed production and oil extraction (Jeftić 1986). Flax plants prefer temperate climates without drastic climatic amplitudes; however, they are also resistant to mild frosts. Although the plants can adapt to a variety of soils and climates, they grow best on well-drained soils. There are both winter and spring varieties of locally grown fiber flax. Autumn varieties are sown in the second half of September, and spring varieties are sown in the middle of March. Like many other crops, flax gives higher yields when sown in autumn; nevertheless, spring varieties were more commonly sown until recently (Jeftić 1986; personal observation). Usually, around 150 kg of flax seeds are broadcast sown per hectare if fiber-flax is grown. Smaller quantities of seeds are sown if oil-flax is grown to allow branching and greater production of seeds. Because of its shallowly developed root system, flax plants have high requirements for water and nutrients; thus, a flax crop is always rotated with other crops and is not sown on the same field for several years. Flax plants cannot compete with weeds, and soil tillage and weeding are necessary for successful crop growth. Before the introduction of modern machinery, flax had to be manually weeded by uprooting while the flax plants were still young. Also, flax plants are very susceptible to insect infestation. Manuring is recommended only before sowing so that development of weeds can be prevented by plowing. In modern times, the most common flax weeds are *Agropyron repens* and *Agrostemma githago* (Jeftić 1986:185–200).

If flax is intended for use as fiber, it is harvested in the earlier phase of maturity, the so-called green phase, typically in June. If flax is intended for oil extraction from seeds, it is harvested a week or two later, in the yellow phase of maturity. Harvested plants are left in the fields for a few hours, and then bound in sheaves 20–40 cm in diameter, arranged in cone-shaped stacks, and left in the fields for several days to dry. When plants are completely dry, the capsules containing the seeds are separated from the stems. Later, the seeds are separated by crushing the capsules. In traditional agriculture, flax was harvested by uprooting, a fairly rapid process (Jeftić 1986; Slavko Borojević, personal communication 1997).

In the northern Balkans, flax was usually grown only in small areas, around 200–300 m^2, solely for domestic needs. Average yields ranged around 4,000 kg/ha before the 1940s (Slavko Borojević, personal communication 1997). Dry stems represented 20–30% of a total yield. Traditionally, retting was practiced in nearby ponds and rivers for 3–4 weeks depending upon water temperature and mineral content. Plant stalks were left in warm waters, and the maceration process was facilitated through the activity of pectin bacteria (e.g., *Clostridum macerans*). Retted stalks, called straw, were dried in the open air. Fibers were separated by a breaking process in which the brittle woody portion of the straw was broken by beating. Then the fiber cells, typically 15–40 mm long and 20–30 microns wide, were separated (Jeftić 1986). Longer fibers were used for weaving yarn and the broken, woody pieces for fuel.

Linen fiber is valued for its strength, luster, durability, and moisture absorbency. Linen fabrics are resistant to the attacks of microorganisms, and linen's smooth surface repels soil. It was widely used in the manufacture of ropes, twines, and bagging materials. Also, fine grades of linen were made into fabrics and laces.

Yields of linen seed ranged from 1,000–1,500 kg/ha. The whole air-dried seeds usually contain 33–43% of oil (by weight). The varieties sown exclusively for seeds are shorter and have more branches than the varieties for fiber; however, there are intermediate forms from which both fiber and seeds are used. Flax seeds were used in antiquity for food, and leftovers from oil extraction can be used as animal fodder (Jeftić 1986).

On the basis of modern analogies (Jeftić 1986), flax fields in the Neolithic period may have been located on nutrient-rich soil that was well aerated, but where ground water was not high. Fields were perhaps situated somewhere between the Opovo settlement and one or more ponds, considering the necessity for transportation of dry and wet plants. The fields had to be well tended and crop rotation might have been practiced. Flax could have been easily rotated with other crops, such as cereals or legumes, identified from the Opovo site. If flax was grown locally in prehistory, it is likely crops were rotated and fields left fallow for several years to prevent soil exhaustion. In that case, flax cultivation required larger areas of arable land.

Because both seeds and linen fabrics were found at Opovo, flax may have been grown for both fiber and seed use. If the flax fields were small, members of one household could have harvested a plot of 200 m^2 in one day. However, for harvesting larger areas, outside help might have been needed. Water was readily available for retting in ponds around the settlement. Retting could have been easily done in August when the water was still warm. The beating and combing of the retted stalks could have been accomplished afterwards by household members. In the retting process, bundles of stacks had to be submerged in water and weighed down, usually with stones or wood. Since there was no nearby source of stones, the Opovo inhabitants may have used wooden logs as weights.

If larger areas of flax were sown, then flax might have been harvested at the same time as cereals and legumes. However, further processing of flax in the late summer and early fall would not have conflicted with the early harvesting and threshing of cereals and legumes, although it might have coincided with water chestnut gathering and processing.

Crop Rotation

Principal crops identified at the Opovo site are emmer and einkorn. Fewer grains of barley and lentil were recovered, in addition to five flax seeds. The present ecological requirements of crops and certain weeds that are intrusions in crops may serve as possible analogies for the ecological requirements of the plants from the past. As mentioned earlier, samples from Feature 41 contained cereals and the seeds of *Polygonum lapathifolium*, which is a common weed in all crops and an indicator of wet places (Čanak et al. 1978). Cereal fields might have had damp areas along the edges. Generally, wheats cannot thrive in places where underground water touches their roots, and some of the rootlets may reach 1.5–2 m deep. The cereals found do not naturally grow in floodplains, although they can thrive in areas around springs. Wheat is generally intolerant of high water table and modern varieties cannot grow well if their roots reach water (Jeftić 1986). Davies and Hillman (1988) have found, however, that some wheat populations can thrive on more humid soils. Finally, legumes, which are even less tolerant to humid soils than cereals, are found along with cereals in the Neolithic sites. At Opovo, in Feature 41 (pit), seeds of lentils were found together with cereals. On those particular fields with damp areas, I assume rotation of legumes and cereals might not have been practiced because legumes, especially lentils, like warm, dry places. Thus, I suggest that some fields in the Opovo environs were left fallow to regain soil fertility. If that was the case, the minimum amount of land needed for cereal and legume cultivation might have been two times larger than if the fields were not left fallow. However, that does not mean that crop rotation was not practiced on other fields or that cereals were not sown after lentils.

***

I have attempted to reconstruct agricultural activities connected with only those crops whose remains were identified in the Opovo archaeobotanical samples. I have discussed in detail archaeological, ethnographic, and historic data concerning cereal yields, use of straw and chaff, plowing, sowing, crop rotation, harvesting, threshing, dehusking, storing of cereals, and the time invested in cultivation of these crops. I have estimated that around 900 kg of wheat might have been necessary for consumption for a household of six persons per year, in addition to 300 kg for the seed stock. This amount of cereal grain could have been obtained from 1–1.5 ha of arable land, probably dispersed in several separate fields. If the soil was manually plowed, members of one household would have needed to work 10–15 days in the fields 1.5 ha large. With an ox-drawn plow, one person could have plowed the same area of land in just a few days. While cereals could have been sawn in a few days, 300 working hours were required for their harvesting and binding. To thresh a postulated 900 kg, additional 60 working hours were necessary. Thus, I suppose that a household of 5–6 members needed additional help for plowing (if animal-drawn plow was not used), for harvesting of cereals, and possibly for threshing. Small amounts of legumes and flax could have been cultivated by one household without external help.

Cereal chaff and straw were used as temper in clay found as house rubble. I suggested that at least 75 kg of straw and chaff might have been added to clay for the wattle-and-daub of an idealized Opovo house. Such a quantity of straw could have been obtained from sowing cereals on a ca. 1,500 m^2 field, if the yield were 500 kg/ha. The area is larger than the existing space between the houses excavated at the Opovo site and between the houses at many other Vinča sites, suggesting that cereals were cultivated outside the settlement, probably in small fields, rather than between the houses.

Some of the weed seeds indicate that some cereals in prehistoric Opovo were autumn sown, while others such as barley, legumes, and flax were probably spring sown. In some fields, cereals were perhaps rotated with other crops such as legumes or flax, while other fields were probably left fallow to regain soil fertility. Flax and legume fields were perhaps located on dry plots, because they are intolerant of high ground water.

Because of the presence of only a few grinding stones and the distance of the Opovo site from rock sources, I suggested that most of the cereal dehusking was performed in wooden mortars, and that the retrieved charred spikelet forks/glume bases from the Opovo pits were likely remains of dehusking thrown into the fire. The absence of grinding stones might indicate that cereals were used to prepare gruels rather than bread made of finely ground flour. Although no grain storage has been excavated at the Opovo site, it is possible that some of the bell-shaped pits were used as grain storage prior to becoming rubbish pits. Also, grain could have been stored in bags or wooden chests.

The unique discovery of two linen fibers in addition to five carbonized flax seeds indicate that flax was grown for fiber and oil use in Vinča culture.

I did not explore all other possibilities of the plants that might have been cultivated before they perished from the archaeological record. Thus, the crops mentioned and accompanying practices reflect neither the full spectrum of plants grown at the Opovo site, nor the total picture of the husbandry of the settlers during the Neolithic period. Perhaps, the Opovo inhabitants grew various sorts of vegetables, spices, or medicinal plants in gardens, whose use is not reflected in the archaeobotanical record. There is, however, clear evidence for wild plant procurement.

CHAPTER 6. WILD PLANT PROCUREMENT

Besides plant remains that undoubtedly point to cultivation of crops such as cereals, legumes, and perhaps flax, other remains indicate procurement of wild plants. In the Opovo samples, stones, seeds, and achenes of various fruits were found. In addition, numerous water chestnut fruit fragments were also recovered, and some of the seeds classified among weedy herbaceous plants might have been intentionally gathered. Modern plant use and present-day and archaeological distribution of each identified taxon have been covered in previous chapters (see Chapter 2). In this chapter, I reconstruct activities regarding wild plant procurement, and discuss the buffering systems used when crop yields were low.

Fruits

The most abundant fruit remains in Opovo were of Cornelian cherries (*Cornus mas*), considering that there is only one stone per drupe. Cornelian cherry fruit stones (119 fragments including whole stones) were identified in ten features and in several non-feature in the Opovo samples (Figure 2.13); thus, I suggest that they were gathered from the wild during several seasons. Several generations could have collected fruits from the same trees or bushes since the trees can live for over 200 years. Cornelian cherry fruits were probably gathered in the fall from trees or bushes around the edges of the oak deciduous forests (Figure 6.1). It is possible that some of the intentional woodland clearing around the Opovo site actually enhanced the growth of these heliophilous and thermophilous trees, but I do not believe Cornelian cherry were intentionally planted or transplanted because they grow very slowly. Cornelian cherries seem to have been systematically procured by the Opovo inhabitants, and possibly stored for later use, either dried or as preserves.

The fruits of blackberries (17 seeds of *Rubus* sp.) and strawberries (70 seeds of *Fragaria* cf. *vesca*) were also found (Figures 2.17 and 2.18). These fruits were probably gathered around the Opovo settlement during several seasons. The blackberries could have been gathered in the late summer, probably from bushes or thorny vines growing on the forest edges and along the roads or fields. They could have been eaten fresh as well as dried and stored for later use.

Strawberries are the earliest fruits in the region and they can be gathered by late spring. The fruits could have been collected from open forests or forest clearings near the Opovo settlement. The forest clearings and local soil disturbance could have reduced the competition from other plants and stimulated asexual propagation of perennial strawberry plants. It is impossible to determine whether the strawberry plants were transplanted to gardens near the settlement in the nitrogen-rich soil, or whether the fruits were only gathered from the wild.

Figure 6.1. Modern *Cornus mas* (Cornelian cherry) tree with ripe fruits (after Šilić 1988b).

Elderberries (*Sambucus ebulus*) and elder (*Sambucus nigra*) are also gathered from the wild in the late summer and early fall. Six carbonized and four uncarbonized seeds were found in the Opovo samples (Figures 2.14 and 2.15). The fruits were probably gathered from the forest edge or near the roads. Perhaps forest clearings and humanly created openings enhanced the growth of some elderberry trees or bushes. Because only a few seeds were recovered from the Opovo samples, I cannot evaluate how extensively elderberries were used by the inhabitants.

Only six grape seeds (*Vitis vinifera* cf. *sylvestris*) were found from one Opovo feature, and they were uncarbonized (Figure 2.16). Grapes were probably gathered in the fall in the alluvial forest, perhaps along the Tamis River. The berries of wild grapes are rather small and the seeds are big; it is possible that in some instances the seeds were preserved uncarbonized in the pit. The extent of grape use is difficult to perceive because only a few seeds were found. As Kroll (1991b:165–171) suggests, perhaps not many grape fruits were left for human harvest because of competition with birds and other wild animals, capable of reaching the grapes in the high trees of dense riverine forests.

The Opovo inhabitants gathered several fruits such as Cornelian cherries, blackberries, and strawberries. They also collected the fruits of elderberries and grapes. All these fruit-bearing plants were probably nearby the settlement and easily gathered within one-hour walking distance or less. I would suggest that the fruits were intentionally gathered because all the identified fruits ripen before or after the cereal harvest, and were probably not unintentionally harvested with the cereal crops, even if some of them were present in the field edges. The Opovo inhabitants perhaps gathered the fragile berry and other fruits into baskets or wooden containers because in cloth or skin bags the fruits would have been smashed. Fruits brought to the settlement that were not immediately consumed could have been dried either in the open in protected sheds, or above house fires, and stored for later use.

Weedy Herbaceous Plants

As mentioned earlier, this category is the most ambiguous because some of the plants could have been weeds in crops, while others could have been intentionally gathered and used. While *Lolium* (2 seeds) and *Bromus* (19 seeds) were most possibly weeds in crops, Solanaceae (262 seeds), *Polygonum* (66 seeds), *Galium* (5 seeds) and *Chenopodium* (20 seeds) species retain the dual status. Except for the Solanaceae and *Lithospermum officinale* (5 seeds) which might have been gathered for medicinal use, other plants could have been gathered for the use of their green parts as vegetables, or in the case of *Chenopodium* for the starchy seed content. Although there are many seeds in each Solanaceae berry, 240 carbonized and 20 uncarbonized Solanaceae seeds were recovered from 10 out of 13 pits and from non-feature samples, thus pointing to some kind of non-accidental, systematic use of these plants in Opovo (Figure 2.22–2.25). All these plants, if not weeds in the crops, could have thrived in the vicinity of the settlement or around the houses and were easily accessible to the Opovo inhabitants. There are no pure caches of any of these seeds recovered at the Opovo site, except for numerous Solanaceae seeds found and distributed in most of the pits; thus, there is not enough evidence to suggest any kind of tillage of these plants. Most of them can be ruderal plants, and require no special management, so their procurement by the Opovo settlers would have been quite easy. In Feature 48, and in several samples from other features, Solanaceae seeds were not present with cereal grains or cereal byproducts, indicating that some of the Solanaceae might have been intentionally collected or even planted for medicinal or other uses (see Chapter 2). As mentioned in the previous chapters, there is archaeological, historic, and ethnographic evidence of eating berries of *Physalis alkekengi* (Figure 6.2) and smoking *Hyoscyamus niger* (Figure 6.3) seeds in pipes for toothache (Lempaiäinen 1992). Also, fumes of burning seeds to apparently relieved itching in the eyes caused by worms (Sezik et al. 1997). There is ethnographic evidence of eating Solanaceae plants as greens in Africa (Fiona Marshall, personal communication 1997).

Figure 6.2. Modern *Physalis alkekengi* (Chinese lantern) fruit (Photo courtesy of Annette Hoeggemeier).

Water Chestnuts

Water chestnuts (*Trapa natans*) were almost ubiquitously present (375 fragments) in all the pits of the three building horizons, and in non-feature samples in the Opovo site (Figure 2.19). The various parts of the hard, outer, water chestnut shell were found carbonized, indicating that the fruits were peeled and starchy parts eaten by humans. The Opovo inhabitants seem to have systematically collected the fruits for food from the stagnant waters around the settlement in late summer and early fall. The fruits were probably collected in baskets or leather bags, either by standing in the shallow water, or possibly from boats in deeper waters. Large quantities of fruits could have been brought into the settlement in baskets. I doubt that cloth bags were used to carry water chestnuts because the fruits are very spiky and wet, and would have easily torn any kind of woven fabric. There is no evidence at Opovo of

Figure 6.3. Modern *Hyoscyamus niger* (henbane) unripe fruit (Photo ©Dieter Wilken).

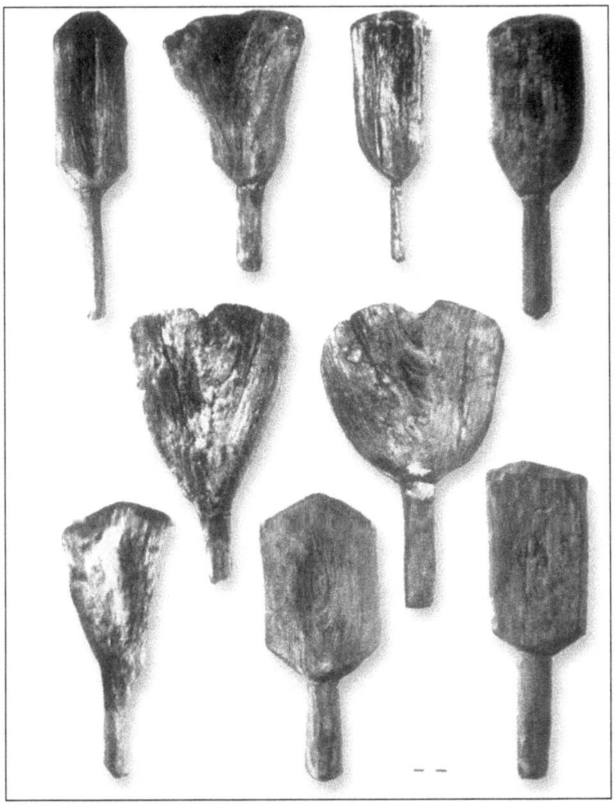

Figure 6.4. Wooden mattocks with inserted *Trapa natans* L. endocarps from the peat bog site Sarnate, Latvia (after Vankina 1970).

storage or the use of wooden mattocks for water chestnut processing as has been documented for eastern Baltic sites, i.e. the Sarnate site in Latvia (Figure 6.4), dated to 2700–2400 b.c. (Vankina 1970).

The breakage pattern of the water chestnuts in the Opovo samples makes it appear that a rather dull instrument was used to peel the fruit because there are no clear, sharp cuts left on the water chestnut shell fragments like those on the ones cleaned with a knife (Figure 6.5). Processing of water chestnuts seems rather laborious work, but it required fewer operations than processing of cereals, for example. The fruits can be eaten fresh, although they can be also roasted or boiled for several hours. It is also possible that once released, fleshy fruits were dried and later pounded into flour and saved for later use, a method that was documented ethnographically (Janković 1952:307–309). Unpeeled fruits preserve well and dried fruits are easier to peel than fresh ones. The gathering of water chestnuts in the Opovo site was not incidental or opportunistic, but intentional and probably systematic. Stands of water chestnut plants could have been abundant in the local waters around the settlement and might not have required special managing (see Figure 4.8), such as fencing areas for protection from animals, a practice common in wetland agriculture in China since Neolithic times (Hui-Lin Li 1983:43). However, the Opovo inhabitants could have encouraged pigs,

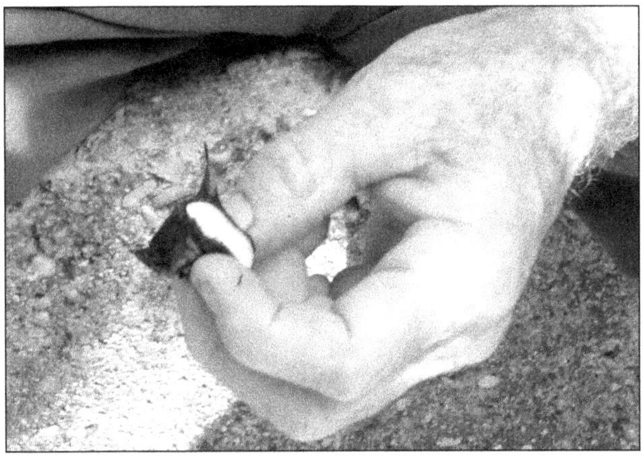

Figure 6.5. *Trapa natans* (water chestnut) fruit, peeled with a sharp knife by Marko Strugar Dodoši, Montenegro, 2001.

whose bones have been identified, to eat water chestnuts in the shallow waters. The consumption of water chestnuts added carbohydrates to the diet of the Opovo settlers, who presumably consumed quantities of wild animal meat. It is impossible to say whether the appearance of water chestnut remains mainly in layer 8 of the various pits points to feasting, or perhaps to a bad year when cereal yields were low. Historically, water chestnuts were considered famine food or food gathered in regions with low agricultural potential, such as mountainous parts of Montenegro. However in Asia, water chestnuts are often eaten with other vegetables and consumed in addition to rice.

Bad Years

I tried to determine from the meteorological data the frequency of bad years that might have caused serious damage to crops and necessitated more intensive wild plant procurement (including gathering of water chestnuts) and hunting. As mentioned earlier, there is no precise information about the climate of Opovo during the Late Neolithic period in the fifth or fourth millennium B.C. (see Chapter 4). Instead, I relied on analogies based on the modern and historic climatic data from the region and assumed a similar frequency of dry and wet years in prehistoric times. Because of the lack of water management, the area was more susceptible to flooding in the past, although the effects of floods were probably less severe than today.

Floods and Droughts

According to the historical records for the period from 989 to 1865, there were 83 catastrophic droughts. During the thirteenth century, not a single catastrophic drought was recorded, whereas in the eighteenth century, 29 years were extremely dry. Two of the driest years were 1147 and 1271 in which famine occurred. During 1473–1474, a drought was so extreme that the Danube could be crossed by foot (Stojšić and Miljković 1992:192).

The Austro-Hungarian records indicate that the regions most vulnerable to flooding were the valleys of the rivers Tamis and Begej, and Opovo is located close to the Tamis river. Despite canal and river bank constructions that started in the seventeenth century, that area of Banat was exposed to flooding in 1919, 1924, 1932, 1939, 1941, 1942, 1956, 1965, 1970, and 1975. In these years, the annual precipitation was above average, and in years for which exact data are available, the heaviest rains fell in May and June (Stojšić and Miljković 1992:203; Savezni Hidrometeorološki Zavod, *Godišnjak* 1985). Statisticians have computed that in the last 100 years in the Voyvodina region, there were only 17 good years for crops. There were 51 dry years, and 32 years were wet (*Dnevnik*, 13 June 1997:7). Therefore, we can generalize that approximately every second year, crop yields were affected by the unfavorable climatic conditions. Similar frequency is characteristic for eastern Hungary where the rivers Tisa, Maros, and Körös floods are common (Chapman 1997).

Based on historical data, if we assume that wet and dry years appeared also during prehistory, we may assume that at every several years either a flood or a drought threatened the Opovo environs. Because the relief around the Opovo site is rather flat and uniform with elevations between 70–90 m above sea, droughts probably affected the surface of the area rather uniformly. Modern water management systems caused a decrease in the underground water table, which is reflected in the salinization of the lower lying parts, and higher water levels in major streams. If, however, the water table was higher and if there was more tree vegetation around the crops as suggested earlier, droughts might have been less severe than in historic periods. While larger areas were susceptible to seasonal inundation, the Opovo settlement and some of the fields on hillocks might have remained unflooded during less severe floods. The Opovo farmers could probably choose to farm on drier areas or on in a more fertile terrain with higher undergorund water or prone to flooding.

As mentioned earlier, a household had to store about 1,200 kg of cereals for a year's consumption and for the next year's sowing (see Chapter 5). If the cereal yields were extremely bad due to drought and the harvests were not more than twice the amount of sown seed, then an additional 600 kg of cereal grain had to be procured from elsewhere. Cereals and legumes are bulky foods and large quantities are difficult to transport between settlements. In that case, an area larger than 1.5 ha had to be cultivated per household to reduce the risk from climatic disasters. Possibly in those years, the Opovo inhabitants had to rely more on barley and legumes, which were identified in the Opovo samples in small quantities and which are more resistant to drought than wheats. In good years, legumes and some of the barley could have been used as fodder, as shown ethnographically (Halstead and Jones 1989:41–55). When the years are extremely dry, then the distribution and production of water chestnut fruits are severely affected; most fruits (seeds) cannot sprout in the spring if left more than a few days outside water (Janković 1958:43). Perhaps in those years, carbohydrate intake would have been significantly lower, unless the Opovo inhabitants had enough stored food from the previous years.

Droughts generally might not have been as severe in the past during the occupation of the Opovo site as in the present, because during the Atlantic the climate was more temperate. In soils that retain more moisture, crop yields might have been less affected than in dry elevated fields. If everyone's fields were similarly affected by drought because of the similar terrain, than the sharing of crops among households or even neighboring communities might not have been viable; some other type of buffering or exchange would have been necessary among inhabitants from more distant places.

In the case of serious flooding, some fields on 70–80 m elevations might have remained unflooded and those crops could have shared. Since legumes prefer dry and warm places, it is more likely that some cereals survived that could have been shared. In addition, more waterfowl and fish were available. The flooding of the rivers facilitated dispersal of water chestnuts, and the fruits of these plants might have been even more abundant around the Opovo site. Water chestnuts could perhaps have supplemented or substituted for cereal intake. So, if the water chestnut fragments found in the Opovo pits represent more intensive gathering after a bad crop harvest, then that year was probably wet.

Buffering

Detailed climatological data and excellent preservation of organic material from the Swiss lake dwellings along with the precise dendrochronological dating (±20 years) allowed the reconstruction of the Neolithic and Bronze Age ecology and economy of these settlements (Schibler et al. 1997). The authors documented increased deer hunting and seed/fruit gathering, e.g., of turnip (*Brassica rapa*), hazelnut (*Corylus avellana*), fat hen (*Chenopodium album*) and intensified flax cultivation. They explain the increase as a consequence of deterioration of climate in the Neolithic lakeside settlements of the 37th and 36th centuries in the northern Alpine (cold phase ca. 3650 B.C. onwards). Hunting and gathering were buffering mechanisms for coping with the food crises (Schibler et al. 1997).

There is no evidence yet that the Opovo site was occupied during a phase of climatic deterioration that could have caused food crises. Most of the synchronization among the Vinča culture settlements is based on pottery typology and not refined chronometric dates. It is possible that the Opovo settlers faced poor agricultural years for which we do not have synchronous evidence of intensified wild plant gathering and hunting at other Vinča sites.

Wild Plant Gathering

To buffer a subsistence agricultural system, a wild resource must fulfill at least four essential requirements: 1) it must occur in dense patches and be suitable for intensive harvesting; 2) it must have a high potential for storage; 3) it must exhibit a pattern of inter-annual variability that is largely independent of agricultural production; and 4) it must have a pattern of seasonal availability compatible with the agricultural cycle and with buffering agricultural shortages (O'Shea 1989:59).

All these requirements are met by water chestnut procurement. Water chestnuts occur in dense stands in many ponds and canals of the Voyvodina region, and that was probably the case during prehistory. Water chestnuts are storable, as shown by ethnographic evidence (see Chapter 2). They are harvested in the late summer and fall, following cereal and legume harvests and before fall plowing. Therefore, large quantities of water chestnuts could have been gathered when crop yields were low and in smaller quantities when cereal yields were high.

For the Opovo case, I would suggest that water chestnut harvests were regular but probably more intensive in years that were bad for crops. Most community members might have been involved in water chestnut gathering given the ubiquity of the water chestnut fragments.

I do not know whether Opovo settlers were considered specialists in water chestnut exploitation and if they exchanged them for other commodities or labor with neighboring communities. Water chestnuts are a bulky food if not processed, and are difficult to transport. In addition, most of the neighboring communities could have gathered them in ponds or rivers closer to their own settlements. I would suggest that water chestnut procurement was probably incorporated into regular crop production, and served occasionally as a buffering mechanism in years of scarcity.

Perhaps the location of the Opovo site was attractive for the first settlers because wild resources were available before the first crop harvests provided substantial supplies of food. Wild plant procurement, including water chestnut gathering, could have continued once the crop fields were established. It was fully compatible with the agricultural cycle and could have supplemented cereal food and buffered agricultural shortages. Thus, the Opovo settlers could have regularly procured wild plants without major labor allocation or reorganization.

Hunting

The large quantities of wild animal bones found at the Opovo site demonstrates the importance of hunting (Russell 1993), atypical for other Vinča sites. Red deer (*Cervus elaphus*) is the most abundant taxon in the Opovo samples, followed by wild boar (*Sus scrofa*). Hunting and killing of roe deer do not show specific seasonal patterning, while red deer seem to be concentrated in summer and fall, when they would be in peak condition. Today, roe deer (*Capreolus capreolus*) are known for readily eating cereals and leaving forest hideouts during the day and night, and they usually graze ca. 11 times during 24 hours (Novaković 1996). Thus, roe deer could have exposed themselves prehistoric hunters during the day and could have been killed when encountered. Since red deer, on the other hand, are nocturnal animals and leave forests less frequently; special hunts might have been organized during the mating period (in July and August in the lowland region). Roe deer might have been hunted year round because the are solitary animals that unite in small herds only just before winter, whereas red deer live in sexually segregated herds throughout the year except during the mating period. The mating period in the lowland regions is in late August and early September, when male deer bellow loudly and could have been tracked more easily. In the bellowing territory, a red deer guards his female herd and fights other males. Not uncommonly, one of the males dies in the fight (Novaković 1996). It might have been the case that prehistoric hunters took advantage of this pattern by collecting killed or wounded males ("scavenging"). Hence, more red deer could have been perhaps killed or collected on one occasion than roe deer. If such a scenario were plausible for the Opovo site, then roe deer might have been hunted more frequently by individuals, considering the recurrent exposure of the animals and their light body weight (not more than 30 kg). On the other hand, as Russell (1993) suggests the red deer would have been hunted in groups in summer/fall when they are fully developed, considering that the weight of male red deer ranges between 120–150 kg.

If red deer hunting began in the early summer, although only indicated as summer/fall in Russell's analysis (1993), then the deer could have been killed while crop raiding. It is known locally that deer invade cereal fields, and this was probably true during prehistoric times as well. This would suggest that the Opovo cereal fields were located on the edges of forests so that deer were attracted to the fields to graze. Therefore, the presence of the numerous red deer, roe deer, and wild boar bones was a result of cereal fields that attracted these animals and the sown crops had to be protected against them. The abundance of deer bones could be regarded as a consequence of defensive measurements and not solely as an unrelated procurement of wild animal meat. Likewise, there is ethnographic and archaeological evidence of game procurement as garden hunting among horticulturists (Linares 1976; Neusius 1996:273–276). The hunting of the roe deer, red deer, wild boars, and hares may all point to a game procurement strategy such as garden hunting. If this were the case in the Opovo site, one could ask why similar practices did not occur in other Vinča settlements surrounded by forests. In these settlements, cereals were identified (Hopf 1974) but instead of wild animals, bones of domestic animals were predominant (Bökönyi 1974).

The increase of wild animal bones and change in domestic versus wild animal ratio was attested at the Zurich lake side settlements were explained as a consequence of climatic change. Schibler et al. (1997) suggest that more meat, particularly of red deer was consumed to compensate for the decline in plant resources due to climatic deterioration between the 38th and 36th centuries B.C. A similar shift in animal species frequency, characteristic for the 40th century B.C., is also justified by a climatic deterioration. Lack of intensification in animal stock keeping between 4300–3300 B.C. is explained by the insufficient quantities of animal fodder (leaf hay and twigs) to feed the cattle during the winter months (Schibler et al. 1997).

Another possible explanation is that natural conditions varied among different geographical regions and that the Opovo environment was still pristine and uninhabited when the Late Vinča people settled, and there was still plenty of game browsing in the local forests. The abundance of game, like the availability of wild plants, could have attracted the Late Vinča settlers to the Opovo environs. Crop cultivation could have then only enhanced "field hunting". On the other hand, if red deer hunting began in the late summer/fall, after cereal and legume harvests, it would not have been a consequence of killing the animals while crop raiding but rather of unrelated hunting of mature males and/or collecting the victims of male fights. Late summer/fall hunting would not have coincided with the main agricultural activities including cereal and legume harvests but would have complemented water chestnut gathering.

***

Wild plants were often neglected in archaeobotanical reports dealing with Neolithic sites, because cereal and legume seeds were mostly retrieved and identified from the Balkan sites. Because of the paucity of cereal remains at the Opovo site, which is not typical for the Late Neolithic sites in the Balkans, a special emphasis was placed on the identification of non-crop plant remains in the Opovo archaeobotanical samples. Such a detailed analysis revealed that plants were gathered from the wild and processed in the Opovo settlement. Besides gathered Cornelian cherries, grapes, strawberries, blackberries, and elderberries, some of the weedy herbaceous plants such as *Chenopodium* and *Polygonum* might not have been only weedy intrusions but intentionally gathered for some other purpose. Some of the carbonized and uncarbonized Solanaceae seeds (which were difficult to identify to the exact taxon) might have been used medicinally along with recovered uncarbonized seeds of *Lithospermum officinale*, which might have been used as ornaments, because of their attractive ivory color and hard structure. In addition to their utilitarian values, some of the plants might have been used in rituals for their supposed "magical" powers.

Of special importance for the Opovo site are water chestnuts, which have been recovered in rather large quantities from the Opovo samples. Because the Opovo area was prone to floods and droughts until modern times, carbohydrate-rich water chestnuts may have been eaten to supplement cereal intake. Water chestnuts might have been collected more intensively during wet years, because the droughts also negatively affect water chestnut sprouting and dispersal.

Opovo fields were probably located close to forests. Location of the fields close to the forests might represent a trade-off for better soils, despite the danger of crop raiding. While some of the roe deer, red deer, and wild boar were killed during spring and summer while raiding crops, those killed in the fall were probably not "garden hunted". Hunting during fall did not coincide with cereal, legume, and flax harvests, but could have complemented water chestnut gathering.

Perhaps when the Late Vinča people settled down in the still pristine and uninhabited Opovo environment, there was still plenty of game browsing and there were available wild plants thriving in local forests. These settlers probably took advantage of the wild resources before the first crops started to yield. After the establishment of the Opovo settlement and crop cultivation, the settlers might have continued to gather fruits, medicinal plants, and water chestnuts to complement their diet. In years when crop yields were low, they might have gathered water chestnuts to buffer agricultural shortages.

A combination of crop cultivation and wild plant procurement on the one hand, and hunting and raising domestic animals, on the other, points to a mixed strategy that might be labeled "*terra et silva*" (land and woods). Montanari (1994:27) refers to this Latin phrase used frequently in the documents from the sixth to tenth century "as a sort of hendiadys that represented interplay of cultivated and uncultivated spaces." Montanari depicts the mixture and copenetration of these spaces involving a panoply of productive activities such as grain cultivation, horticulture, hunting, fishing, free pasturing, and gathering. These diverse activities also fostered dietary diversity that through a combination of environmental and social factors spread to all social levels. The abundance of woods and pastures was generally unlimited, and the relation between property and production did not forbid the use of open spaces during the early Medieval Period in many parts of Europe. The woods were so important that "forest famine" was depicted as an equally serious as "agricultural famine" according to historical records; for example, good weather for spawning fish and abundant acorns necessary for fattening pigs were as important as good weather for harvest and vintage (Montanari 1994).

However, more recent ethnographic data seems to contradict the historical evidence of *terra et silva* practice that I suggest for the Opovo case. Subsistence data from 200 societies show that societies either depend heavily on agriculture or engage in it to a very small degree. These data indicate that agriculture forms between 5–45% of subsistence only in a small percentage of societies (Hunn and Williams 1982:6, cited in Zvelebil 1996). Nevertheless, Zvelebil (1996) argues for the existence of mixed hunting-farming groups in some parts of the Baltic region from prehistory until the Middle Ages. He also mentions the evidence for reverse transition from farming to foraging in south central Sweden in the mid third millennium b.c.

Thus, it seems that the term *terra et silva* can be appropriately applied to the Opovo diversified mixed subsistence strategy.

CHAPTER 7. LAND USE AND SOCIAL DIFFERENTIATION

Theories about communal ownership and land tenure versus private ownership have been the focus of many historians, rural sociologists, anthropologists, and economists. Some of the issues raised about landownership and tied to socio-political organization allow me to suggest various possible models in relation to the prehistoric evidence (Allen 1992; Dahlman 1980, cited in Fleming 1985; Ellickson 1993; Kerridge 1992; McCay and Acheson 1987; Neeson 1993, cited in Turner 1994; Ostrom 1990, cited in White and Runge 1994; Runge 1981).[1]

Different field systems have been studied archaeologically, mainly in the British Isles (Bowen and Fowler 1978; Bradley 1978; Fleming 1979; 1985, 1989, Fowler 1981, 1983), as well as in other parts of northwest Europe (Barker 1985; Bradley 1978). Prehistoric land boundaries or so-called "Celtic fields" or coaxial field systems have been discovered in Britain and Ireland, and discussed by Fleming (1982, 1985, 1989), Fowler (1981, 1983), and others. A coaxial field system is defined as a series of fields laid out according to a single dominant axis of orientation, in such a way that most of the major boundaries are parallel to one another. They manifest as soil marks on air photographs, as lynchets or ruined walls in the field, or as patterns still visible on maps of recent or present-day field boundaries. In the British Isles, they appear to have been laid out and used from the late third millennium b.c. onwards, and they are widely distributed. The only systematic investigations of prehistoric field systems in the central Balkans so far have been done by Chapman et al. (1987; Chapman and Shiel 1993) in Dalmatia. Although remains of prehistoric coaxial field systems have not been discovered in the Balkans, the lack does not necessarily negate their existence. I believe that on the basis of archaeological analogies between similar subsistence patterns, in addition to ethnographic and historic evidence for settlement patterning, household size, plant and animal remains, some inferences could be derived for field systems in the prehistoric Balkans.

Landownership is often connected with social stratification and/or ranking (e.g., Bintliff 1982). I usually use the term social stratification without intending to equate the term social stratification with state societies, or the term ranking with chiefdoms (see Fried 1967). I employ both terms to describe the differences that might have existed among households in the Vinča society and not to describe age and sex inequalities. Age and sex inequalities might have cut across other social privileges and perhaps might have been common for all households. As Hastorf (1993:222) correctly indicates, land-use is "determined by many factors including demographics, historic boundaries, spiritual landscapes, ancestral connections, environmental conditions, as well as by the way people have chosen to organize their production".

On the basis of archaeobotanical data and the proximity of houses revealed at the Opovo site, I discuss possible locations of the ancient fields of the Opovo households. On a more theoretical level, citing examples from anthropological literature, I explore different types of landownerships that could be suggested for the Opovo case. On the examples from archaeobotanical literature, I examine the importance of collective labor, feasting, intra-site differentiation, crop purity, and special plants for social stratification at the Opovo site and Vinča culture.

Houses

The excavations at the Opovo site revealed four complete houses and part of a fifth house (Figure 1.9). The houses were built close together, and are interpreted as residences (Stevanović 1996, 1997a, Tringham 1994). On the basis of house sizes, it was suggested that a family was the basic household unit in the Vinča culture (Chapman 1981; Tringham and Krstić 1990). Probably small families inhabited the Opovo houses, because none of them exceeds ca. 40 m^2 in area. Stevanović and Tringham, who studied the architectural remains at the Opovo site in detail, argue that the houses at the Opovo site, and at other Vinča culture sites, were intentionally burned by separate conflagrations for non-utilitarian reasons at the end of the house-life cycle (Stevanović 1996, 1997a; Stevanović and Tringham 1997; Tringham 1994). Life continued in preexisting houses and in the newly built ones. Given the proximity of the houses at Opovo and other Vinča sites, it seems that the house burning would need to be carefully scheduled and probably communally sanctioned, as argued by Whittle (1996:107). Regardless of the cause for their destruction, Vinča houses were replaced at the Opovo site probably over a 200 year period.

The Opovo houses from the last building phase copied the orientation of the houses build in the previous phases. For Whittle (1996:105–107) this evidence indicates strong conformity to communal practice rather than individual ownership for tenure of building plots. For Chapman (1989:33–53; 1997:131–149), the fact that houses were often built directly above those where the ancestors lived is a phenomenon of ancestral ideology embedded in the structure of the Middle and Late Neolithic tell settlements.

Bailey (2000) argues that generally in the Balkans after 6500 B.C., the more substantial surface level structures with multiple rooms and activity areas indicate more exclusive units based on individual buildings and mark the evidence of the household as fundamental unit of social and economic decisions. He suggests that membership of household to village was fixed to physical structures perhaps through rituals involving representational material culture such as figurines. Cooperation of labor or sharing of resources would not have extended, on one level beyond the physical dimensions of the household, and on other levels beyond the physical environment. However, Bailey (2000) admits variation among the sites.

Fields

The excavations at the Opovo site and other Vinča sites did not reveal traces of prehistoric fields or gardens. The houses revealed were built only a few meters apart and surrounded by many pits and hearths (Figure 1.9); there was not sufficient space for a having a garden plot next to the house, i.e., "house-and-garden" complex as suggested by Chapman (1990:38) for the early occupation at Selevac. Perhaps, in the case of Opovo, small vegetable gardens could have been located close to the clusters of houses, but there was certainly a lack of arable land for cereal cultivation between the houses. I have suggested previously that approximately 1.5 ha of land scattered in small fields might have been under cereal cultivation to meet the needs of one family (see Chapter 5). Furthermore, I estimated that in order to produce enough straw to add to daub for the construction of one house, a surface of 1,500 m^2 had to be under cereal cultivation. The quantity of cereal produced per household must have exceeded the quantity of vegetables grown, considering the amount of cereal needed for food and amount of chaff used for house daub. Ethnographic and historical evidence from the region implies that wheat and barley were not the kinds of plants easily grown together with other vegetables in gardens, but rather were grown in fields (Jeftić 1986; Slavko Borojević, personal communication 1997). Moreover, wild emmer and wild einkorn, when growing in primary habitats, commonly form vast, dense stands resembling unfenced fields, and do not resemble any kind of garden (personal observation; Hillman and Davies 1992).

However, Jones (1992:133–143) suggests that the cultivation regime for cereals and pulses, respectively, at the Bronze Age site of Assiros may have resembled intensive gardening rather than the extensive field cultivation characteristic of present-day Amorgos. The seeds of weed class Chenopodiatea were more common in the samples from the Bronze Age than in the ethnographic samples from Amorgos, while in the ethnographic samples, the weeds from the class Secalinitea were prevalent. The weeds of the Chenopodiatea class are considered the weeds of hoed gardens and ruderal places (Ellenberg 1979). Chenopodiatea weeds often accompany spring crops, although some of the plants from this class can accompany both winter and spring crops. Additionally, studies of the pulse cultivation on the island of Evvia, Greece, showed that Chenopodiatea species are associated with small gardens within the village, while Secalinitea species are associated with large fields at some distance from the village. These studies support the earlier statement that intensive cultivation on a garden scale is a possible interpretation of the distinctive weed flora for ancient cereal and crops cultivation (Jones et al. 1999).

Halstead (1996:296–306) uses archaeobotanical data by Jones (1992) in explaining the development of agriculture and pastoralism in the Greek Neolithic. He suggests that small holdings were often hand cultivated under an intensive "horticultural" regime, which would include rotation of cereals and legumes, intensive weeding, and manuring. The several lines of evidence that he mentions for a horticultural regime are: the absence of evidence for plow use; the labor-intensive growing of lentils; absence of large-scale forest clearance, help from small-scale sheep husbandry in maintaining soil fertility; and effective isolation of domestic animals, which prevents them from breeding with their larger wild relatives. In the case of more pervasive crop failure, Halstead (1996) suggests that livestock, particularly sheep, might have offered a significant alternative source of food and ensured a reliable supply of energy in the late Neolithic Greece.

At the Opovo site, clearance of the supposed mixed-oak forest might have not been dramatic. There is not much evidence of mixing wild and domestic populations of animals (Russell 1993). The bones of a few domestic pigs do not indicate breeding with wild populations, but do indicate that they were probably kept within the settlement limits. Although both aurochs (wild cattle) and cattle are common in the northern Pannonian Plain (Bökönyi 1974), the forests that presumably existed and the marshy area in the Opovo environs are not a primary habitat for cattle. There is indication that the Opovo cattle was mixed with wild cattle. Cattle bones were most abundant among the bones of domestic animals; however the wild fauna dominated the Opovo zooarchaeological assemblage (Russell 1993).

Although a model of a horticultural regime suggested by Halstead (1996) seems possible for Neolithic Greece, a horticultural regime seems less plausible for central Balkans and the Vinča culture in general, where the legumes and ovicarpids are scarce, and where wild resources could have buffered crop failures.

It is difficult to imagine how the area that I suggested for the Opovo case of ca. 1.5 ha of cereals per household (even in small plots) could have been weeded. This would have been even more difficult if mixed einkorn and emmer crops were grown together as indicated in several archaeological sites (Hopf 1974). During tilling, it might have been difficult to differentiate seedlings of weedy grass (Poaceae) plants from the cereal crops. Experimentally, Reynolds (1979) had shown that tilling and sowing the cereals in rows produces very high yields even without the use of manure. Although sowing of cereals in rows and extensive tilling is absent from the ethnographic present and past in the Balkans and Near East, it still could have been practiced in the past. In prehistory, the parcels under cereal cultivation were probably small and could have been hand-tilled if plows did not exist (see Bogaard 2004).

Cereals were probably cultivated in small parcels and they might have received attention different than later crops grown more extensively on larger fields using plows. Influenced by ethnohistoric record of exclusive broad sowing of cereals in the Balkans and Near East and the absence of weeding wheat, I cannot envision how cereals could have been tended and weeded like other crops that were grown even on a smaller scale, such as legumes or flax. For the Opovo archaeobotanical samples in which the Chenopodiatea seeds were also dominant weeds (e.g., Feature 41), the newly cultivated land and nutrient-rich chernozem soil could have encouraged the development of these weeds in the Opovo fields. As to the absence of group of weeds that belong to Secalinitea, they might not have been so aggressive yet at the beginning of the agricultural period as they became in later periods, such as in Greece during the Bronze Age.

In the Opovo case, however, lentils were found only in small quantities. Furthermore, lentils are far less ubiquitous in the Vinča culture sites than on the Greek Neolithic sites, and are present in smaller quantities than are cereals. Based on ecological requirements of both the crops and the identified weeds, I think that some Opovo parcels where cereals were grown were not suitable for lentil cultivation because of the high soil humidity. Flax cultivation requires a long period of soil regeneration, and perhaps those plots were left fallow before other crops were sown and/or rotated.

Bailey (2000:148) realizes the great regional variation among the sites in the Balkans, and suggests that there is no unilinear sequential development from garden horticulture to swiddening to larger field cultivation. He emphasizes the importance of organizational requirements and social consequences of the cultivation of the new species rather then their nutritional advantages.

The excavations at the Opovo site and other Vinča sites did not reveal traces of prehistoric fields, but the remains of cultivated crops, including cereals, legumes, and flax. Based on the ecological requirements of the identified plant species, it is likely that some fields were located on drier and some on wetter soils, and that perhaps the inhabitants practiced crop rotation and leaving fields fallow. On the basis of this evidence, in addition to ethnographic and historic sources, I have suggested that there were probably small, dispersed fields around the Opovo settlement.

Landownership and Land Rights

On the basis of house sizes as discussed above, it was suggested that a family was the basic household unit in the Vinča culture (Chapman 1981; Tringham and Krstić 1990). If we assume that a household was the basic economic unit and that one family cultivated crops in several dispersed fields (see Chapter 5), then we must ask question who owned these fields. At least two different types of landownership could be suggested for the Opovo settlement: 1) the land was communal property of a sub-group that settled in the Opovo site, or 2) the land was privately owned by households. Here, I do not draw a distinction between ownership and effective control over the land, assuming that those who owned the land could decide how the land was used and how the produce was disposed of. Neither do I claim that a house can be necessarily equated with one household and one family.

If we assume the land was entirely communal, more communication and negotiation were needed to coordinate activities, such as sowing, tilling, and harvesting of crops on dispersed plots, than if the land was managed by separate entities. It seems that communal land management would require social organization on the community level to regulate rights for land use on the one hand, and to coordinate and synchronize activities on the other hand. My concern is the extent of communal organization needed if common fields existed. Labor allocation and the timing of sowing and harvesting needed to be planned to coordinate the whole community. Although the sowing of cereals requires little labor, harvesting is highly labor intensive, especially in areas where summer rains are common. The harvest must be performed in a very limited time span. Cooperation would have been needed for tilling crops, especially without plows. In the case of communally owned arable land, and hence communally harvested crops, some central redistribution would be needed to share harvests among households.

Likewise, Winterhalder and Goland (1997:123–160), from their evolutionary ecological perspective, suggest that a few high-yield domesticates will face selection pressures to risk management within the households. The input costs of plant management, processing, and storage are costly in agricultural societies. On the basis of ethnographic data from the Andes and historical data from Medieval England, they computed that it is more beneficial and less catastrophic if families have scattered fields than nucleated ones. According to the authors, the free-rider problem during periods of long production and denser population would make inter household pooling less attractive.

Some authors in development literature (see Axelrod 1984) question the validity of the free rider-problem and application of the tragedy of the common paradigm. For example, White and Runge (1994:1–42) demonstrate that common property and collective action functioned in the cooperative watershed management in rural Haiti. They conclude that collective action institutions are appropriate for solving various issues such as gaining prestige or short–term gains or building labor debts. In their case of watershed management, both private incentives respecting individual parcels and public incentives for transboundry problems are involved. The authors define the following theoretical components of voluntary contribution to a public good: the nonseparable costs of watershed management, the critical role of expectations and contingent upon choice, the collective consequences of individual behavior, the redistributive function of shared responsibilities, and the survival capacity of the watershed management institution (White and Runge 1994:6).

The first component of nonseparable costs of watershed management does not apply to the suggested dispersed parcels of arable land in general, and to the Opovo case. The gains of one farmer do not necessarily hinge upon the actions of the others, although the principles of reciprocity and labor exchange could have existed. Also, White and Runge (1994) predict that smaller groups rather than large ones would be more appropriate for solving assurance problems, which they define as difficulties in acquiring information about the likely behavior of other group members. Because of the heterogeneity of the parties' interests, large groups would probably break into smaller ones in order to eliminate or purge noncooperative members from the group.

In the Opovo settlement, where probably few households existed once, small household groups, affiliated perhaps through kin ties, could have cooperated in managing arable land. However, if the household is defined as a basic production unit, then each household that worked certain plots could have claimed rights to these plots, at least as long as they kept cultivating them. Hence, the land would have been owned by the community, although each household could have been responsible for the cultivation of its own fields.

Fleming (1979, 1982, 1985, 1989) has argued that the Bronze Age land division at Dartmoor (Great Britain) did not reflect individual landownership, but rather communal. For an individual or a small sub-group of a local population, an attempt to control additional land without access to additional supplies of labor would have been pointless; ineffective, unnecessary, expensive, and strongly opposed by others (Fleming 1989:68). Control of products of their labor could have been more effectively exercised through exchange of bride wealth or some other kind of exchange. Therefore, the subsections of the population would have found it more worthwhile to make a small investment in the control and defense of land administered by the community. As Dahlman (1980:129) suggests, "collective property rights and decision–making can be quite consistent with private wealth maximization and under certain conditions with respect to transaction and decision costs, such an arrangement will be superior to private ownership and control." On the other hand, in the case of collective land for hunting, woodland browsing, grazing, wood collecting, etc., the costs of maintaining and policing individual boundaries would be much higher for individuals than for the community, especially in cases when it was necessary to build fences or other barriers (Fleming 1985). I would add that collective use of water resources such as wells, rivers, and ponds was also common.

In the present context, rights in land are defined similarly to definitions used in anthropological development literature, and are partially applied to prehistoric cases. Land rights consist of claims that are socially recognized and enforceable by external legitimized authorities (e.g., councils of elders) either at the village or clan level. The perceived rights would include managing of the field parcels and the agricultural products by households, but a clan or village community owns and controls the land, for which there is a wealth of ethno–historic evidence in Asia and Africa (see Agarwal 1994; Bassett and Crummey 1993). Thus, these rights would have locational and temporal restrictions; they accrue only for a person's lifetime or lesser period, and they are located in the village where a person is residing.

However, communal ownership does not necessarily imply communal cultivation. There are advantages to having members of a kin–structured group controlling a large territory. One advantage is that continued access to land affords much better protection than vesting ownership rights in many fixed parcels of land. If arable land is periodically available for reallocation, the community may be able to match land and labor resources more evenly than might be the case within available parcels with fixed boundaries. Fleming (1985, 1989) mentions evidence for former land–use systems of Scotland, Ireland ("runrig"), and Medieval England, including the rotation of parcels or strips of land among tribesmen or tenants, as well as substantial rights of "common" grazing.

If the right to clear and use a certain plot is established, it could lead to social stratification. Thus, a bigger family may or may not be advantageous, taking into account a number of unproductive members of the family. Those household heads with larger families can call on a larger workforce for agricultural and construction tasks, in turn leading to larger surpluses recycled to kinsfolk through feasting and ceremonies. The differential production of household surpluses can lead to a wider chain of alliance networks through surplus exchange, given checks and balances against excessive accumulation. Such expansion of family power would be unlikely to last for more than two or three generations unless institutionalization of such structures were achieved (Chapman 1991:87). Bintliff (1982:100–106) has persuasively argued that land ownership and control of agricultural products, rather than control of trade, transformed wealthy peasants into local political elites.

Another possibility is that the inherited right to cultivate land might not necessarily mean land ownership. If the land, for some reason, was not cultivated, the community could have decided to expropriate the right, a practice not unknown in the ethnographic present. For example, Fleming (1989:72) mentions that groups of commoners would reach a decision through a sort of parliament as at the St. Kilda, or old men's court at County Clara, or headmen "kings" who acted as communal coordinators and arbitrators, and not through a rise in power via accumulation of wealth through exchange. This model entails a durable system of communal land holding and can be proposed for Neolithic communities.

In contrast to having only rights to cultivate the land, the second type of landownership that could be suggested for the Opovo settlement is that each household owned its own fields. If ancestral ideology had been embedded in the settlements whereby the replacement of the houses was a common practice, as suggested by Chapman (1981, 1992, 1998), possibly the ancestral ideology was also reflected in land inheritance (Srejović 1984; Tringham et al. 1992). Moreover, Chapman (1981) interprets the incised pottery signs at the Vinča sites (e.g., Banjica), as signs that suggest owner's marks. The contexts of Banjica signs suggest family land ownership and property inheritance, since the same incised sign was found on pots in two superimposed house floors, built ca. 50–100 years apart. Gimbutas (1991) and other authors would argue that the incised signs on the pottery and other Vinča artifacts are actually the signs of the Old European script and a form of sacred writing (cf. Gimbutas 1991).

Collective Labor

If the household is defined as the primary level of organization, then secondary and tertiary organizational levels may also be recognized. The secondary level is frequently described by sociologists as a cooperation among households located in the same neighborhood (Fleming 1985). These groups, whose size is appropriate for pooling labor and resources and sometimes for making economic decisions, are too small to make arrangements for territorial defense. Despite the lack of territorial defense, the groups have invested too much in a specific area to abandon it as hunter-gatherers do. The third level of organization on a larger scale for territorial defense is not detectable from European ethnography (Fleming 1985:132), but could detected in archaeological records in form of defensive ditches and remains of palisades. However, my own focus was on the primary and secondary levels of organization because they are more closely connected with agricultural practices than with larger defensive systems.

An important function of the second level of organization was to provide a pool of labor for various activities. Many ethnographers, social anthropologists, and rural sociologists have discussed collective labor (e.g., Johnson 1982, Olson 1971). Mutual obligations generated by the pooling of labor or the flexible use of labor counteract the inevitable mismatch, at the family level or even at the household level, between the need for labor and its availability. For Opovo prehistoric agriculture, I estimated that to harvest the suggested 1.5 ha, 300 hours would have been necessary; to hand till the same area, 600 hours would have been required, unless plows were used. Although each household would have had the same labor demands at approximately the same time, it seems reasonable to suggest (on the basis of ethnohistoric analogies) that in the Opovo case, households exchanged labor for harvesting, hand tilling, and perhaps threshing.

This type of cooperation was also needed for forest clearance, fencing, and house building. For example, Startin (1978:157) concluded that the task of building LBK houses did involve communal effort, but that not a complicated social hierarchy. The LBK houses are in general considerably larger than the Vinča houses, particularly the Opovo ones. As mentioned in Chapter 6, in Voyvodina, a group of eight people was involved in building a mud house (*pisé*) 50 m^2 in area, over a period of 6–10 days (Ginder 1996). The construction of an Opovo house might not have involved a large-scale communal effort, but probably did involve additional labor outside one household. In the construction of the Voyvodina mud houses, not a single man was a professional mason and no outside organization was necessary to conduct construction. Thus, similar to Startin's suggestion for LBK houses, I would suggest that a social hierarchy was not necessary for construction of the Opovo wattle-and-daub houses.

Feasting

Gatherings and feasting during and after communal activities play an important role in exchange of labor; probably that was the case during prehistory as well. Dietler (1990) suggests that drinking ceremonies and feasts can be important means of mobilizing labor in societies lacking coercive political authority, assuming that labor was not a marketable commodity in prehistory. Hayden (1992) accepts Dietler's (1990) argument that early grains were originally more important in brewing than in subsistence during prehistory. This argument could probably be used for wheat and barley, but no beer is made from pulses.

In the study of the political inequality before the Inka in the upper Montaro Valley of Peru, Hastorf (1993) suggests that the main artifact differences between the elite and non-elite statuses were in the sphere of food preparation, storing, and consumption. The author proposes that the elite gave more feasts in exchange for allegiance and authority. The best evidence was the increase of maize, used for brewing beer, and elite having larger cooking bowls during Wanka I period (a.d. 900–1300).

Russell (1993) analyzed animal bones from the Opovo site and suggests that feasting occurred at Opovo. She bases this conclusion on the presence and condition of bones discarded in the pits that may have resulted from feasting. For example, groups of bones from similar large cuts of meat but from different individual animals appear to have been roasted and carved. Red deer, wild boar, and domesticated cattle were the main animals served during feasts, implying that the degree of meat sharing depended on the size of the animal and not on its domestication status. According to Russell (1993:428), evidence for feasting at Opovo was found in pits: Features 26, 30, 31, 41, 46, and 50, and somewhat tenuously in Features 48, 43 and 45 (Figure 3.13). Russell states that the sharing of meat does not seem to be asymmetrical, i.e., good cuts of meat were distributed to different households. However, she argues that prestige was gained more through distribution rather than possession, making it quite prestigious to host a feast. She further argues that—although the Vinča culture was "not yet solidified to a truly stratified system" (Russell 1993:475)—competitive feasting was a feature of inter-household competition at Opovo in the late Neolithic of southeast Europe, such competition resulting from power imbalances and overall status competitions in Vinča culture.

In addition, on the basis of functional analysis of pottery, it has been suggested that there was a decline over time in the importance of storage in ceramic containers and an increase of serving vessels, thus implying feasting and other forms of conspicuous consumption (Kaiser in Tringham et al. 1992:377).

I have tried to use Russell's (1993) suggestions to extract archaeobotanical evidence of feasting at the Opovo site, although it might be more difficult to detect feasting and intra-site variation among households by using plant remains than by using animal remains. In the plant remains, I find no pattern like the ones found in animal bones and, therefore, I can provide no archaeobotanical evidence for feasting at the Opovo site (Figure 3.13). I believe that the sharing of meat from large animals could have occurred in several households, perhaps after a successful hunt or after crop-raiding animals were killed.

The evidence of accumulated late summer-fall debris in pits could perhaps point to feasting. As discussed in previous chapters (see Chapters 2–5), all the plants identified at the Opovo site were gathered or harvested from early summer to fall, but they could have been consumed year round. Unfortunately, we do not have evidence of strictly seasonal food that could have been consumed only during a very limited period. Moreover, the seasonal pattern of animal bone deposition is not corroborated by the deposition of plant remains. In the two pits where wheat spikelet forks/glume bases are the most abundant category, there are no remains of animals killed during the summer. This might further suggest that bones from consumed animals were not thrown in the pits along with cereal byproducts. The most appropriate plant remains for seasonal determination in the Opovo site are water chestnuts, which are gathered during the late summer and fall. It could be suggested that water chestnuts that were gathered during late summer and fall were additional foods offered during feasts. Water chestnut fragments were found in the layers where either no animals were killed during the summer months or fall, or where animals were killed during summer-autumn. Because fall is a season of water chestnut collecting, water chestnuts might have been consumed with meat in some instances. Afterwards the peelings of water chestnuts were discarded together with bones in some pits, while not in others.

Cereals are often mentioned in connection with beer brewing, another aspect of evidence for feasting. One example of such evidence is Hastorf's (1993) study of agriculture and political inequalities in pre-Inka Peru. She found evidence that maize was used more through the Late Intermediate Period (A.D. 1000 to 1460) than earlier because it was used for brewing beer to accompany feasts through which the hosts gained prestige and access to mountain spirits. In the Old World, evidence for beer brewing can be detected in the archaeological record when, for example, caches of barley grains are found similar in size and equally sprouted, suggesting that they may have been charred during malting (Hillman 1984). Beer could have been made from cereals at Opovo, but no sprouting grains were found. There is, however, clear evidence that cereal grains were dehusked for consumption, and I suggest that the recovered quantities of spikelet forks and glume bases represent several short episodes during which handfuls of einkorn grains were dehusked, a step that is not necessary for beer brewing. The wheat spikelet forks are the most abundant category of plant remains at Opovo, both in the pits where animal bone evidence indicated feasting (e.g., Feature 31) and in the features representing everyday activities (e.g., Feature 52). Cereal grains were found in many pits and were primarily deposited as food refuse.

On the basis of the analysis of plant remains, it seems the Opovo inhabitants relied on locally available food resources. There were no remains of plants from beyond Opovo environs that would point to special use in feasting. There is no pattern between deposited wild and/or domestic animals and plant remains to provide archaeobotanical evidence for feasting at the Opovo site (Figures 3.11–3.13). However, feasting, including the sharing of meat from large wild and smaller domesticated animals, might have taken place at the end of agricultural activities in the late summer-fall, and during special occasions and rituals.

Intra-site Differentiation

On the basis of plant taxa recovered from various samples, I discuss Opovo land and colonization of "marginal lands".

It can be postulated that weed flora accompanying crops varied according to the different and permanent field locations and that plant sample composition might reflect use of different plots. Some of the pits may contain a specific weed assemblage reflecting the cultivation of a particular field during a longer period by a specific household (taking into account that different weed taxa accompany different crops). Similar analyses were undertaken by Jones (1992:133–143), where she compared ethnographic and archaeological samples of cereal and legume processing stages in Greece. She was also able to discriminate between two types of soils used for cultivation, derived on the limestone bedrock and on the schist. At the Opovo site, however, the soil around the settlement is predominantly chernozem over loess substrate. Presently there are saline soils in the Opovo region, but these have developed more recently due to the lowering of water table.

Most of the identified weed seeds from Opovo belong to the class Cheopodiatea, which includes weeds of arable land and of ruderal plants. After detailed identification of all small seeds including possible weeds of arable land and their ecological requirements, I found no specific plant indicator species repeatedly occurring in the same pit that could point to permanent cultivation of one field, presumably by members of one household. Even if the same fields were permanently cultivated, the practice of crop rotation and leaving the fields fallow might have obscured the typical weed assemblage that would have been present in the specific soil type. Crops can also be harvested from various plots and then threshed and processed at the same time, obscuring differences between individual plots. Moreover, the taphonomy of the Opovo pits reflects different depositional processes rather than cultivation of the fields in individual plots.

Not a single storage space or large concentration of grains was recovered at the Opovo site; thus, there is no evidence that could ascertain that some houses stored more grain than others. The fact that no stored grain was found in or outside the excavated Opovo houses does indicate that either all grain was used up, or that it was removed before the abandonment of the houses, unless postdepositional processes uniformly destroyed all the plant remains. It can be argued that the absence of storage at the Opovo site was not incidental but related to different organization of production and communal storage in the Opovo case. At other Vinča sites, where there was evidence of grain storing (see Chapter 5), grain was found either within or outside the individual residential structures in pits or silos; thus, I am assuming that communal or centralized storage was not typical for the Vinča culture, including the Opovo site.

At the Opovo site, no structures or areas of specialized production were uncovered; however, that might be due to the limited excavation area. The Opovo houses did not have complex internal divisions like those present at southern Vinča culture sites (e.g., Vinča-Belo Brdo, Gomolava), although one of the Opovo houses contained an oven and food preparation facilities, and the other had a second story interpreted as a storage space (Tringham et al. 1992). Among the few houses that were excavated in Opovo, none resembled a shrine nor had artifacts that could have been interpreted as prestige items. Commonly, Vinča religion is interpreted as household religion due to presence of fine pottery and figurines in all domestic contexts, including middens (Chapman 1981; Srejović 1984). In a few sites, bucrania were found (e.g., Vinča-Belo Brdo, Gomolava), which in some instances (e.g., Jakovo-Kormadin) were interpreted as possible shrines (Jovanović and Glišić 1960). Perhaps, production in the Vinča culture generally was not specialized to the extent that it was alienated from household food production, at least in the smaller settlements, such as the Opovo site.

Crop Purity

At the Opovo site, crop remains were primarily found in refuse pits. Composition of plant assemblages varies among the pits, so it is impossible to ascertain whether there were pure stands of crops. In the example of the Selevac site, Chapman (1990:32–38) suggests that creation of fields facilitates the separation of crops into pure stands, and that there is evidence for increased purity of grain crops in the Late Neolithic (Kosse 1979; Renfrew 1973). These changes would characterize Chapman's (1990) "Agricultural model" in contrast to his "Horticultural model". He describes the Horticultural model as the practice of rebuilding houses in the same area with fixed gardens in between them, and later, shifting locations of the gardens. In contrast, he defines the Agricultural model as rebuilding houses at various locations and practicing crop rotation in fields outside the settlement. Horticulture would imply the use of hoe and mattock technology, while agriculture would necessitate the use of the ard. (However, the evidence for plow technology is still extremely scarce from the Balkan Neolithic.) According to Chapman (1990:38), in the second half of the Selevac occupation, yields of fixed gardens utilized for several decades may have declined and the relocation of the house-and-garden complex was unavoidable. Gardens were separated from houses, and the new strategy offered greater flexibility and potential for expansion. Citing Goody (1976), Chapman (1990:38) suggests that it is likely that the new land-use strategies were associated with changes in kinship, descent rules, and patterns of landholding, with a possible shift from matrilineal descent to male-dominated kinship and patrilineal descent in the Late Neolithic.

Crop purity was used by Dennell (1978) as the evidence for large-scale field cultivation in Bulgaria in the sixth millennium B.C. On several Neolithic sites from Bulgaria samples of large quantities of a single species seeds were recovered, the largest from the site Slatina where over

200 kg of charred wheat, barley, and pulses were identified from six storehouses (Bailey 2000:145). According to Dontscheva (1990), one container was dominated by 99% barley seeds, others contained a mixture of *Triticum monococcum*, *Triticum boeticum*, and *Triticum dicoccum* seeds, in addition to lentil (*Lens* sp.) seeds. The combination of different cereals found in the same storages was interpreted as several cereals sown together (Dontscheva 1990).

On several Vinča sites, the seeds of emmer and einkorn were found together (Hopf 1974). So, it is difficult to show archaeobotanical evidence for crop purity in fields. For cereals, it is easier to assess the crop-processing stage of a sample or possible harvesting technique employed than to assess crop purity of fields as a result of agriculture as opposed to horticulture. Chapman (1990:39) cites Rappaport's (1968) ethnographic accounts of horticulture and stresses the multiplicity of plants growing together, i.e., in "maslin" fashion. Chapman (1990:39) argues that the use of fields facilitates the separation of crops into pure stands. However, he does note that cereal crop purity may have resulted from crop processing or dietary requirements. There is ethnographic and historic evidence of agricultural practices in which maslins of various types of cereals and pure "monocrops" are deliberately and separately grown on different plots in one area. Ethnographic evidence from Greece, Moldavia, and the former Yugoslavia shows that in addition to more or less pure monocrops, deliberately mixed maslins (as classified by peasants) were grown in order to exploit the advantages of maslin components which can grow under different conditions (Jones and Halstead 1995; Janushevich 1976; Hegediš and Čobanović 1991). The composition of these maslins can be highly variable. On the Greek island of Amorgos (Jones and Halstead 1995), it was impossible to identify the deliberate cultivation of a maslin from monocrops archaeobotanically, as there are no fixed ratios between different components of maslins. Monocrops and maslins contained a low level of contamination by other cultigens, and as Dennell (1974) noted, these contaminant seeds were mainly introduced with seed stock, not through crop rotation.

When a large concentration of a single crop is discovered in a household and/or site, the same crop must be ubiquitously present in all other crop-plant assemblages on the site, including agricultural waste, in order to prove monoculture specialization (Bakels 1996). Even when such evidence exists, as in case of the Late Neolithic site of Makriyalos in Greece, where einkorn chaff ubiquitously dominates in crop assemblages in all refuse pits, its presence is explained as a small-scale cultivation of crops, einkorn in particular (Valamoti 2004).

The utilization of fixed gardens would certainly diminish the yields of any crops sown in the same plot for several decades. However, experiments at Butser farm demonstrated that even after 15 years of soil use, and even without manuring, wheat yields were still rather high (Reynolds 1981, 1992). Also, ethnohistoric examples in traditional farming provide the evidence for cropping arable land longer than a few decades without intensive manuring or the use of special technical equipment. My speculative assumption is that information (and/or experience) about crop rotation, leaving fields fallow, and manuring might have been passed around together with the introduction of wheat, barely, and legumes.

The Role of Special Food Plants

The other line of evidence that can be used in detecting inter-household differentiation in subsistence based on the plant remains would be to evaluate the importance of special plants in the nutritional regime. Perhaps the best way to detect social differences in prehistory would be by the presence of rare or exotic (luxury) plants that could not grow locally, and by the evidence of storing large quantities of food by particular households.

Scholars of antiquity often mention classical sources in which barley is described as food for common people and slaves, while wheat bread was commonly consumed by the middle- and upper classes (White 1976). Likewise, archaeobotanists have attempted to discover whether social differentiation can be detected from Medieval sites in Europe (Grieg 1988; Haaster 1994; Paap 1984). In most cases, the archaeobotanists analyzed samples taken from urban medieval cesspits. Usually, the appearance of seeds of exotic fruits in the archaeobotanical record was used as distinguishing criteria for indicating different status. Often the seeds of exotic fruits, such as figs, appeared everywhere in northern European towns in the late Middle Ages. Also, the spectrum of the identified plant foods was not substantially different between the houses of poor and rich. It seems that imported foods were given to the poor for special occasions and that by the late Medieval period, imported food was not limited to the higher classes. However, it seems that consumption of various kinds of meats varied more among the households of different status than did consumption of plant foods. Thus, Greig (1988:127) mentions evidence (Donaldson et al. 1980) that food remains of a clearly high-status context at Barnard Castle, County Durham, display a great range of animals, including birds and fish, but only four different plants.

Hingh and Bakels (1996:117–120) carried out analyses for the early Medieval site of Serris-Les Ruelles in northern France. The authors studied samples both from the manor complex and from village structures dating to the seventh and eighth centuries in order to find "socio-botanical differences". On the basis of historical records, they assumed that wheat remains would be prevalent in the manor complex and rye remains in the village structures. However, there were no striking differences between the two contexts, and naked wheat was the most common cereal in both cases. In the samples from the manor house, the following most important cereal was rye, and in the samples from the village, barley. The authors conclude that although there is evidence for social differentiation on the basis of other archaeological material, it was not detectable in plant remains. They suggest that food preparation and consumption was very similar for the lord and the peasants in a small, closed society such as the villa-type because the peasants produced both for themselves and the lord.

More recently, Bakels and Jacomet (2003) investigated the archaeobotanical evidence of luxury food in Central Europe during the Roman occupation. On the basis of data compiled from a 180 sites dating from the first until the fourth century AD, the authors concluded that initially the new "Roman foods" were found primarily in military camps, but soon afterwards in civil settlements but not yet rural sites. The so called "luxury plants" were found everywhere within the Roman Empire and seemed to be part of the Roman lifestyle. Some plants that could grow in the local environment, e.g., walnuts, apples, cherries, probably garlic and herbs, were incorporated in local agriculture and horticulture. Those which could not remained "luxuries", e.g., rice, chickpeas, pistachio, and pomegranate, disappeared in Central Europe after the collapse of Roman Empire.

In the examples from the Roman and Medieval Europe, it seems apparent that luxury plants foods quickly become available to all social strata, especially, the plants that could be locally grown.

In the search for palaeoethnobotanical evidence of social stratification, Hastorf (1993) has noted that despite the fact that there were differences between the elite and non-elite in the pre-Inka Peru, all the wild and domesticated animals, as well as crops were found in compounds of both groups during Wanka II Period (starting ca. A.D. 1300). Thus, the differences are not archeologically reflected in the farming, processing, and grinding tools. However, the elite had more maize, camelid meat, metal jewelry, and spindle whorls. The observed differences were due to the elite having access to more domestic labor. The insufficiency of more pronounced differences, Hastorf (1993) explains by the absence of specialized production and centralized distribution.

A similar type of study was conducted by Lepofsky et al. (1996:31–62), who tried to reconstruct prehistoric socioeconomies from the site Keatley Creek in British Columbia. They analyzed and compared palaeoethnobotanical and zooarchaeological data from the floors of a small, medium-sized, and large housepit. The authors hypothesized that differences in residence structure size would correlate with economic status, and that larger housepits would exhibit greater internal differentiation than the smaller structures. The results of both faunal and floral analysis did indicate that density and diversity vary with housepit size, and that more diverse activities took place in the largest structure. However, the archaeobotanical analysis showed no evidence of a regular, repeated pattern that would indicate distinct domestic subgroups in any of the housepits. In contrast, the results of the archaeozoological analysis did show a difference in activities related to the processing and consumption of animals among houses. It seems that animal processing activities and food storage were more communal in the medium-and small-sized houses than in the large one. The archaeobotanical results obtained from the analysis of British Colombia housepits seemed more promising for detecting social differentiation than did archaeobotanical data from European medieval sites.

At the Opovo site, it was impossible to conduct an analysis similar to the one done by Lepofsky et al. (1996) that might have provided evidence for intra-structure differentiation. In the case of the Opovo site, where houses were also interpreted as residential structures, house size did not vary much. Unfortunately, in Opovo, plant remains were not preserved within the house structures and most data concerning food processing and consumption came from pits.

A comparable type of plant analysis that employed differentiation between farmsteads and the center of the Moundville chiefdom (A.D. 900–1250) was done in southeastern North America. Based on analysis of samples from various features and different chronological phases of the Moundville society, Scarry and Steponaitis (1997:107–122) have found differences in the abundance of hickory and acorn nutshells on the one hand, and the abundance of corn cupules, on the other. The authors state that while corn production increased in the later phases of the occupation, the procurement of nuts decreased. Scarry and Steponaitis (1997) argue that people at the farmsteads were producing and processing more corn than their predecessors in the villages, and also more corn than the elite at the Moundville center. Also, shelled kernels and processed nut oil and meal were sent to Moundville as tribute. The residents of the farmsteads continued to process nuts as their predecessors had done, but nut procurement diminished in the Moundville center. The authors suggest that in the later phases of the occupation, most crops were grown in the farmstead fields but also in communal fields which were supervised by chiefs.

I attempted to apply an approach analogous to the one used by Scarry and Steponaitis (1997) to the Opovo plant remains, taking into account that the Opovo site was occupied for a shorter period of time than the Moundville sites, and that all the samples were from one site only. At Opovo, cereal remains were the most abundant category among the crops, and their abundance could be compared to the abundance of corn kernels at the Moundville sites. On the other hand, the presence of hickory and acorn nutshell fragments might be compared to the abundance of water chestnut fragments at the Opovo site. It may be hypothesized that pits with predominant cereal remains could be associated with wealthier households or once producing for the elite. Pits with mostly water chestnut remains could be associated with poorer households that relied mostly on wild resources. Regardless of the difficulty of associating certain pits with specific houses at the Opovo site, the plant remains do not point to such conclusions. In the pits where water chestnuts were found, cereal remains, either in the form of spikelet forks/glume bases or seeds were also found, often in large quantities. Unfortunately, it is impossible to compare the Opovo plant assemblage with other Vinča sites because Opovo is the only site where large number of water chestnuts (starchy wild fruits) fragments were identified.

Einkorn, emmer, and lentils that were already cultivated in the Starčevo and later Vinča culture and were present in the region for at least two millennia (Table 2.5); therefore, they cannot be considered exotic or luxury foods. Plants that could not thrive locally were not discovered at Opovo or any other Vinča site that would imply some sort of distance trade of plants.

Social Ranking

Chapman is one of the few archaeologists who has attempted to explain social ranking in the Vinča culture. He states that the cultivation of higher yield wheat and barley, intensive bovine husbandry, and ard (plow) agriculture were geared to sufficient productivity to ensure a "ritual surplus" (Chapman 1981). The homogeneity of the Vinča culture rested on the establishment of social ranking, in turn stimulated and consolidated by a system of ritually prescribed values, activities, and attitudes whose ramifications touched every facet of behavior. These forces of systematic control supported cultural equilibrium for more than a millennium. In sum, two mechanisms of cultural integration can be observed in the Vinča societies: ritually prescribed behavior and social ranking. In the case of Serbia and Voyvodina, intensive cultivation of fertile soil for two millennia would have increased the tendency towards differential soil exhaustion, already more advanced on marginal soils. Lower agricultural yields would have affected two sensitive areas: the degree of settlement nucleation and the availability of "ritual surplus" (Chapman 1981).

Of particular interest are relations that Chapman (1981) has found between the degree of agricultural potential at the regional level, and the presence of prestige and ritual items and fine ceramics in the Vinča culture. He found that arable potential and occurrence of the items are significantly related. In settlements in the less fertile valleys of the rivers Tisa, Tamis, and Sava, there would be fewer items related to prestige and rituals than in the settlements in more fertile valleys south of the Danube. However, in an area of medium arable potential, in which the Opovo site would be included, the focus would be on prestige items rather than on ritual of fine wares, suggesting that prestige is not defined solely by agricultural productivity. Chapman (1981:70–71) explains such a phenomenon in the Pannonian Plain as an absence of several key resources in the plain (e.g., stone), and the greater need for exchange than in the area south of the Danube, abundant in various ores. Adopting the argument proposed by Chapman, Russell (1993) suggests that the Opovo people settled in the area for the exchange opportunities between the Tiza culture people to the north and the Vinča culture inhabitants to the south. Travel could have been possible along the Tisa and Tamis rivers and their tributaries but restricted through the marshes and wetland forest. According to Russell (1993), the Opovo dwellers were willing to accept smaller herd sizes and grain yields because of the rich exchange opportunities in this location. Perhaps the Opovo environment was not chosen for its agricultural potential, because there were marshy areas and forests that had to be cleared. Depending on the density of the forest and marshland cover and frequency of the floods, such a landscape might not have been easily accessible for gift exchange or long-distance trade, either.

Chapman (2000) argues that it is likely that craft and long-distance specialists emerged in the fifth millennium B.C. in the Balkans. Kaiser and Voytek (1983) note that evidence of social stratification does not appear in southeast Europe until the late Chalcolithic/early Bronze Age, although they document intensification in resource acquisition, production, and consumption of lithics and ceramics for two Vinča sites, Selevac and Gomolava. Greenfield (1991) suggests that specialization is limited to a few sites which are situated upon resources unavailable elsewhere (e.g., copper ores, obsidian, and spondylus shell), but that there is no specialization of commonly available resources by specialists in southeast Europe during the Late Neolithic. Similarly, Evans (1978) writes that it was not until the Chalcolithic (Eneolithic) period that craft specialization became an important feature of the sociocultural system in the eastern Balkans.

The different use of space and density of built houses between flat sites and tell sites has been recently examined in terms of spatial organization and agricultural production and land ownership (see Bailey 1999; Chapman 1991, 1998; Halstead 1994, Halstead and Frederick 2000; Kotsakis 1999). In two articles, Chapman (1989, 1991) discusses different land-use for tell sites vs. flat sites (extended) in the Balkans. Within the Vinča culture, both tell sites (e.g., Gomolava) and flat sites (including Opovo) were occupied during the late Neolithic period. According to Chapman, one of the variables differentiating flat and tell sites was probably land-holding patterns. In the north and west Balkans, the typical house-and-garden unit of the less-compact flat sites prevented communal ownership of all productive land. Chapman asks whether the development of large villages reflected the success of some lineages through the exploitation of lineage labor at the expense of lesser lineages.

The rare appearance of tell sites outside the original area of tell domination provides evidence for possible inter-village ranking. However, according to Chapman (1989, 1991), there is no evidence for inter-polity domination at the village level in the Balkan Neolithic and Copper Age. In the flat sites (extended villages) of the west and north Balkans, the ratio between built versus unbuilt space is 1:8 or more, according to Chapman (1989:38; 1991:91). Thus, each household would have a minimum of 500 m^2 for garden cultivation or pasture. Chapman cautiously states that although horticulture did not meet all subsistence needs, this form of cultivation would have made a substantial contribution. It is true that houses in the flat sites are less restricted by the available space, and that small gardens might have existed near the houses. Nevertheless, the excavated houses at Opovo and many other Vinča sites were built close to each other, often only a few meters apart, so it seems unlikely that every household would have had space for a garden-and-house complex. Garašanin (1979:199) has also pointed out that the density within Vinča settlements was high and that arable land had to be located close to the settlements. He also suggests that several Vinča settlements that were excavated near the type-site Vinča-Belo Brdo were contemporaneous, and that their surrounding territories might have been divided between the settlements.

I have suggested that ca. 1.5 ha (15,000 m^2) of land had to be under cereal cultivation to meet one household needs of six persons (see Chapter 5). Even if the yields were high, it is unlikely that less than one hectare of arable land would have been sufficient, not including the fields that might have been left fallow or the pastureland. If we assume, for the moment, that 500 m^2 of land were cultivated per household as suggested by Chapman (1989, 1991), that area represents only ca. 5% of the estimated 15,000 m^2 area that had to be under cereal cultivation to meet the needs of one household. Thus, I would suggest that a combination of horticulture and agriculture could have been practiced. In the gardens, herbs, medicinal plants, and some legumes and/or flax might have been grown, but probably not cereals. Hence, the inhabitants of both tell sites and flat sites had to cultivate crops in small fields in addition to gardens, which might have been close or further away from their houses. The cultivation of small dispersed fields, crop rotation, and fallow fields might have been plausible agricultural strategies for both tell and flat settlement inhabitants; thus, the types of land-holding might not have been different between the two types of settlements. The physical locations of fields must have been different between tell and flat sites, considering the limited space of the tell settlements. Although settlement size and internal planning varied between both the flat and tell sites, land-holding customs might have been similar among the contemporaneous settlements where the inhabitants of the same cultural group lived, assuming that a family was the basic production unit in both types of settlements. Differential success of some lineages through the exploitation of lineage labor might have been possible both in the tell and larger flat sites, despite the restricted settlement space in the tell sites, as long as arable land and/or other resources were accessible.

Because of inadequate number of detailed archaeobotanical investigations, it is difficult to determine differences in possible crop specializations between tell and flat sites of the Vinča culture. Wherever flotation has been employed, the main crops are einkorn and emmer, including the tell site of Gomolava (Vinča-Pločnik layers), the large flat site Selevac, and the small flat site Opovo (Table 2.5). More diverse plants were recovered from the tell site Gomolava (van Zeist 2001/2002) than at the flat sites of Selevac and Opovo (Table 2.5). The difference in archaeobotanical record between Vinča layers of the tell Gomolava and the Opovo site (unless attributed to the different individuals analyzing plant remains) may reflect different plant material deposited and recovered, different preservation conditions, different length of occupation, and different intensity of crop cultivation or macro regional differences between the two sites (Borojević 1988). Similarly, Valamoti (2004) was not able to distinguish flat from tells sites on the basis of present plant species or plant parts at the three tell sites and two flat sites in northern Greece dated to the late Neolithic.

Storage of grains is often discussed as an indicator of surplus and intensification of production (see Bakels 1996). Stored glume wheats were found both at tell sites, e.g., Vinča and flat sites, e.g., Staro/Selo Selevac, and not at the flat site Opovo. At the Vinča sites, there is no evidence of central grain storage yet. Houses in all three sites were destroyed in fires, the source of carbonization, and thus the likelihood of preservation of carbonized stored grain was similar at all three sites. It can be argued that at the Opovo site, stored grain was absent because storage was outside the revealed houses and not subject to fires, and therefore not preserved and/or recovered. This argument would make a case for different storage techniques between different Vinča sites, reflecting different storage practices in different regions. Another possible explanation is that at small sites such as Opovo, which was not occupied for more than 200 years, there was a small scale cultivation of cereals. Such cultivation did not generate large quantities of grains that had to be stored in special storage pits, while in the larger and longer occupied sites (either tell or flat sites) more cereals were produced, stored, and likely to get preserved.

It remains uncertain whether the settlers of small flat sites such as Opovo had to produce surplus crops for exchange of deficient items or ritual services, or if they could have offered game meat, antler, labor, or other services in exchange, if necessary. The Opovo settlers were probably not the major producers of cereals in the region. The production was perhaps undertaken to satisfy household demands without generating surplus.

In the search for of social differentiation, Chapman (1992:305–317) examines the uses of social space in the article "Arenas of Social Power: The Case of Serbian Prehistory". He adopts Foucault's concept (1984) that "control over space lies at the heart of social power" and combines it with Moore's (1981, 1983) notion that information underpins the configuration of social relations. Thus, he defines the term "arenas of social power" as the nexus of meanings, activities, and prior knowledge. In his preliminary application of the arenas of social power to the model of Serbian prehistory, Chapman points out that the rare use of a combination of symbolism of the living and the dead can be the most effective means for ideological control, that is, in the domestic and mortuary spheres. Different tensions were identified in the balance between these two spheres that resulted in significant changes in the use of social arenas. For the Vinča culture, Chapman argues that the problem of communal versus group or household accumulation was the main contradiction within the social order. The existence of large settlements consisting of well-built houses with an increased concentration of household possessions created tension between the accumulation of wealth and an egalitarian ethos. As a response to social restrictions versus group or household accumulations, a novel social arena of social differentiation within mortuary practices becomes apparent.

Though numerous Vinča settlements have been discovered, there are only two known cemeteries, Botoš and Gomolava, in addition to a few graves from other sites. In total, there are some 50 graves attributed to the Vinča culture in general (Garašanin 1979). The Gomolava cemetery is located within the tell next to the settlement, while the Botoš cemetery was presumably located close to the flat (unexcavated) settlement. At Opovo, partially burnt and unburnt human bones were found in the area of concentrated rubble (Russell 1993).

On the basis of typological analysis of grave goods, the Botoš cemetery is dated to the earlier period of the Vinča culture, the Vinča-Tordoš phase (Brukner et al. 1974), while the cemetery at Gomolava is dated to the later Vinča-Pločnik phase (Brukner 1988), and was perhaps used when the Opovo settlement existed. If we consider that the dating of the both cemeteries is accurate, then the Botoš cemetery is several hundred years older than the one excavated at Gomolava. There is little evidence for change in social differentiation spheres from the domestic to the mortuary, as argued by Chapman (1992), because both spheres seem to have existed during both the early and the late phase of the Vinča culture.

Similarly to Chapman (1992), Halstead (1995b; 1996) also sees rising tensions between the Neolithic household on the one side, and the village community on the other. The tensions are reflected in the architectural isolation of houses, symbolic emphasis on the human figure, and decorative elaboration of vessels suitable for the provision of hospitality in the Greek Neolithic. Because the household units become more visible in the archaeological record beginning with the Neolithic period, archaeologists have begun to pay more attention to these tensions. I believe that the tensions — be they within household, family, or even individuals, on the one hand, and within the community on the other — could have been present both before and during the Neolithic period. Evidence for these tensions is the existence of strict social rules that restrict individual selfish behavior wherever communal activities and property exist, regardless whether the communities are composed of hunters and gatherers, or farmers. As Price and Feinman (1995:4) have noted: "Every human population that we know exhibits aspects of egalitarian and nonegaliterian relations."

Cooperation is perhaps easier in smaller communities where there is more limited scope for self-interest at any one time than in larger ones, and in that respect Chapman (1992) and Halstead (1996) are right because in the Neolithic period there is evidence of large sedentary settlements where the plurality of self-interests could have been problematic. Group fission as a solution to these problems, in sedentary communities, is more costly considering invested labor in fields, houses, heavy tools, as discussed by Kaiser and Voytek (1983) for the large Vinča settlements. Group fissioning is certainly less likely to occur in permanently settled communities than among more mobile groups, but it is precisely the group fission that is used to explain the development of many emerging late Neolithic settlements. From the processualist perspective, Tringham (1994) explains the establishment of the small hamlet of Opovo as a group fission and development away from any growth of centralized organization. A small newly established, kin-based domestic group would maintain much of the social and economic structure of the larger Vinča settlements south of the Danube, but would resist and attempt to change the traditional dominance pattern by inhabiting agricultural marginal areas and by hunting red deer. From the feminist standpoint, Tringham explains that in Opovo, each household would have different relations with the larger villages south of the Danube, and that "there would have been rapid decrease in pooling of labor and extra-kin support especially for the female members of the households" (Tringham 1994:190). From a similar perspective, Russell (1998) argues that where labor is the limiting factor in production, bridewealth would be vital for securing rights of the labor of women and their children, and that cattle were likely used as bridewealth in Neolithic Europe, including the Opovo site.

***

Food has both nutritive and symbolic value, and in archaeology it is often easier to access its practical value. Plant remains are not always the most appropriate material for determining social differences in food consumption. The preparation and consumption of meat may leave more evidence. In contrast, culinary peculiarities of preparing vegetable food (including staple items) may have little chance of being preserved in the archaeological record. There is also another possibility, that within a small group, social differences are not reflected by the vegetable food that members of the group consume. Perhaps, for men, women, and children of all social strata, vegetable food was simply more accessible than meat at some times. Although diets do differ in their details, the bulk of food usually comes from locally available meat and vegetable sources, not from rare and exotic plants. In more complex societies, peasants provided staples for the urban centers and the elite. Even if the elite controlled food production, the producers could have kept some of the vegetable food for themselves. However, some of the specialists, non-food producers who had to exchange their products or services for food, could have eaten differently than the elite and food producers. It seems that social stratification does not equally affect every aspect of human behavior. The type of food characterizing distinct strata might have been similar, although the quality and the quantity might have been different. In some cases, however, the elite could have had a monopoly over the production of certain crops, as shown by Hastorf's (1993) convincing evidence for elite manipulation of corn production and differential consumption of beer made from maize. The quantity of the food might be detected archaeologically by the evidence of large storage places, but the quality of food would be difficult to test archaeologically. Based on modern and historic evidence, the culture of food, especially vegetable food, differs sometimes more among groups of various ethnic and religious backgrounds and locally available resources rather than on social status within a particular group. If wheat and barley were indeed introduced from the Near East to Europe at the beginning of the Neolithic, then their use might have been limited to people who either brought them or had first access to these newly acquired staples. But by the fifth millennium B.C. when the Opovo site was occupied, the cultivation of cereals was already well established in Vinča culture and initially different "food cultures" might not have persisted. However, the tradition of exploitation of locally available starchy food had not ceased, as demonstrated by the presence of water chestnuts in the Opovo samples.

To summarize, if there were any kind of social differentiation, it was not reflected in the plant food at the Opovo site. The available archaeobotanical evidence does not demonstrate differential plant consumption among households. In the case of plant remains from the site, the differences in plant distribution observed among features, mostly pits, were due to various taphonomic processes and plant processing activity areas rather than social stratification.

It is, however, a different question whether various households had different access to land and labor, and whether some farmers were more skilled in production. Perhaps, these differences could have existed among various Vinča households even if the land remained communal and households had a right to the land they cultivated. These differences could have been reflected archaeologically in the building of larger houses and the possession of fine ware and prestigious items. If rights to land were not perceived on a communal basis, then they might not have led to social stratification.[2] A limitation on land tenure might not have been imposed on animal herding, for example. Then, any agricultural surplus might have been exchanged for animals, unless the herds were communal property as well. Hence, one questions how long community control should have persisted without becoming hereditary and institutionalized, especially in some long standing Vinča settlements. It is uncertain whether these rights were strongly entrenched in community tradition, preventing a system of land inheritance, or whether some achieved accomplishments became ascribed.

The other possibility is that households had the security of land tenure and that individuals could not only cultivate land, but also transfer it to another holder.[3] Nevertheless, communal restrictions may have applied to privately owned land; for example, forests and hunting grounds and water resources might have remained communal. Thus, this type of land tenure also involved combination of private and communal incentives. In addition, the "sale" of ancestral land to strangers might have been socially disapproved by kin or community because arable land has symbolic values, as documented ethnographically (e.g., Luo in Kenya, Shipton 1989; tribes in Asia, Agarwal 1994). Land negotiations and transactions among households were probably limited to the community level regardless whether it was matrilocal or patrilocal. In many rural areas of the world, including the former Yugoslavia, most land transactions take place within a village.

It seems that continual land tenure might have given more incentive for agricultural production, as argued by new institutional economic theorists (e.g., North 1989). Because wealth and agricultural productivity are related, this type of continual land tenure system might have led to more obvious social stratification, at least within the community sphere. However, agricultural productivity might have been intensified for other reasons, e.g., population growth.

Inequality, power, wealth, and other characteristics vary in degree along the continuum among societies, as noted by Price and Feinman (1994) in their book about social inequalities. It remains unclear whether the observed differences in house size, the rare occurrence of prestige items in the Vinča houses and graves, or the even less pronounced differences among the few excavated Opovo houses were so great that they could be interpreted as social stratification.

CHAPTER 8. SUMMARY AND CONCLUSIONS

The study of plant remains from the late Vinča culture site of Opovo (4700–4500 B.C.) on the Pannonian Plain provides important information about different aspects of plant uses, taphonomic processes, ancient vegetation patterns, agricultural practices, wild plant procurement, and subsistence strategies during the Late Neolithic period. I examined plant remains from the Opovo site to determine whether archaeobotanical data corroborate faunal data and provide evidence for the permanent occupation of the site, for the change of plant composition and abundance through time, and for intra-site differentiation among households. On the basis of this information and by using ethnohistoric analogies, I attempted to reconstruct field locations and propose models of land use that could have existed during the Vinča culture period. On a more theoretical level, I have discussed the importance of communal versus private land tenure and its consequences for possible social stratification in the Vinča culture.

Systematic recovery and flotation of ca. 3,000 liters of sediment provided archaeobotanical data from 15 features consisting mostly of pits but also the floors of two houses (House 4 and 6), an oven floor, a bedding trench, a posthole, and non-feature contexts (Tables 2.1–2.3; Figures 3.11–3.13). Remains of cereals, legumes, oil-fiber plants, fruits, water chestnuts, and weedy herbaceous plants were recovered from the Opovo archaeobotanical samples. Most of the plant remains were carbonized, although a number of small seeds were found both in carbonized and mineralized form. Unusual preservation of mineralized seeds (n=41) could be explained by the presence of burnt house rubble that capped many pits and prevented disintegration of these uncarbonized seeds.

Identified Plants and Their Use

The identified plant remains from the Opovo site (Table 2.5) fall into the typical spectrum of plant species identified from the Vinča sites and other Late Neolithic/Chalcolithic (Eneolithic) sites of the region (Figure 2.32).

Triticum dicoccum (emmer) grains and *Triticum monococcum* (einkorn) spikelet forks were the most abundant among cereal remains at the Opovo site. Both types of wheats represent glume wheats ubiquitously present on many Neolithic sites in the Balkans. Emmer seeds (n=732) were found in all pits and on the floor of House 4. The largest quantity of emmer grains (n=626) was recovered from a pit, Feature 41. Einkorn grains (n=136) were found in nine pits and also on the floor of House 4. However, only 26 *Hordeum vulgare* (barley) grains were recovered, all from one pit, Feature 41. Einkorn spikelet forks (n=978) were identified from six pits, being most abundant in the pits from the northern part of the excavation block, especially in Feature 52 (Figures 3.11–3.13).

Although cereal storage was not recorded at the Opovo site, the ubiquity of the cereal remains, including both grains and spikelet forks/glume bases, provided evidence for farming and cereal preparation at the site. I have suggested that the quantities of spikelet forks recovered represent short episodes of dehusking einkorn grains, a necessary step in preparation of glume wheats intended for human consumption. Cereal chaff and straw were regularly used as temper for house daub, and in small quantities as temper for coarse pottery. There is no available evidence for beer brewing at Opovo and because no concentrations of equally sprouted cereal grains were found.

In addition to cereals, a small number of legume seeds (n=51) were recovered, mostly *Lens culinaris* (lentils) and *Vicia* sp. (vetches). Lentils were probably intentionally grown and eaten by humans; vetches might have been weedy intrusions in crops or grown as fodder.

Despite the fact that only five *Linum usitatissimum* seeds were found, their presence in samples, in addition to the discovery of a cloth and a cord made from flax, indicates that flax might have been grown for both seeds and fibers. Seeds could have been used for oil extraction or medicinally, while fibers were used for making linen fabrics and cords.

In total, 142 fruit seeds and stones were recovered from many Opovo samples. They include stones of *Cornus mas* drupes (Cornealian cherry), and seeds of *Sambucus ebulus*, *Sambucus* sp. (elderberries), *Vitis vinifera* cf. *sylvestris* (wild grapes), *Rubus* sp. (blackberries), and *Fragaria* cf. *vesca* (wild strawberries), which were gathered from the wild for human consumption and perhaps for medicinal purposes. Elderberry fruits might have been used as a source for black dye.

The discovery of 374 carbonized fragments of *Trapa natans* (water chestnut) fruits from eight features indicates that the fleshy fruits were released and eaten by humans either fresh, cooked, or dried and ground into flour. The Opovo inhabitants could have encouraged pigs, whose bones were identified, to eat water chestnuts brought directly to the site and/or from the shallow waters in the vicinity of the site.

Seeds of weedy herbaceous plants (n=379) were recovered from almost all features. Some of the identified small seeds such as *Bromus*, *Polygonum*, *Lolium*, *Galium*, and perhaps *Chenopodium* sp. species could bee classified as weedy intrusions in cereal crops. Solanaceae seeds (n=262), which were recovered in substantial numbers in samples together with or without cereal remains, were either weeds or were intentionally gathered as fruits or for medicinal purposes. *Lithospermum officinale* (white gromwell) plants were perhaps gathered for medicinal purposes.

Besides their utilitarian values, at least some of these plants might have had magical powers and been used in various rituals.

Taphonomic Processes

Examination of taphonomic processes and the mode of arrival of the identified plants facilitated the identification of the features' functions. None of the pits was identified as a primary grain silo. Previously burnt plant remains were thrown into garbage pits, as indicated by various non-cereal remains, bones, and other artifacts found in pits. It is likely that many carbonized plant remains were swept as ashes into pits after the plant parts were burnt in fires. Cereal grains were dehusked and perhaps ground in wooden mortars and then made into gruels and porridges because not many grinding stones were found. Since the pits in the northern part of the excavation block contained the highest number of spikelet forks, it appeared that byproducts of cereal processing were thrown into those pits. In contrast, most of the cereal grains and water chestnut fragments were in the pits in the central and southern parts of the excavation block, indicating that refuse of cooking and eating was dumped into the pits in the southern part of the site. Carbonized water chestnut hard shells were deposited in the pits after their fleshy fruits were released.

On the basis of plant remains, I have attempted to determine the use-life of features and seasonality patterns. It is very difficult to infer seasonality of various features, considering that the site was occupied for some 200 years and refuse pits were used for long periods. Additionally, in continental Europe harvesting and gathering season of all crops and wild plants lasts between late May and early October. Only storable food can be consumed during winter. If deposition of stored food is not taken into account, all the Opovo features exhibit similar seasonality patterns, from late spring until early fall. There are no clear patterns in density distribution or in combination of plant taxa from different building horizons (Figure 3.12).

Ancient Vegetation

The ecological requirements of several identified plants – such as Cornelian cherry, flax, legumes, and to some extent cereals (wheat and barley) indicate that there must have been areas that were not wet or damp. Also, the omnipresent wood charcoal in archaeobotanical samples shows that wood was used extensively and was deposited in refuse pits along with other plant remains. Fragments of *Quercus* (oak), *Ulmus* (elm), *Cornus* (Cornelian cherry), and Salicaceae (willow/poplar) wood were identified from several charcoal specimens (Table 4.2). My results are supported by multiple lines of evidence which suggest that during the Neolithic period, the Opovo region had mixed-oak forests, marshland, and perhaps some steppe in the dry areas. Russell's (1993) analysis of animal bones implies the existence of mixed forest as the primary habitat of red deer, roe deer, and wild boar, all of whose bones were identified in large numbers in the Opovo pits. Moreover, analyses of wood impressions in the house daub have shown that some of the wood used in the house constructions was cut from large tree trunks (Stevanović 1997a). An additional line of evidence was provided by lithic analysis which demonstrated that many of the tools were used for woodworking, especially during the earliest period of site occupation (Voytek in Tringham et al. 1985, 1992). All these independent studies have convincingly shown that there were forests in Opovo environs during the Neolithic period and that perhaps some had to be partially cleared for crop cultivation (Figure 4.1).

Agricultural Activities

Farming at the Opovo site included cultivation of einkorn, emmer, barley, legumes, and flax. Although many of the identified weedy herbaceous seeds belong to the weeds of`spring-sown crops, the presence of *Bromus arvensis*, *Galium aparine*, and *Lolium temulentum* seeds, which are weeds of winter crops, may indicate that some cereals might have been sown in fall, while others were sown in spring. Einkorn was perhaps sown in fall, while emmer, barley, lentils, and flax, which are less resistant to cold winters, might have been sown in spring. Some einkon and emmer could have been sown in a maslim fashion. Flax and legume fields were probably located on dry plots as these crops are intolerant of high ground water. Weeds, like *Polygonum lapathifolium,* are indicators of dump places, and their seeds were identified together with cereal remains. In those fields, cereals were perhaps not rotated with legumes or flax, which cannot tolerate high underground water.

Based on the ecological demands of crops and their weeds and ethnographic analogies, I suggested that crop rotation and fallowing were practiced, because some the identified crops tolerate better high underground water than others, and some crops exhaust soil more then others. In traditional agriculture, crop rotation and fallowing usually imply the existence of separate parcels of arable land in various locations. With this dispersed field system, the crops in various locations are not likely to be equally affected in the case of crop failure.

On the basis of ethnohistoric data, I estimated that ca. 900 kg of wheat per year might have been necessary to feed a family of five or six members. An additional 300 kg might have been needed for the seed stock. This amount of cereal grain could have been obtained from ca. 1.5 ha of arable land.

The Opovo site is considered a rather small Vinča site, covering an area about 5 ha. Assuming that population was ca. 50–250 people at one time, sufficient arable land on the higher elevations around the settlement was available for crop cultivation.

If we assume that 1.5 ha of land scattered in small fields was manually tilled (since there is no direct evidence for animal-drawn plows) then each member of the household would have had to have spent 10–15 days per growing season in the fields. Cereals could be sown in just a few days; however, an additional 300 working hours were needed for harvesting and binding, and 60 working hours were necessary for threshing cereals. Thus, some households probably needed additional help for manual tilling, harvesting and perhaps threshing of cereals. Small legume and flax fields could have been cultivated by members of one household.

Grains could have been stored either underground or above ground, but there is no evidence of cereal storage at the Opovo site. The cereal supplies might have been depleted before the houses were abandoned, or else the cereals could have been stored in containers made of perishable material that left no traces.

Straw and chaff were used as temper for house daub at the Opovo site. On the basis of the modern use of straw and chaff temper for building mud (pisé) houses in the Voyvodina region, I calculated that at least ca. 75 kg of straw and chaff was necessary for tempering daub for an idealized Opovo house measuring ca. 4 × 8 × 2 m. An estimated area of ca. 1,500 m^2 had to be under cereal cultivation to yield sufficient chaff and straw to temper the clay for plastering a house built in the wattle-and-daub technique (Tables 5.4 and 5.5). This large area of land per household was not available between the houses within the Opovo village limits, nor between houses at many other Vinča sites. Therefore, small fields of these sizes had to be located outside the settlements. The interhouse space at other sites might have been used for vegetable gardens where horticulture was practiced. Since emmer and einkorn grains and spikelet forks were ubiquitously present in archaeobotanical samples, and since large amounts of cereal chaff were used as temper in clay for building of the houses, I suggest that crops were cultivated in loco in the small fields around the site which was occupied year round. Seasonality estimates for animals killed from various layers of several pits also show that the Opovo site was occupied year round (Russell 1993).

Wild Plant Procurement

Because wild plants were often neglected in archaeobotanical reports dealing with the Balkan sites, I placed special emphasis on the identification of non-crop plant remains from the Opovo samples. The discovery of water chestnut remains, in addition to fruit seeds and stones at Opovo, demonstrates the continuing procurement of wild plants for human nutrition. Besides fruits of Cornelian cherry, grapes, strawberries, blackberries, and elderberries, some of the weedy herbaceous plant such as chenopods, knotweeds, and nightshades might not have been weedy intrusions but rather gathered for other uses. In addition to cultivating crops, the Opovo settlers gathered fruits, medicinal plants, and water chestnuts to complement their diet.

Modern and historical data about the unfavorable years for agriculture in the region indicate that on the average, either floods or droughts affected crop yields every two to four years. I postulate that a similar frequency of wet and dry years might have occurred in prehistory. At the Opovo site, water chestnuts might have been gathered more intensively during the bad years to offset the deficiency in carbohydrate food supply. They were also gathered more in wet years since water chestnut dispersal is affected by droughts. Water chestnuts are storable, as shown by ethnographic evidence. They are harvested in the late summer and fall, that is, after cereal and legume harvests and before fall plowing/tilling. Water chestnut procurement was probably incorporated into the regular crop production of Opovo settlers, serving occasionally as a buffering mechanism in years of scarcity.

Mixed Subsistence

The analysis of plant remains, in addition to Russell's (1993) analysis of animal bones, allow me to suggest the use of the phrase *terra et silva* as an appropriate way to describe the agro-gathering subsistence strategy of Late Neolithic Opovo. The Latin phrase *terra et silva* appears frequently in the documents from the sixth-to-tenth century A.D. and represents an interplay of cultivated and uncultivated spaces (Montanari 1994:27). However, this mixed strategy is not seen as characteristic for Vinča culture, which is usually described as a typical agricultural society. Detailed ecofactual analysis at the Opovo site has shown that it is an excellent example of both the diverse subsistence strategies used within a general cultural context of the Vinča culture, and of the use of so-called marginal lands. Activities that incorporated farming, herding, hunting, and wild plant procurement, fostered by environmental and social factors, resulted in a mixed economy. On the basis of the Opovo architectural and artifactual evidence characteristic of the Vinča culture, I suggest that the Opovo settlers practiced "broad spectrum" subsistence. Such a proposition was suggested as one of the possible interpretations by Tringham et al. (1985) in their first article about the Opovo site.

Land Use and Social Stratification

I suggest two plausible models of land ownership for ancient Opovo: either arable land was communally owned or land was owned by households (or certain individuals within a household). For both models, I believe that hunting grounds, including woods and water resources, were communally owned. Considering the pronounced importance of hunting and wild plant procurement at the Opovo site, strong cooperation in managing communal land must have been practiced, regardless whether arable land was communally or privately owned.

For the first model, the analysis of plant remains indicates that crop rotation and fallowing could have been practiced on separate dispersed fields. Arable land was probably divided into numerous parcels scattered around the Opovo settlement. Such a division of arable land places constraints on the synchronization of cooperative work. Thus, I suggest that if a household is defined as a basic production unit, each household could have claimed rights to certain plots as long as they kept cultivating them. This type of land division would have allowed incentives for household (individual) optimization, while eliminating the "free-riders" in crop cultivation. In addition, the practice would have permitted some socioeconomic mobility, but the inequality would not necessarily lead to enduring social stratification.

I propose that the entire community (village or clan) might have owned the land, but that each household was responsible for cultivating its own fields. Although each household could perhaps cultivate enough cereal crops for their needs, cooperation was necessary during harvest time and for tilling, especially if animal-pulled plows were not used. Teamwork was also necessary for clearing of forest, building of houses, fencing fields, and hunting large game.

The second plausible model is that the rights to cultivation became hereditary and that the parcels of arable land were actually available for private transactions among various households. Even if individual households owned the land, local restrictions may have prohibited the "sale" of ancestral land to foreigners, so that all transactions would have been limited to village members. Such security might have increased incentives for a higher labor investment in the arable land, and might have resulted in agricultural surplus and wealth. Thus, inherited wealth and perhaps associated prestige, might have become ascribed and institutionalized. This type of land tenure could have led to enduring inequality and social stratification.

The differences in plant remains at the Opovo site did not show any distributional pattern of gathered plants versus crops, since both types of plants occurred together in several pits, regardless of the animal bone context. Furthermore, there were no exotic/rare plants and no concentrations of stored crops. There was no available evidence that cereals were used for brewing beer that might have been served during feasts. However, there was clear evidence that cereal grains were dehusked and perhaps later ground for food. Precise taphonomic analysis of the plant remains did not show any differences among pits that could be explained as changes in subsistence pattern or food consumption during the occupation of the Opovo site. Rather the patterns resulted from different plant processing and deposition areas within the site (Figures 3.11 and 3.12). Russell (1993) argues that the concentration of larger bones of both domestic and wild animals found in some pits may have resulted from roasting and carving larger pieces of meat, and may be interpreted as the refuse of feasting and may reflect social differentiation among households at the Opovo site (Figure 3.13). Because bones of red deer, roe deer, wild boar, and domestic cattle were found together, it seems that the size of animals was the main factor in sharing, rather than domesticatory status. According to Russell (1993), feasting was important for certain households for gaining prestige through distribution and not through domination. Feasts facilitated exchange at the Opovo site, which was located on the border of the Vinča and Tisza cultures. Control of land and herd sizes was more important for gaining prestige in the larger Vinča sites such as Vinča-Belo Brdo, Banjica, Gomolava, and Selevac than in Opovo. At Opovo, social storage gained through gift exchange would have been of greater importance than hoarding (Russell 1993).

Due to the different types of consumption and preservation of vegetable food versus meat (bones), plant remains might not always be the right archeological material for discerning competitive feasting and social stratification. I suggested that people were not necessarily stratified according to vegetable (staple) food, although some individuals might have eaten more copious meals than others—evidence likely not to be preserved in the archaeological record.

The few houses revealed within the excavation block at Opovo did not vary much in size and internal space. There were was also no concentration of rare and exotic items either within the houses or pits. Thus, on the basis of the artifactual analysis, and on the basis of plant remains, the differences between the households (of the houses that were excavated) were not pronounced; perhaps, the land tenure did not lead to enduring inequality and social stratification.

Implications

The Opovo settlement may represent an example of a regional variation that existed among the Vinča cultural variants. On a larger scale, the Opovo site can serve as an example of a small Late Neolithic settlement that practiced a mixed economy, including small-scale farming and procurement of wild plants and animals, at least a millennium after the establishment of agriculture in the central Balkans.

Since the Opovo site was settled for the first time in the late Vinča period, it does not appear to represent a settlement established by the relicts of local hunter-gatherer groups who finally adopted agriculture. Identified artifacts and the establishment of permanent settlements indicate the presence of Neolithic cultures rather than Mesolithic traditions.

The Vinča culture is traditionally characterized only by the cultivation of cereals and legumes and the raising of domestic animals. Emmer, einkorn, barley, and pulses thought to be introduced from the Near East to Europe were actually identified at Opovo and the majority of other Vinča sites. During the Neolithic period, new cereal habits might have developed that were more resistant to the temperate climate of the northern Balkans. Agriculture in floodplains of the major rivers might not have been possible in the northern Balkans, since floods typically occur in the late spring and early summer. The period between spring sowing and early summer floods is not long enough to allow for the maturation of crops. Therefore, cereal fields were probably located on more elevated terrain (levees) around Neolithic settlements.

Besides Opovo, many Vinča settlements have been excavated, including tell and flat sites, some of them on a scale large enough to allow some insight about settlement patterns, house sizes, and storage facilities. At some Vinča sites, there are differences in house sizes and internal divisions. The differences in the house-sizes might be explained as differences in the number of household members who needed more space and/or who were able to mobilize more labor for the construction of the larger houses. Some household members might have been more skillful than others. Also, some of the houses might have had more ceremonial character than others.

In several Vinča settlements, silos or pithoi containing cereal grains have been excavated in or around the houses, indicating that most staples were stored individually, and not on the communal level. There is no evidence of excess storage space, or of central storage that would indicate the grains were controlled by the whole community, or by groups that had preferential access to agricultural staples. Nevertheless, social sharing of food in times of need and/or festive occasions might have been a social obligation. Gathering from the wild could have been a common practice in providing dietery diversity, as well as medicinal and ritual plants. When crop yields were low, Opovo inhabitants migh have intensively gathered fruits containing carbohydrate-rich fruits such as water chestnuts. Such a practice might have been more widespread in newly settled regions that were abundant in wild plant and animal resources before the full establishment of farming, as shown in the Opovo example.

It is possible that in areas where a mixed economy was practiced, the exploitation of wild resources was not limited by the property rights of certain individuals within one settlement and that an egalitarian ethos might have existed until the Bronze Age. On the other hand, it can be argued that in agriculturally marginal lands where there is higher risk of crop failure, there is greater emphasis on overproduction and exchange to buffer risks, leading to social stratification and creating an elite. Ordinary households would have depended on the elite for survival in bad years, as Halstead (1994:195–219) suggested for the Late Neolithic–Early Bronze period in Thessaly. Because the Pannonian Plain has rather uniform topography without mountains, with major variations in the level of underground water, natural disasters such as floods and especially droughts might have similarly afflicted many households within one village or even among several villages in one area. Exchange of some other kinds of products for subsistence goods might not have been possible for many settlements within a region, although labor or services might have been exchanged for staples.

In the central Balkans during the Late Neolithic there is evidence of larger settlements and colonization of marginal lands. There are also some differences between grave goods of various individuals in a limited number of graves uncovered from the two Vinča cemeteries. All these differences imply some sort of inequality among various individuals during the Vinča culture period. Whether or not the observed differences were so prominent that they could be interpreted as social stratification depends on the interpretive model rather than on institutionalized wealth differences. Such differences are not evident in the archaeobotanical material and yet to be documented in archaeological material.

Through the detailed study of plant macro remains from the Opovo site, and by incorporating the faunal, architectural, and artifactual information, I attempted to contribute to better understanding of the human-plant interrelationship in the context of the prehistoric site of Opovo and its environs during the late Neolithic period in the Pannonian Plain and central Balkans.

NOTES

Chapter 1

[1]This is only a narrow selection, primarily in English, from the vast literature about the Vinča culture, and it does not include many other works published in Serbo-Croatian and other languages.

[2] When using radiocarbon dates, I refer to the abbreviation "b.c." for uncalibrated dates and to the abbreviation "B.C." for calibrated dates, or I cite directly from the primary sources.

[3] The Opovo Archaeological Project was carried out under the auspices of the Joint U.S.-Yugoslav Fund for Cooperative Scientific, Cultural, and Technical Research, which was sponsored by the Smithsonian Institution, the Self-Management Interest Community for Scientific Work of the Province of Voyvodina (1983–1989), and the University Research Expeditions Program of the University of California. The late Professor Bogdan Brukner (who passed away in January, 2006) from the University of Novi Sad in the former Yugoslavia and Professor Ruth Tringham from the University of California, Berkeley, were the project's principal investigators. The late Ljubomir Bukvic, a local archaeologist and the curator at the National Museum of Pančevo, was in charge of the organizational logistics of the excavations. The brief description of the Opovo site is primarily based on the combined information from two articles published by the investigators of the site (Tringham et al. 1985, 1992).

[4] ©Creative Commons License: "This work is licensed under the Creative Commons Attribution-Non Commercial-Share Alike License. To view a copy of this license, visit http://creativecommons.org/licenses/by-nc-sa/2.0/ or send a letter to Creative Commons, 559 Nathan Abbott Way, Stanford, California 94305, USA."

[5] DZ quantifications defined by J. Watson (cited in Russell 1993:107). J. Watson (1979:129) introduced the concept of *diagnostic zone* "where any zone of a bone can be treated as a diagnostic entity in its own right and used for MNI counts". She further defines a number of diagnostic zones for each bone, as long as the same or equivalent zone is present in all the species under consideration, e.g., the left ulna (proximal articulation).

Chapter 2

[1] *Cornus mas* whole stones from Feature 30 are from layer 14, and two fragments from flotation are from layer 11. Material designated to layers 5 and 6 probably dates to the use of Feature 30, BH 3 (Ruth Tringham, personal communication 1995).

[2] Application of Elleneberg's (1979) indicator values made on field observations (e.g., shade loving species) and its use for the interpretation of archeological weed floras was recently questioned by Charles et al. (1997) because the approach cannot provide answers to the environmental factors underlying species distribution. Charles et al. (1997) now apply the Functional Interpretation of Botanical Surveys (FIBS) method developed by the Unit of Comparative Plant Ecology (UCPE). The FIBS method is used to investaige ecological processes of species distribution within wide range of habitats and to help reconstruct past agricultural systems. This method was not applied in the analysis of the Opovo weed flora since it was not available during the initial plant analysis and, more importantly, because the number of identified weed seeds from the Opovo samples was rather small. Bogaard's book *Neolithic Farming in Central Europe* (2004) was published as the manuscript of this book was submitted to the press.

Chapter 3

[1] To avoid confusion between house number assignments from the previous publications (Tringham et al. 1985, 1992; Russell 1993; Borojevic 1998) and the new one, here I adopted the new numeration of the houses (Stevanovic 1997a; Tringham personal communication 2004). For assignment of the features to the building horizons, I used the latest list "Sequence of Building Horizons at Opovo-Ugar Bajbuk (1983–1989)" based on personal communication with Ruth Tringham (2004) and the plan of building horizons published by Stevanovic (1997a: Figure 6). In a short archaeological description of the features, I combined the available information from all these sources and from my personal observations during field work, without citing all the mentioned sources every time.

Chapter 4

[1] Pollen analyses from the Danube's Iron Gate area do not seem relevant for this study due to the very different altitudes and geographic locations where pollen samples were taken, as well as the thermophilous character of vegetation of the gorges. For synthesis of pollen analyses from the Iron Gates, see Radovanović (1996).

[2] Schwarzerde soil is described as fairly thick, black soil rich in mull (fermenting) humus and lime, secondarily developed out of a peaty wet soil. (Fink 1964b, cited in Havinga 1972:320).

[3] Chernozem soil is described as a blackish soil, rich in mull (fermenting) humus highly saturated with calcium ions and with intense biological activity. It represents climax soil in loess or clayey sediments in a more arid, continental climate (Havinga 1972:320).

[4] The highest level is presented by "formations", e.g. "water plants" or "forest". Further, the formations are divided into "classes" such as *Alneta glutinose*. Classes are divided into "orders" (e.g. *Fagatalia sylvaticae*, mesophylic deciduous forests). Orders are then divided onto "alliances" (e.g. *Alno-Padalia* growing on higher but still damp habitats). Furthermore, each alliance is divided further into numerous "associations" which form the smallest phytosociological unit.

Chapter 7

[1] This is only a very narrow selection from the literature about land tenure with which I became acquainted while writing this book.

[2] Here, land rights are treated as a term abstractly attached to a household and not to one or more individuals. The question of who in the household possessed these rights or who inherited the land would be a completely separate topic.

REFERENCES

Agarwal, B.
1994 Gender and Command over Property: A Critical Gap in Economic Analysis and Policy of South Asia. *World Development* 22:1455–1478.

Allen, R. C.
1992 *Enclosure and the Yeomen: The Agricultural Development of the South Midlands 1450–1850.* Clarendon Press, Oxford.

Andrič, M.
2001 The Holocene Vegetation Dynamics and the Formation of Neolithic and Present-day Slovenian Landscape. *Documenta Praehistorica* 28:133–175.

Arnaudov, N.
1949 Predistoričeski rastitelni materiali. *Annuaire de l'Université de Sofia, Faculté des Sciences* 45:68–81.

Ataman, K.
1992 Threshing Sledges and Archaeology. In *Prehistoire de l'Agriculture: Nouvelles Approches Experimentales Ethnographique*, edited by P. C. Anderson, pp. 305–319. Monographie du CRA No. 6, Paris.

Axelrod, R.
1984 *The Evolution of Cooperation.* Basic, New York.

Baczyńska, B. and M. Lityńska-Zając,
2005 Application of Lithospermum officinale L. in early Bronze Age medicine. *Vegetation History and Archaeobotany* 14: 77-80.

Bailey, D. W.
1999 What is a Tell? Settlement in Fifth Millennium Bulgaria. In *Making Places in the Prehistoric World: Themes in Settlement Archaeology*, edited by J. Brück and M. Goodman, pp. 95–111. University College London Press, London.
2000 *Balkan Prehistory: Exclusion, Incorporation and Identity.* Routledge, London.

Bakels, C. C.
1988 On the Location of the Fields of the Northwestern Bandkeramik. In *Archeologie en Landschap* (Festschrift H. T. Waterbolk), edited by M. Bierma, O. H. Harsema and W. van Zeist, pp. 49–57. Biologisch-Archaeologisch Instituut, Groningen.
1992 Research on Land Clearance during the Early Neolithic in the Loess Regions of the Netherlands, Belgium, and Northern France. In *Evaluation of Land Surfaces Cleared from Forests by Prehistoric Man in Early Neolithic Times and the Time of Migrating Germanic Tribes,* edited by B. Frenzel, pp. 47–55. Gustav Fisher Verlag, Stuttgart.
1996 Growing Grain for Others or How to Detect Surplus Production? *Journal of European Archaeology* 4:329–336.

Bakels, C., and S. Jacomet
2003 Access to luxury foods in Central Europe during the Roman Period: the Archaeobotanical Evidence. *World Archaeology* 34:542–557.

Barker, G.
1985 *Prehistoric Farming in Europe.* Cambridge University Press, Cambridge.

Bassett, T. J., and D. E. Crummey (editors)
1993 *Land in African Agrarian Systems.* The University of Wisconsin Press, Madison.

Beijerinck, W.
1976 *Zadenatlas der Nederlandsche Flora: Ten Behoeve van de Botanie, Palaeontologie, Bodemcultuur en Waren-Kennis.* Backhuys and Meesters, Amsterdam.

Benac, A.
1978 Les thèse fondamentales sur l'origine du neolithique dans les Balkans et les régions avoisinantes. *Godišnjak Centra za balkanološka ispitivanja* XVI:9–21. Sarajevo.
1990 Vinča et les cultures contemporaines à l'ouest de la Yugoslavie. In *Vinča and Its World,* edited by D. Srejović and N. Tasić, pp. 17–23. Serbian Academy of Sciences and Arts, Belgrade.

Beug, H-J.
1986 Vegetationsgeschichtliche Untersuchungen über frühe Neolithicum im Unterreichsfeld, Landkreis Göttingen. In *Anthropogenic Indicators in Pollen Diagrams*, edited by K. E. Behre, pp. 115–125. Balkema, Rotterdam.

Bintliff, J.
1982 Settlement Patterns, Land Tenure and Social Structure: A Diachronic Model. In *Ranking, Resource and Exchange: Aspects of the Archaeology of the Early European Society,*

edited by C. Renfrew and S. Shennan, pp. 106–111. Cambridge University Press, Cambridge.

Bittman, F. and D. Kučan
2004 Erste archäobotanische Untersuchungen in Okolište 2002. http://www.jungsteinSITE.de, 24. Januar 2004. Electronic document, accessed June 1, 2005.

Blamey, M., and P. Blamey
1984 *Fruits, Nuts and Berries and Conspicuous Seeds.* Harper Collins Publishers, London.

Bökönyi, S.
1974 *History of Domestic Mammals in Central and Eastern Europe.* Akadémiai Kiadó, Budapest.
1990 Tierknochenfunde der neusten Ausgrabungen in Vinča. In *Vinča and Its World*, edited by D. Srejović and N. Tasić, pp. 49–54. Serbian Academy of Sciences and Arts, Belgrade.

Boardman, S., and G. Jones
1990 Experiments on the Effects of Charring on Cereal Plant Components. *Journal of Archaeological Science* 17:1–11.

Bogaard, Amy
2004 *Neolithic Farming in Central Europe. An archaeobotanical study of crop husbandry practices.* Routledge, London.

Bohn, U., G. Gollub, and C. Hettwer
2000 *Karte der natürlichen Vegetation Europas.* (Map of the Natural Vegetation of Europe). Bundesamt für Naturschutz, Bonn.

Borojević, K.
1988 Differences in Plant Macro Remains from the Neolithic Level at Gomolava and the Neolithic Site of Opovo. In *Gomolava: Hronologija i stratigrafija u praistoriji i antici Podunavlja i jugoistočne Evrope*, edited by N. Tasić and J. Petrović, pp. 109–115. Arheološki institut, Beograd i Vojvodjanski muzej, Novi Sad.
1990 Analize paleobotaničkih makroostataka sa arheoloških lokaliteta u Srbiji. Unpublished M.A. thesis. Department of Archaeology, University of Belgrade, Yugoslavia.
1998 The Relations among Farming Practices, Landownership, and Social Stratification in the Balkan Neolithic Period. Ph.D. Dissertation, Washington University, St. Louis. University Microfilms, Ann Arbor.

Borojević, S.
1995 *Zapisi o poljoprivredi* (Remarks about Agriculture). Srpska akademija nauka i umetnosti, ogranak Novi Sad.

Bottema, S.
1974 *Late Quaternary Vegetation History of Northwestern Greece.* University Press, Groningen.

Bottema, S., and B. S. Ottaway
1982 Botanical, Malacological and Archaeological Zonation of Settlement Deposits at Gomolava. *Journal of Archaeological Science* 9:221–246.

Bowen, H. C., and P. J. Fowler (editors)
1978 *Early Land Allotment.* BAR International Series 48, Oxford.

Bown, D.
1995 *Encyclopedia of Herbs and Their Uses.* Darling Kindersley, London.

Bradley, R.
1978 Prehistoric Field Systems in Britain and North-West Europe: A Review of Some Recent Work. *World Archaeology* 9:265–280.

Braun, T.
1995 Barley Cakes and Emmer Bread. In *Food in Antiquity I,* edited by J. Wilkins, D. Harvey, and M. Dobson, pp. 25–37. University of Exeter Press, Exeter.

Braun-Blanquet, J.
1951 *Pflanzensoziologie: Grundzüge der Vegetationskunde.* 2nd ed. Springer, Wien.

Brockman, H.
1914 *Vergessene Nutzpflanzen.* Wissen und Leben, Zürich.

Brothwell, D. R., and P. Brothwell
1969 *Food in Antiquity.* Praeger, New York.

Brukner, B.
1988 Die Siedlung der Vinča-Gruppe auf Gomolava (Die Wohnschicht des Spätneolithikums und Frühäneolithikumus-Gomolava Ia, Gomolava Ia-b und Gomolava Ib) und der Wohnhorizont des äneolithischen Humus (Gomolava II). In *Gomolava: Hronologija i stratigrafija u praistoriji i antici Podunavlja i jugoistočne Evrope*, edited by N. Tasić and J. Petrović, pp. 19–39. Arheološki institut, Beograd i Vojvodjanski muzej, Novi Sad.

Brukner, B., B. Jovanović, and N. Tasić
1974 *Praistorija Vojvodine*. Institut za izučavanje istorije Vojvodine, Novi Sad.

Čajkanović, V.
1985 *Rečnik srpskih narodnih verovanja o biljkama* (Dictionary of Serbian Folk Beliefs about Plants), edited by V. Djurić. Srpska književna zadruga i Srpska akademija nauka i umetnosti, Beograd.

Čanak, M., S. Parabućski, and M. Kojić
1978 *Ilustrovana korovska flora Jugoslavije* (Illustrated Weed Flora of Yugoslavia). Matica srpska, Novi Sad.

Cârciumaru, M.
1973 Analyse pollinique des coprolites livres par quelques stations archaeologiques des deux bords du Danube dans la zone des 'Portes de Fer'. *Dacia* 17:53–60.
1978 L'analyse pollinique des coprolithes de la station archeologique de Vlasac. In *Vlasac*, Vol. II, edited by D. Srejović and Z. Letica, pp. 31–34. Serbian Academy of Sciences and Arts, Belgrade.
1996 *Paleoetnobotanica. Studii in preistoria şi protoistoria Romaniei*. Istoria agriculturii din Romania, Bucureçti.

Champier, L.
1952 A porpos de l'oeuvre de Robert Gradmann. Methods de recherche en geographie agraire. *Annales Universitatis Saraviensis* 2:190–203.

Chapman, J.
1981 *The Vinča Culture of South-East Europe*. BAR International Series 117, Oxford.
1983 The Secondary Products Revolution and the Limitations of the Neolithic. *Bulletin of the Institute of Archaeology* 17:107–122. London.
1989 The Early Balkan Village. In Neolithic of Southeastern Europe and its Near Eastern Connections. *Varia Hungarica* II:33–53.
1990 The Neolithic in the Morava-Danube Confluence Area: A Regional Assessment of Settlement Pattern. In *Selevac: A Neolithic Village in Yugoslavia*, edited by R. Tringham and D. Krstic, pp. 13–45. University of California, Los Angeles.
1991 The Early Balkan Village. In *Social Space: Human Spatial Behaviour in Dwellings and Settlements*, edited by O. Grøn, E. Engelstad, and I. Lindblom, pp. 79–100. Odense University Press, Odense.
1992 Arenas of Social Power: The Case of Serbian Prehistory. *Zbornik Narodnog muzeja* 14:305–317. Beograd.
1997 Prehistoric Landscapes in Eastern Hungary. In *Landscapes in Flux*, edited by J. Chapman and P. Dolukhanov, pp. 137–162. Colloquia Pontica 3. Oxbow Books, Oxford.
1998 Objectification, Embodiment and the Value of Places and Things. In *The Archaeology of Value: Essays on Prestige and the Processes of Evaluation,* edited by D. Bailey with the assistance of Steve Mills, pp. 106–131. BAR International Series 730, Oxford.
2000 *Fragmentation in Archaeology: People, Places and Broken Objects in the Prehistory of South-Eastern Europe*. Routledge, London.

Chapman, J., and R. S. Shiel
1993 Social Change and Land Use in Prehistoric Dalmatia. *Proceedings of the Prehistoric Society* 59:61–104.

Chapman, J., R. S. Shiel, and S. Batović
1987 Settlement Patterns and Land Use in Neothermal Dalmatia, Yugoslavia: 1983–1984 Seasons. *Journal of Field Archaeology* 14:123–146.

Charles, M., G. Jones, and J. G. Hodgson
1997 FIBS in Archaeobotany: Functional Interpretation of Weed Floras in Relation to Husbandry Practices. *Journal of Archaeological Science* 24:1151–1161.

Clason, A. T.
1979 The Farmers of Gomolava in the Vinča and La Tene Period. *Palaeohistoria* 2:42–81.
1988 The Equids of Gomolava. In *Gomolava: Hronologija i stratigrafija u praistoriji i antici Podunavlja i jugoistočne Evrope*, edited by N. Tasić and J. Petrović, pp. 99–105. Arheološki institut, Beograd i Vojvodjanski muzej, Novi Sad.

Columella, Lucius Junius Moderatus
1941 *On Agriculture*. Book I, II, III. Translated by Harrison Boyd Ash. Harvard University Press, Cambridge.

Culiberg, Metka, and Alojz Šercelj
1981 Pollen Analysis of the Sediments of Plitvička Jezera (Lakes of Plitivce). *Acta Botanica Croaticae* 40:147–154.

Dahlman, C. J.
1980 *The Open Field System and Beyond*. Cambridge University Press, Cambridge.

Davies, M. S., and G. C. Hillman
1988 Effects of Soil Flooding on Growth and Grain Yield of Populations of Tetraploid and Hexaploid Species of Wheat. *Annals of Botany* 62:597–604.

Demoule, J-P., and C. Perlès,
1993 The Greek Neolithic: A New Review. *Journal of World Prehistory* 7:355–416.

Dennell, R.
1974 The Purity of Prehistoric Crops. *Proceedings of the Prehistoric Society* 40:132–135.
1978 *Early Farming in South Bulgaria from the VI to the III Millennia B.C.* BAR International Series (Supplementary) 45, Oxford.
1992 The Origins of Crop Agriculture in Europe. In *Origins of Agriculture: An International Perspective*, edited by C. W. Cowan and P. J. Watson, pp. 71–100. Smithsonian Institution Press, Washington.

Dietler, M.
1990 Driven by Drink: The Role of Drinking in the Political Economy and the Case of Early Iron Age France. *Journal of Anthropological Archaeology* 9:352–406.

Dnevnik (Daily Newspaper)
1997 Ukroćena panonska voda (Tamed Pannonian water). 13 June 1997:7, Novi Sad (Vojvodina).

Dobrohotov, V. N.
1961 *Semena sornih rastenij* (The Seeds of Weed Plants). Izd. seljskohazjastvenoj literaturi, zurnalov i plakatov, Moskva.

Donahue, R. E.
1992 Desperately Seeking Ceres: A Critical Examination of Current Models for the Transition to Agriculture in Mediterranean Europe. In *Transitions to Agriculture in Prehistory*, edited by A. B. Gebauer and T. D. Price, pp. 73–80. Prehistory Press, Madison.

Donaldson, A. M., Jones, A. K. G., and D. J. Rackham
1980 Barnard Castle, Co. Durham. A Dinner in the Great Hall: Report on the Contents of a Fifteenth-Century Drain. *Journal of the British Archaeological Association* 133:86–96.

Dontscheva, E.
1990 Plant Macorest Research of Early Neolithic Dwelling in Slatina. *Studia Praehistorica* 10:86–90.

Ellenberg, H. von
1979 *Zeigwerte der Gefäßpflanzen Mitteleuropas* (2. verbesserte und erweiterte Auflage). Verlag Erich Goltze KG, Göttingen.

Ellickson, R. C.
1993 Property in Land. *The Yale Law Journal* 102:1315–1400.

Ern, H.
1990 Threatened Wetland Ecosystems in the Floodplains of the River Sava and Its Tributaries (Northern Yugoslavia). In *Vegetation and Flora of Temperate Zones*, edited by U. Bohn and R. Neuhausl, pp. 51–59. SPB Academic Publishing, The Hague.

Ertug-Yaras, F.
1997 *An Ethnoarchaeological Study of Subsistence and Plant Gathering in Central Anatolia.* Unpublished Ph.D. dissertation. Department of Anthropology, Washington University, St. Louis.

Evans, R.
1978 Early Craft Specialization: An Example from the Balkan Chalcolithic. In *Social Archaeology: Beyond Subsistence and Dating*, edited by C. L. Redman, M. J. Berman, E. V. Curtin, W. T. Longhorne, N. M. Versaggi and J. C. Wanser, pp. 113–129. Academic Press, New York.

Fernald, M., and A. Kinsey
1958 *Edible Wild Plants of Eastern North America.* Harper Brothers, New York.

Fink, J.
1964 Die Boden Niederösterreichs. *Jahrb. Landesk. Niederösterreich* 36:965–988.

Fischer, E., and M. Rösch
2004 Archäobotanische Untersuchungen. In Vorbericht über die rumänisch-deutschen Prospektionen und Ausgrabungen in der befestigten Tellsiedlung von Uivar, jud. Timiş, Rumänien (1998–2002), edited by W. Schier and F. Draşovean. *Praehistorische Zeitschrift* 79:145–230.

Fleming, A.
1979 The Dartmoor Reaves: Boundary Patterns and Behaviour Patterns in the Second Millennium bc. *Proceedings of the Devon Archaeological Society* 37:115–131.
1982 Social Boundaries and Land Boundaries. In *Ranking, Resources and Exchange*, edited by C. Renfrew and S. Shennan, pp. 52–55. Cambridge University Press, Cambridge.

1985 Land Tenure, Productivity, and Field System. In *Beyond Domestication in Prehistoric Europe: Investigations in Subsistence Archaeology and Social Complexity*, edited by G. Barker and C. Gamble, pp. 129–143. Academic Press, London.
1989 The Genesis of Coaxial Fields Systems. In *What's New? A Closer Look at the Process of Innovation,* edited by S. E. van der Leeuw and R. Torrence. Unwin Hyman, London.

Foucault, M.
1984 Interviews. In *The Foucault Reader*, edited by P. Rabinow. Penguin, London.

Fowler, P. J.
1981 Wildscape to Landscape: "Enclosure" in Prehistoric Britain. In *Farming Practice in British Prehistory*, edited by R. Mercer, pp. 9–55. Edinburgh University Press, Edinburgh.
1983 *The Farming of Prehistoric Britain.* Cambridge University Press, Cambridge.

Fried, M. H.
1967 *The Evolution of Political Society*. Random House, New York.

Fritz, G. J.
1986 *Prehistoric Ozark Agriculture: The University of Arkansas Rockshelter Collections.* Ph.D. Dissertation. Department of Anthropology, University of North Carolina, Chapel Hill, University Microfilms, Ann Arbor.
1987 The Trajectory of Knotweed Domestication in Prehistoric Eastern North America. Paper (poster) presented at the 10th Annual Ethnobiology Conference, Gainesville.

Garašanin, M.
1951 *Hronologija Vinčanske grupe* (Chronology of the Vinča Group) Arheološki Seminar, Ljubjana.
1961 Die Siedlung der Starčevo Kultur in Nosa bei Subotica und das Problem der neolithischen Lehmschermen, pp 303–307. *Fifth Congress ISPP*, Hamburg 1956.
1979 Centralnobalkanska zona. In *Praistorija jugoslavenskih zemalja II: Neolit*, edited by A. Benac, pp. 33–79. Akademija nauka i umetnosti Bosne i Hercegovine, Sarajevo.
1980 Les origines du néolithique dans le bassin da la Méditeranée et dans le sud-est européen. In *Problèmes da la néolithisation dans certaines régions de l'Europe*. Polska Akademija Nauk, Prace komisiji archeologicznej 21:57–72.
1984 Uvod. Vinča i Vinčanska kultura u neolitu jugoistočne Evrope (Introduction. Vinča and the Vinča Culture in the Neolithic in southeast Europe). In *Vinča u praistoriji i srednjem veku*, pp. 1–11; 57–65. Srpska akademija nauka i umetnosti, Beograd.

Gaul, J. H.
1948 *The Neolithic Period in Bulgaria.* American School of Prehistoric Research Bulletin No. 16, Cambridge.

Gigov, A.
1959 Dosadašnji nalazi o postglacijalnoj istoriji šuma Srbije (The present results about the finds of postglacial history of forests in Serbia). *Zbornik radova Instituta za Ekologiju i Biogeografiku* 7:3–26. Beograd.
1964 Typen der Pollen-Diagramme am Gebiet Jugoslawiens im Laufe der Nacheszeit. *Advancing Frontiers of Plant Sciences* 9:9–14. New Delhi.

Gimbutas, M.
1976 *Neolithic Macedonia.* Institute of Archaeology, University of California, Los Angeles.
1991 *The Civilization of the Goddess,* edited by Joan Marler. Harper Collins Publishers, San Francisco.

Ginder, E.
1996 *Vojvodjanske seoske kuće od naboja*. Kulturno-istorijsko društvo "Proleće na Čenejskim salašima" PČESA, Novi Sad.

Glišić, J.
1968 Ekonomika i socijalno ekonomski odnosi u neolitu Podunavsko-pomoravskog basena (Economy and socio-economic relations during the Neolithic in the Danube-Morava Basin). In *Neolit Centralnog Balkana*, pp. 21–61. Narodni muzej, Beograd.

Goody, J.
1976 *The Development Cycle in Domestic Groups*. Cambridge University Press, Cambridge.

Goren-Inbar, N., C. S. Feibel, K. L. Verosub, Y. Melamed, M. E. Kislev, E. Tchernov, and I. Saragusti
2000 Pleistocene Milestones on the Out-of-Africa Corridor at Gesher Benot Ya'aqov, Israel. *Science* 289:944–947.

Goren-Inbar, N., G. Sharon, Y. Melamed, and M. Kislev
2002 Nuts, Nut Cracking, and Pitted Stones at Gesher Benot Ya'aqov, Israel. *Proc. Natl. Acad. Sci. USA*. Feb 19; 99:2455–2460.

Gradmann, R.
1901 Das mitteleuropäische Landschaftsbild nach seiner geschichtlichen Entwicklung. *Geographische Zeitschrift* 7:361–377.
1933 Die Steppenheidetheorie. *Geographische Zeitschrift* 39:265–278.

Greenfield, H. J.
1986 *Paleoeconomy of the Central Balkan (Serbia): A Zooarchaeological Perspective on the Late Neolithic and Bronze Age (4500–1000 BC).* BAR, International Series 304, Oxford.
1991 Kula Ring in Prehistoric Europe? A Consideration of Local and Interregional Exchange During the Late Neolithic of the Central Balkans. In *Between Bands and States*, edited by S. Gregg, pp. 287–309. Center for Archaeological Investigations, Occasional Paper No. 9, Southern Illinois University, Carbondale.

Gregg, S. A.
1988 *Foragers and Farmers*. The University of Chicago Press, Chicago.

Greig, J.
1988 Plant Resources. In *Countryside of Medieval England*, edited by G. Astill and A. Grant, pp. 108–127. Basil Blackwell, Oxford.

Greig, J., and J. Turner
1974 Some Pollen Diagrams from Greece and their Archaeological Significance. *Journal of Archaeological Science* 1:177–194.

Gross, E., S. Jacomet, and J. Schibler
1990 Stand und Ziele der wirtschaftsarchäologischen Forschung an neolithischen Ufer- und Inselsiedlungen im unteren Zürichseeraum (Kt. Zürich, Schweiz). In *Festschrift Hans R. Stampfli Schibler*, edited by J. Schibler, J. Sedlmeier, and H.-P. Spycher, pp. 77–100. Verlag Helbing & Lichtenhahn, Basel.

Grüger, E., and H. J. Beug
1988 Botanical Investigations at Divostin and Grivac. In *Divostin and the Neolithic of Central Serbia.* Ethnology Monographs No. 10, edited by A. McPherron and D. Srejović, pp. 415–418. Department of Anthropology, University of Pittsburgh.

Gyulai, F.
1993 *Environment and Agriculture in Bronze Age Hungary*. Archaeolingua. Budapest.

Haaster, H. van
1994 Plant Resources and Environment in Late Medieval Lübeck. In Urban-Rural Connections: Perspectives from Environmental Archaeology. *Environmental Archaeology* 12:79–84.

Halstead, P.
1987 Traditional and Ancient Rural Economy in Mediterranean Europe: Plus Ça Change? *Journal of Hellenic Studies* 107:77–87.
1994 The North-South Divide: Regional Paths to Complexity in Prehistoric Greece. In *Development and Decline in the Mediterranean Bronze Age*, edited by C. Mathers and S. Stoddart, pp. 195–219. Sheffield Archaeological Monographs 8, Sheffield.
1995a Plough and Power: The Economic and Social Significance of Cultivation with the Ox-Drawn Ard in the Mediterranean. *Bulletin of Sumerian Agriculture* 8:11–22.
1995b From Sharing to Hoarding: The Neolithic Foundations of Aegean Bronze Age Society. In *Politeia: Society in the Aegean Bronze Age*, edited by R. Laffineur and W-D. Niemeier, pp. 11–21. Universite de Liege, Liege.
1996 The Development of Agriculture and Pastoralism in Greece: When, How, Who and What? In *Origins of the Spread of Agriculture and Pastoralism in Eurasia*, edited by D. R. Harris, pp. 296–310. Smithsonian Institution Press, Washington, D. C.

Halstead, P., and C. Frederick (editors)
2000 *Landscape and Land Use in Postglacial Greece.* Sheffield Academic Press, Sheffield.

Halstead, P., and G. Jones
1989 Agrarian Ecology in the Greek Islands: Time Stress, Scale and Risk. *Journal of Hellenic Studies* 109:41–55.

Hanf, M.
1990 *Ackerunkräuter Europas mit ihren Keimlingen und Samen.* BVL Verlagsgesellschaft mbH München-Wien-Zürich.

Hansen, J. M.
1991 *The Paleoethnobotany of Franchthi Cave.* Indiana University Press, Bloomington.

Harlan, J. R.
1967 A Wild Wheat Harvest in Turkey. *Archaeology* 20:197–201.
1981 The Early History of Wheat: Earliest Traces to the Sack of Rome. In *Wheat Science: Today and Tomorrow,* edited by L. T. Evans and W. J.

Peacock, pp. 1–19. Cambridge University Press, Cambridge.

Hartyányi, B.
1988/89 Növényleletek a Battonya-Parázstanyai neolitikus lakótelepen (Pflanzenfunde der neolitischen Wohnsiedlung bei Battonya – Parázs-Gehöft). *Magy. Mezög. Múz. Közlem*, pp. 39–67. Budapest.

Hartyányi, B., and S. M. Máthé
1980 A Berettyóújfalu-Szilhalom-i későneolitikus lakótelep 1976-ban feltárt növényleletei (An die Oberfläche 1976 Gebrachte pflanzliche Überreste einer Wohnsiedlung in Szilhalom bei Berettyóújfalu aus der späteren Zeit des Neolithikums). *Magy. Mezög. Múz. Közlem*, pp 125–145. Budapest.

Hartyányi, B., and G. Nováki
1975 Samen- und Fruchtfunde in Ungarn von der Neursteinzeit bis zum 18. Jahrhundert. *Agrártörténeti Szemle. Historia Rerum Rusticarum*, XVII. évfolyam (Supplementum). A Magyar Tudományos Akadémia Agrártörténeti Bizottságának folyóirata, Budapest.

Hastorf, C. A.
1993 *Agriculture and the Onset of Political Inequality before the Inka*. Cambridge University Press, Cambridge.

Havinga, A. J.
1972 A Palynological Investigations in the Pannonian Climate Region of Lower Austria. *Review of Paleobotany and Palynology* 14:319–352.

Hayden, B.
1992 Contrasting Expectation in Theories of Domestication. In *Transitions to Agriculture in Prehistory*, edited by A. B. Gebauer and T. D. Price, pp. 11–20. Prehistory Press, Madison.

Hegediš, A., and K. Čobanović
1991 *Demografska i agrarna statistika Vojvodine 1767 1867* (Demographic and Agricultural Statistics in Vojvodina 1767–1867). Filozofski i Poljoprivredni fakultet Univerziteta u Novom Sadu, Novi Sad.

Hegi, G.
1965 *Illustrierte Flora von Mittel-Europa*, Band V/2. Teil. Carl Hanser Verlag, München.

Heun, M., R. Schafer-Pregl, D. Klawan, R. Castagna, M. Accerbi, B. Borghi, and F. Salamini
1997 Site of Einkorn Wheat Domestication Identified by DNA Fingerprinting. *Science* 278:1312–1314.

Hillman, G. C.
1981 Reconstructing Crop Husbandry Practices from Charred Remains of Crops. In *Farming Practice in Britain Prehistory,* edited by R. Mercer, pp. 123–162. University Press, Edinburgh.
1984 Interpretation of Archeological Plant Remains: The Application of Ethnographic Models from Turkey. In *Plants and Ancient Man,* edited by W. van Zeist and W. A. Casparie, pp. 1–39. Balkema, Rotterdam.
1991 Phytosociology and Ancient Weed Floras: Taking Account of Taphonomy and Changes in Cultivation Methods. In *Modeling Ecological Change,* edited by D. R. Harris and K. D. Thomas, pp. 27–40. Institute of Archaeology, University College, London.

Hillman, G. C., and M. E. Davies
1992 Domestication Rate in Wild Wheats and Barley under Primitive Cultivation: Preliminary Results and Archaeological Implications of Field Measurements of Selection Coefficient. In *Prehistoire de L'Agriculture:Nouvelles Approches Exprimentales et Ethnographiques,* edited by P.C. Anderson, pp.113–158. Monographie du CRA No. 6, Paris.

Hingh, de A., and C. Bakels
1996 Paleobotanical Evidence for Social Difference? The Example of the Early Medieval Domain of Serris-Les Ruelles, France. *Vegetation History and Archaeobotany* 5:117–120.

Hopf, M.
1958 Neolithische Getreidefunde aus Bosnien und der Hercegovina. *Glasnik Zemaljskog muzeja u Sarajevu* (nova serija) 13:97–103.
1961 Untersuchungsbericht über Kornfunde aus Vršnik. *Zbornik na Štipskiot Naroden Muzej* 2:41–45.
1967 Untersuchungsbericht über die botanischen Reste aus der neolithischen Ansiedlung in Gornja Tuzla. *Glasnik Zemaljskog muzeja Bosne i Hercegovine* 21/22:169–171.
1973 Frühe Kulturpflanzen aus Bulgarien. *Jahrbuch des Römisch-Germanischen Zentralmuseums Mainz* 20:1–47.
1974 Pflanzenreste aus Siedlungen der Vinča-Kultur in Jugoslawien. *Jahrbuch des Römisch-Germanischen Zentralmuseums Mainz* 21:1–11.
1978 Frühe Kulturpflanzen in Südeuropa. *Berichte der Deutschen Botanischen Geselschaft* 91:31–38.

Horvat, I., V. Glavač, and H. Ellenberg
1972 *Vegetationskarte von Südosteuropa.* Erläuterungen zur Karte 1:2 Millionen. Gustav Fischer Verlag, Stuttgart.
1974 *Vegetation Südosteuropas.* Gustav Fischer Verlag, Jena.

Hubbard, C. E.
1968 *Grasses: A Guide to Their Structure, Identification, Uses, and Distribution in the British Isles.* Penguin Books, England.

Hubbard, R. N. L. B.
1980 Development of Agriculture in Europe and the Near East: Evidence from Quantitative Studies. *Economic Botany* 34:51–67.

Hui-Lin, Li
1983 The Domestication of Plants in China: Ecogeographical Considerations. In *Origins of Chinese Civilizations*, edited by D. N. Keightley, pp. 21, 43–46. University of California Press, Berkeley.

Hunn, E. S., and N. M. Williams
1982 Introduction. In *Resource Managers: North American and Australian Hunter-Gatherers*, edited by N. M. Williams and E. S. Hunn, pp. 1–16. Westview Press, Boulder.

Jager, K-D., and R. Neuhausl
1992a Koinzidenzen zwischen neolithischer Siedlungsverteilung und aktueller Vegeta tionsdifferenzierung. In *Das Neolithikum in Mitteleuropa: Kulturen - Wirtschaft - Umwelt vom 6. bis 3. Jahrtausend v. u. Zeit.*, Übersichten zum Stand der Forschung, edited by J. Preuss. Berlin (in press). Cited in: Jager, K-D. and R. Neuhausl, 1992b.
1992b Interactions Between Natural Environment and Neolithic Man in Central Europe: An Investigation Based on Comparative Studies on Vegetation and Settlement with Special Emphasis on the View of Natural Science. In *Evaluation of Land Surfaces Cleared from Forests by Prehistoric Man in Early Neolithic Times and the Time of Migrating Germanic Tribes*, edited by B. Frenzel, pp. 75–81. Gustav Fisher Verlag, Stuttgart.

Janković, M. M.
1952 Wirtschaftliche Bedeutung der Wassernuß (*Trapa natans* L.) in Jugoslavien. *Archiv für Hydrobiologie* 47:307–309.
1958 *Ekologija, rasprostranjenje, sistematika i istorija roda* Trapa L. *u Jugoslaviji* (Ekologie, Verbreitung, Systematik und Geschichte der Gattung *Trapa L.* in Jugoslawien). Societe Serbe de Biologie: Editions Speciales 2, Beograd.
1973 Fam. *Trapaceae* Durmotier. In *Flora SR Srbije*, edited by M. Josifović, pp. 35–46. Srpska akademija nauka i umetnosti, Beograd.

Janushevich, Z. V.
1976 *Kulturnije rastenija jugo-zapada SSSR po paleobotaniceskim isledovanijam.* Kisinev.
1978 Prehistoric Food Plants in the South-West of Soviet-Union. *Berichte der Deutschen Botanischen Gesellschaft* 91:59–66.

Járai-Komlódi, M.
1966 Adatok az Alföld negyedkori klíma- és vegetáció-történetéhez I (Quaternary Climatic Changes and Vegetational History of the Great Hungarian Plain) I. *Bot. Közl.* 53:191–201.
1968 The Lateglacial and Holocene Flora of the Hungarian Great Plain. *Ann. Univ. Sci. Budap. Rolando Eotvos Nomin., Sect. Biol.* 9–10:199–225.

Jeftić, S.
1986 *Posebno Ratarstvo I, II* (Crop Science). Naučna knjiga, Beograd.

Jennings, D. L.
1976 Raspberries and Blackberries. In *Evolution of Crop Plants,* edited by N. W. Simmonds, pp. 251–254. Longman, London.

Johnson, G. A.
1982 Organizational Structure and Scalar Stress. In *Theory and Explanation in Archaeology*, edited by A. C. Renfrew, M. J. Rowlands and B. Abbot Segraves, pp. 389–421. Academic Press, London.

Jones, G.
1984 Interpretation of Archaeological Plant Remains: Ethnographic Models from Greece. In *Plants and Ancient Man*, edited by W. van Zeist and W. A. Casparie, pp. 46–61. A. A. Balkema, Rotterdam.
1992 Weed Phytosociology and Crop Husbandry. *Review of Paleobotany and Palynology* 73:133–43.

Jones, G., and P. Halstead
1995 Maslins, Mixtures and Monocrops: On the Interpretation of Archaeobotanical Crop Samples of Heterogeneous Composition. *Journal of Archaeological Science* 22:103–114.

Jones, G., A. Bogaard, P. Halstead, M. Charles, and H. Smith
1999 Identifying the Intensity of Crop Husbandry Practices on the Basis of Weed Floras. *The Annual of the British School at Athens* 94:167–189.

Jones, G., S. Valamoti, and M. Charles
2000 Early Crop Diversity: A "New" Glume Wheat from Northern Greece. *Vegetation History and Archaeobotany* 9:133–146.

Josifović, M.
1970–1977 *Flora SR Srbije* (Flora of SR Srbia), Vol 1–2 (1970); Vol. 3–4 (1972); Vol. 5 (1973); Vol. 6 (1974); Vol. 7 (1975); Vol. 8 (1976); Vol. 9, Supplement (1977). Srpska akademija nauka i umetnosti, Odeljenje prirodno-matematičkih nauka, Beograd.

Jovanović, B.
1982 *Rudna Glava: najstarije rudarstvo bakra na centralnom Balkanu* (Rudna Glava: the Oldest Copper Mining in Central Balkans). Muzej rudarstva i metalurgije, Bor i Arheološki institut, Beograd.
1984 Naselje Vinčanske culture. Stratigrafija. (The Vinča settlement: stratigraphy). In *Vinča u praistoriji i srednjem veku.* (Vinča exibition catalog), pp. 23–34. Galerija srpske akademije nauka i umetnosti, Beograd.

Jovanović, B., and J. Glišić
1960 Eneolitsko naselje na Kormadinu kod Jakova. *Starinar* N.S. 11:113–142.

Kaczanowska, M., and J. K. Kozłowski
1990 Chipped Stone Industry of the Vinča Culture. In *Vinča and its World*, edited by D. Srejović and N. Tasić, pp. 49–54. Serbian Academy of Sciences and Arts, Belgrade.

Kaiser, T.
1984 *Vinča Ceramics: Economic and Technological Aspects of Late Neolithic Pottery Production in Southeast Europe.* Unpublished Ph.D. Dissertation, Department of Anthropology, University of California, Berkeley.

Kaiser, T. and B. Voytek
1983 Sedentism and Economic Change in the Balkan Neolithic. *Journal of Anthropological Archaeology* 2:323–353.

Kalis, A. J.
1988 Zur Umwelt des frühneolithischen Menschen: Ein Beitrag der Pollen Analyse. *Forsch. ber. Vor-und Frughesch. Ba-Wu* 31:125–137.

Karg, S.
1996a Bizarre Früchte aus dem Wasser. Am Federsee wurde eine vergessene Nutzpflanze wiederentdeckt. *Schönes Schwaben* 7:8–11.
1996b *Aus Pflahlbauers Pflanzenwelt.* Trapa natans - *Die Wassernuß.* Exibition Guide Württemberg Landesmuseums. Stuttgart.

Karg, S. and J. Müller
1990 Neolithische Getreidefunde aus Pokrovnik, Dalmatien. *Archäologisches Korrespondenzblatt des Römisch-Germanischen Zentralmuseums Mainz* 20: 373–386.

Katić, P., D. Djukanović, and P. Djaković
1979 *Klima SAP Vojvodine* (The Climate of Socialistic Autonomous Province of Vojvodina). Poljoprivredni fakultet, OOUR Institut za ratarstvo i povrtarstvo, Novi Sad.

Katz, N. J., S.V. Katz, and M. G. Kipiani
1965 *Atlas i opredelitelj plodov i semjan, v strecajushija v cetverticnjih otlozenijah SSSR* (Atlas and Keys of Fruits and Seeds Occurring in the Quaternary Deposits of the USSR). Akademija Nauk, Moscow.

Kerridge, E.
1992 *The Common Fields of England.* Manchester University Press, Manchester.

Kislev, M.
1980 Contenu d'un silo a ble de l'epoque du fer ancien. In *Tell Keisan (1971–1976): Une cite phenicienne en Galilee,* edited by J. Briend and J. B. Humbert, pp. 361–379. Orbis Biblicus et Orientalis. Series Archaeologia 1, 361–379, Taf. 139–140. Fribourg.

Kosse, K.
1979 *Settlement Ecology of the Koros and Linear Cultures in Hungary.* BAR International Series 64, Oxford.

Kotsakis, K.
1999 What Tells Can Tell: Social Space and Settlement in the Greek Neolithic: In *Neolithic Society in Greece*, edited by P. Halstead, pp. 66–76. Sheffield Academic Press, Sheffield.
2001 Mesolithic to Neolithic in Greece. Continuity, Discontinuity or Change of Course? In *Documenta Praehistorica XXVIII: 8th Neolithic Studies*, edited by M. Budja, pp. 63–75. Filozofska fakulteta, Ljubljana.

Kreuz, A.
1995 On-site and Off-site Data: Interpretative Tools for a Better Understanding of Early Neolithic Environments. In *Res Archaeobotanicae: Proceedings of the 9th International Workgroup for Paleoethnobotany*, edited by H. Kroll and R. Pasternak, pp. 117–135. Institut für Ur- und Frühgeschichte der Christian-Albrecht-Universitat, Kiel.

Kreuz, A., E. Marinova, E. Schäfer, and J. Wiethold,
2005 A comparison of early Neolithic crop and weed assemblages from the Linearbandkeramik and the Bulgarian Neolithic cultures: differences and similarities. *Vegetation History and Archaeobotany* 14 [ISSN: 0939-6314 (Paper) 1617-6278 (Online) DOI: 10.1007/s00334-005-0080-0]

Kroll, H.
1983 *Kastanas: Ausgrabungen in einem Siedlungshügel der Bronze- und Eisenzeit Makedoniens 1975–1979: Die Pflanzenfunde.* Verlag Volker Spiess, Berlin.
1990 Melde von Feudvar, Vojvodina. Ein Massenfund bestätigt *Chenopodium* als Nutzpflanze in der Vorgeschichichte. *Präehistorische Zeitschrift* 65:46–48.
1991a Südosteuropa. In *Progress in Old World Palaeoethnobotany,* edited by W. van Zeist, K. Wasylikowa and K. E. Behre, pp. 161–177. A. A. Balkema, Rotterdam, Brookfield.
1991b Botanische Untersuchungen zu pflanzlichen Grundnahrungsmitteln. Bier oder Wein? In Vorbericht über die jugoslawisch-deutschen Ausgrabungen in der Siedlung von Feudvar bei Mošorin (Gem. Titel, Vojvodina) von 1986–1990: Bronzezeit-Vorrömische Eisenzeit, edited by B. Hansel and P. Medović. *Bericht der Römische-Germanischen Kommission* 72:165–171.
1992 Einkorn from Feudvar, Vojvodina, II: What is the Diffrence between Emmer-Like Two-Seeded Einkorn and Emmer? *Review of Paleobotany and Palynology* 73:181–185.

Küster, H.
1989 Ein Fund von Kornelkirschen in Drama. In Bericht über die bulgarisch-deutschen Ausgrabungen in Drama (1983-1988). *Bericht der Römische-Germanischen Kommission* 70:119–122.
1991 Phytosociology and Archaeobotany. In *Modeling Ecological Change,* edited by D. R. Harris and K. D. Thomas, pp. 17–26. Institute of Archaeology, University College, London.

Kyparissi-Apostolika, N.
2000 The Mesolithic/Neolithic Transition in Greece as Evidenced by the Data at Theopatra Cave in Thessaly. In *Documenta Praehistorica XXVII: 7th Neolithic Studies*, edited by Mihael Budja, pp. 133–141. Filozofska fakulteta, Ljubljana.

Laatsch, W.
1957 *Dynamik der mitteleuropäischen Mineralböden.* 4th ed. Steinkopff, Dresden-Leipzig.

Legge, A. J.
1990 Animals, Economy, and Environment. In *Selevac: A Neolithic Village in Yugoslavia*, edited by R. Tringham and D. Krstić, pp. 215–246. University of California Press, Los Angeles.

Lempaiäinen, T.
1992 Macrofosil Finds of Henbane (*Hyoscyamus niger*) in the Old Settlement Layers in Southern Finland. *Review of Paleobotany and Palynology* 73:227–239.

Lepofsky, D. K., K. D. Kusmer, B. Hayden, and K. P. Lertzman
1996 Reconstructing Prehistoric Socioeconomies from Paleoethnobotanical and Zooarchaeological Data: An Example from the British Columbia Plate. *Journal of Ethnobiology* 16:31–62.

Linares, O. F.
1976 "Garden Hunting" in the American Tropics. *Human Ecology* 4:331–349.

Lisitzyna, G. N., and L. A. Filipovitch
1980 Paleoetnobotaniceskije nahodki na Balkanskom poluostrove (Palaeoethnobotanical Findings in the Balkan Peninsula). *Studia Praehistorica* 4:5–90.

Long, D. J., R. Tipping, T. G. Holden, M. J. Bunting, and P. Milburn
2000 The Use of Henbane (*Hyoscyamus niger* L.) as a Hallucinogen at Neolithic "Ritual" Sites: A Re-evaluation. *Antiquity* 74:49–53.

Lundstrom-Baudais, K., and G. Bailly
1995 In the Celler of a Wine-Maker during the 14th Century: The Archaeobotanical Study of Ilôt Vignier, Besançon (France). In *Res Archaeobotanicae: Proceedings of the 9th International Workgroup for Paleoethnobotany*, edited by H. Kroll and R. Pasternak, pp. 165–195. Institut für Ur- und Frühgeschichte der Christian-Albrecht-Universitat, Kiel.

Maier, U.
1993 Archäeobotanische Untersuchungen. In *Archäeologische und archäeobotanische Untersuchungen in der Goldberg III - Siedlung Alleshausen - Grundwiesen am Federsee, Kreis Biberach*, edited by H. Schlichtherle and U. Maier, pp. 88–93. Archaeologische Ausgrabungen Baden, Württemberg.
1996 Morphological Studies of Free-Threshing Wheat Ears from a Neolithic Site in Southwest Germany, and the History of Naked Wheats. *Vegetation History and Archaeobotany* 5:39–55.

Marinova, E.
2001 Vergleichende paläoethnobotanische Untersuchung zur Vegetationsgeschichte und zur Entwicklung der prähistorischen Landnutzung in Bulgarien. Dissertation zur (Dr. Rer. Nat.) der Mathematisch-Naturwissenschaftlichen Fakultät der Rheinischen Friedrich-Wilhelms-Universität Bonn, Bonn.

Martin, A. C.
1987 *Weeds.* Golden Press, Racine.

Martin, A. C., and W. D. Barkley
1973 *Seed Identification Manual.* 2nd ed. University of California Press, Berkeley.

McCay, M., and M. Acheson (editors)
1987 *The Question of the Commons: The Culture and Ecology of Communal Resources.* University of Arizona Press, Tuscon.

McLaren, F. S., and R. N. L. B. Hubbard
1990 The Archaeobotanical Remains. In *Selevac: A Neolithic Village in Yugoslavia,* edited by R. Tringham and D. Krstić, pp. 247–254. Los Angeles.

McPherron, A., and D. Srejović
1988 *Divostin and the Neolithic of Central Serbia.* University of Pittsburgh, Pittsburgh.

Meurers-Balke, J., and J. Lüning
1992 Some Aspects and Experiments Concerning the Processing of Glume Wheats. In *Prehistoire de l'Agriculture: Nouvelles Approches Experimentales Ethnographique*, edited by P. C. Anderson, pp. 341–362. Monographie du CRA No. 6, Paris.

Miller, N. F.
1991 The Near East. In *Progress in Old World Paleoethnobotany*, edited by W. van Zeist, K. Wasylikowa and K-E. Behre, pp. 133–160. A. A. Balkema, Rotterdam, Brookfield.
1996 Seed Eaters of the Ancient Near East: Human or Herbivore? *Current Anthropology* 37:521–528.

Milojčić, V.
1949 *Chronologie der jüngeren Steinzeit Mittel- und Südosteuropas.* Berlin.

Montanari, M.
1994 *The Culture of Food.* Translated by Carl Ipsen. Blackwell, Oxford.

Moore, J. A.
1981 The Effects of Information Networks in Hunter-Gatherer Societies. *In Hunter-Gatherer Foraging Strategies*, edited by B. Winterhalder and E. A. Smith, pp. 194–217. University of Chicago Press, Chicago.
1983 The Trouble with Know-It-Alls: Information as a Social Ecological Resource. In *Archaeological Hammers and Theories*, edited by J. A. Moore and A. S. Keene, pp. 194–217. Academic Press, London.

Nandris, J.
1970 Ground Water as a Factor in the First Temperate Neolithic Settlement of the Koros Region. *Zbornik Narodnog muzeja* 6:59–69. Beograd.

Neeson, J. M.
1993 *Commoners: Common Right, Enclosure and Social Change in England, 1700–1820.* Cambridge University Press, Cambridge.

Nejgebauer, V.
1952 Činioci stvaranja zemljišta u Vojvodini (The Elements of Soil Development in Vojvodina). *Zbornik za prirodne nauke Matice srpske* 2:63–135. Novi Sad.

Nejgebauer, V., Živković, B., Tanasijević, D., and N. Miljković
1971 *Pedološka karta Vojvodine* (Soil Map of Voyvodina). Institut za poljoprivredna istraživanja, Novi Sad.

Neusius, S. W.
1996 Game Procurement among Temperate Horticulturists: The Case for Garden Hunting by the Dolores Anasazi. In *Case Studies in Environmental Archaeology*, edited by E. J. Reitz, L. A. Newsom and S. J. Scudder, pp. 273–288. Plenum Press, New York.

North, D.
1989 Institutions and Economic Growth: A Historical Introduction. *World Development* 17:1319–1332.

Novaković, V.
1996 *Divlji papkari: Tehnologija gajenja i korišćenja* (Wild cloven-footed animals: technology of raising and utilization). Srbija Šume, Beograd.

Olson, M.
1971 *The Logic of Collective Action.* Harvard University Press, Cambridge, Massachusetts.

O'Shea, J.
1989 The Role of Wild Resources in Small-Scale Agricultural Systems: Tales from the Lakes and the Plains. In *Bad Year Economics: Cultural Responses to Risk and Uncertainty*, edited by P. Halstead and J. O'Shea, pp. 57–68. Cambridge University Press, Cambridge.

Ostrom, E.
1990 *Governing the Commons: The Evolution of Institutions for Collective Action.* Cambridge University Press, Cambridge.

Özdoğan, M.
1993 Vinça and Anatolia: A New Look at a Very Old Problem (or Redefining Vinça Culture from the Perspective of Near Eastern Tradition). *Anatolica* XIX:173–193 (Special Issue on Anatolia and the Balkans Symposium).
1997 The Beginnings of Neolithic Economies in Southeastern Europe: An Anatolian Perspective. *Journal of European Archaeology* 5:1–33.

Pantić, N.
1960 *Paleobotanika* (Palaeobotany). Naučna knjiga, Beograd

Parabućski, S. (editor)
1983 *Vegetation Maps of Voyvodina* (Section 3). Unpublished, on file in the Department of Botany, Faculty of Biology, University of Novi Sad, Yugoslavia.

Parabućski, S., and M. Janković
1978 Pokušaj utvrdjivanja potencijalne vegetacije Vojvodine (An Attempted Determination of the Potential Vegetation of Voivodina). *Zbornik za prirodne nauke Matice srpske* 54:5–20. Novi Sad.

Pearsall, D. M.
2000 *Palaeoethnobotany: A Handbook of Procedures* 2nd ed. Academic Press, San Diego.

Percival, J.
1921 *The Wheat Plant.* Duckworth and Co., London.

Perlès, C.
2001 *The Early Neolithic in Greece.* Cambridge World Archaeology. Cambridge University Press, Cambridge.

Pešić, R.
2001 *Vinčansko pismo i drugi gramatološki ogledi* (Vinča alphabet and other grammatical considerations). Pešić i sinovi, Beograd.

Petersen, P. B., and L. Munck
1993 Whole-Crop Utilization of Barley, Including Potential New Uses. In *Barley: Chemistry and Technology,* edited by A. W. MacGregor and R. S. Bhatty, pp. 437–474. American Association of Cereal Chemists, St. Paul.

Plas van der, L. J., H. Pons, H. Groendijk, G. de Jong, G. J. Krook, and M. Kuiper
1985 Opovo-Reports, Data, Ideas: 1985. Unpublished manuscript, on file at the Department of Soil Science and Geology, Agricultural University, Wageningen.

Polunin, O.
1980 *Flowers of Greece and the Balkans: A Field Guide.* Oxford University Press, Oxford and New York.

Popova, T.
1995 Plant Remains from Bulgarian Prehistory (7000–2000 BC). In *Prehistoric Bulgaria,* edited by D. W. Bailey and I. Panayotov, pp. 193–203. Monographs in World Archaeology, No. 22. Prehistory Press, Madison.

Price, T. D., and G. M. Feinman (editors)
1995 *Foundations of Social Inequality.* Plenum Press, New York.

Pullen, D. J.
1992 Ox and Plough in the Early Bronze Age Aegean. *American Journal of Archaeology* 96:45–54.

Pustovoytov, K., S. Riehl, and S. Mittmann
2004 Radiocarbon Age of Carbonate in Fruits of *Lithospermum* from the Early Bronze Age Settlement of Hirbet ez-Zeraqon (Jordan). *Vegetation History and Archaeobotany* 13:207–212.

Radenković, L.
1996 *Simbolika sveta u narodnoj magiji južnih Slovena* (World Symbolics in Southern Slavic Sorcery). Prosveta, Niš i Balkanološki institut SANU, Beograd.

Radovanović, I.
1996 *The Iron Gates Mesolithic*. International Monographs in Prehistory, Ann Arbor.

Rappaport, R.
1968 *Pigs for the Ancestors*. Yale University Press, New Haven.

Renfrew, C.
1970 The Place of Vinča Culture in European Prehistory. *Zbornik radova Narodnog muzeja* 6:45–57. Beograd.

Renfrew, J. M.
1973 *Paleoethnobotany: The Prehistoric Food Plants of the Near East and Europe*. Methuen, London.
1974 Report on the Carbonized Cereal Grains and Seeds from Obre I, Kakanj, and Obre II. *Wissenshaftliche Mitteilungen des Bosnisch-Hercegowinischen Landesmuseums in Sarajevo* 4:47–53.
1976 Carbonized Seeds from Anza. In *Neolithic Macedonia*, edited by M. Gimbutas, pp. 300–312. Monumenta Archaeologica, Los Angeles.
1979 The First Farmers in South East Europe. In Festschrift Maria Hopf, edited by U. Korber-Grohne, *Archaeo-Physika* 8:243–265. Rheinland-Verlag GMBH, Köln.

Reynolds, P. J.
1979 *Iron-Age Farm: The Butser Experiment*. British Museum Publications, London.
1981 Deadstock and Livestock. In *Farming Practice in British Prehistory*, edited by R. Mercer, pp. 97–123. Edinburgh University Press, Edinburgh.
1992 Crop Yields of the Prehistoric Cereal Types Emmer and Spelt: The Worst Option. In *Prehistoire de l'Agriculture: Nouvelle Approches Experimentales et Etnographiques*, edited by P. C. Anderson, pp. 383–395. Monographie du CRA No. 6, Paris.

Rimantiené, R.
1979 *Šventoji*. Mosklas, Vilnius.

Riviera Núñes, D., and J. Walker
1989 A Review of Paleobotanical Findings of Early *Vitis* in the Mediterranean and of the Origin of Cultivated Grape-Vines with Special Reference to New Pointers to Prehistoric Exploitation in the Western Mediterranean. *Review of Paleobotany and Palynology* 61:205–217.

Rösch, M., and E. Fischer
2000 A Radiocarbon Dated Holocene Pollen Profile from the Banat Mountains (Southwestern Carpathians, Romania). *Flora* 195:277–286.

Rösch, M., O. Ehrmann, L. Herrmann, E. Schulz, A. Bogenrieder, J. P. Goldammer, M. Hall, H. Page, and W. Schier
2002 An Experimental Approach to Neolithic Shifting Cultivation. *Vegetation History and Archaeobotany* 11:143–154.

Runge, C. F.
1981 Common Property Externalities: Isolation, Assurance and Resource Depletion in a Traditional Grazing Context. *American Journal of Agricultural Economics* 63:595–606.

Russell, K. W.
1988 *After Eden: The Behavioral Ecology of Early Food Production in the Near East and North Africa.* BAR International Series 391, Oxford.

Russell, N.
1993 *Hunting, Herding and Feasting: Human Use of Animals in Neolithic Southeast Europe*. Ph.D. Dissertation, Department of Anthropology, University of California, Berkeley. University Microfilms, Ann Arbor.
1998 Cattle as Wealth in Neolithic Europe: Where's the Beef? In *The Archaeology of Value: Essays on prestige and the processes of evaluation,* edited by D. Bailey with the assistance of Steve Mills, pp. 42–54, BAR International Series 730, Oxford.

Sarić, M.
1986 *Flora SR Srbije* (Flora of SR Srbia), Vol. 10, Supplement (1986). Srpska akademija nauka i umetnosti, Odeljenje prirodno-matematičkih nauka, Beograd.

Šarić, T.
1989 *Atlas korova:100 najvažnijih vrsta korovskih biljaka u Jugoslaviji* (Weed Atlas). Svjetlost, Sarajevo.

Savezni Hidrometeorološki zavod
1985 *Godišnjak Saveznog hidrometeorološkog zavoda* 1948–1984. Beograd.

Savezni Zavod za statistiku
1957 *Statistički bilten* (Statistic Bulletin) No. 95, Novembar, Beograd.

Scarry, C. M., and Steponaitis, P. V.
1997 Between Farmstead and Center: The Natural and Social Landscape of Moundville. In

People, Plants, and Landscapes: Studies in Paleothnobotany, edited by K. J. Gremillion, pp. 107–123. The University of Alabama Press, Tuscaloosa.

Scheffer, F., and P. Schchtschabel
1966 *Lehrbuch der Bodenkunde.* 6th ed. Enke, Stuttgart.

Schiemann, E.
1951 New results on the history of cultivated cereals. *Heredity* 5:305-318.

Schibler, J, H. Hüster-Plogmann, S. Jacomet, C. Brombacher, E. Gross-Klee, E., and A. Rast-Eicher
1997 *Ökonomie und Ökologie neolithischer und bronzezeitlicher Ufersiedlungen am Zürichsee.* Mongraphien der Kantonsarchäologie Zürich 20, Zürich.

Schier, W.
1996 The Relative and Absolute Chronology of Vinča: New Evidence from the Type Site. In *The Vinča Culture, Its Role and Cultural Connections*, edited by F. Draşovean, pp. 141–159. National Museum, Timişoara.

Schier, W., and F. Draşovean
2004 Vorbericht über die rumänisch-deutschen Prospektionen und Ausgrabungen in der befestigten Tellsiedlung von Uivar, jud. Timiş, Rumänien (1998–2002). *Prähistorische Zeitschrift* 79:145–230.

Schoch, W. H., B. Pavlik, and F. H. Schweingruber
1988 *Botanische Makrorezte - Botanical Macro-remains - Macrorestes botaniques.* Paul Haupt Publishers, Bern and Stuttgart.

Schröter, C.
1895 Über die Pflanzenreste der neolithischen Landsiedlung von Butmir in Bosnien. In *Die neolithische Station von Butmir bei Sarajevo in Bosnien,* edited by J. Radimsky and M. Hoernes, pp. 37–46. Bosnisch-Herzegovinisches Landesmuseum, Wien.

Sezik, E., E. Yesilada, M. Tabata, G. Honda, Y. Takaishi, T. Fujita, T. Tanaka, and Y. Takeda
1997 Traditional Medicine in Turkey: Folk Medicine in East Anatolia; Erzurum, Erzincan, Agri, Kars, Igdir Provinces. *Economic Botany* 51:195–211.

Sherratt, A.
1981 Plough and Pastoralism: Aspects of the Secondary Products Revolution. *Patterns of the Past: Studies in Honour of David Clark*, edited by I. Hodder, G. Isaac, and N. Hammond, pp. 261–307. Cambridge University Press, Cambridge.
1996 Flying Up with the Souls of the Dead. *British Archaeology*, No. 15, June 1996: Essay. http://www.britarch.ac.uk/ba/ba15/ba15int.html Electronic document accessed December 6, 2004.

Shiel, R. S.
1997 Surviving in the Tisza Valley: Plant's and People's Perspectives in the Environment. In *Landscapes in Flux*, edited by J. C. Chapman and P. Dolukhanov, pp. 181–192. Colloquia Pontica 3. Oxbow Books, Oxford.

Shipton, P.
1989 *Bitter Money: Cultural Ecology and Some African Meanings of Forbidden Commodities.* American Ethnological Society Monograph 1. Washington, D.C.

Sigaut, F.
1975 L'Agriculture et le Feu. *Cahiers des Etudes Rurales* 1. Paris.
1988 A Method for Identifying Grain Storage Techniques and Its Application for European Agricultural History. *Tools and Tillage* 6:3–32.
1992 Rendements, semis et fertilite: signification analitique des rendements. In *Prehistoire de l'Agriculture: Nouvelles Approches Experimentales Ethnographique*, edited by P. C. Anderson., pp. 395–403. Monographie du CRA No. 6, Paris.

Šilić, C.
1988a *Šumske zeljaste biljke.* Svjetlost, Sarajevo.
1988b *Atlas drveća i grmlja.* Svjetlost, Sarajevo.

Skakun, N. N.
1992 Evolution des techniques agricoles en Bulagarie chalcolitique (d'apres les analyses traceologiques). In *Prehistoire de l'Agriculture: Nouvelles Approches Experimentales Ethnographique*, edited by P. C. Anderson, pp. 289–303. Monographie du CRA No. 6, Paris.

Slavnić, Z.
1952 Nizinske šume Vojvodine (Die Niederungswalder der Vojvodina). *Zbornik za prirodne nauke Matice srpske* 2:17–39. Novi Sad.

Sponholz, B.
2004 Geomorphologisch-sedimentologische Untersuchungen. In Vorbericht über die rumänisch-deutschen Prospektionen und

Ausgrabungen in der befestigten Tellsiedlung von Uivar, jud. Timiş, Rumänien (1998–2002), by W. Schier and F. Draşovean. *Prähistorische Zeitschrift* 79(2):145–230.

Srejović, D.
1978 La néolithisation de la région de Portes de Fer. *Godišnjak Centra za balkanološka ispitivanja* XVI:21–31. Sarajevo.
1984 Umetnost i religija. In *Vinča u praistoriji i srednjem veku*, pp. 42–56. Srpska akademija nauka i umetnosti, Beograd.

Srejović, D., and N. Tasić (editors)
1990 *Vinča and its World.* Serbian Academy of Sciences and Arts, Belgrade.

Srpska akademija nauka i umetnosti
1984 *Vinča u praistoriji i srednjem veku.* (Vinča Exibition Catalog). Galerija srpske akademije nauka i umetnosti, Beograd.

Startin, W.
1978 Linear Pottery Culture Houses: Reconstruction and Manpower. *Proceedings of the Prehistoric Society* 44:143–159.

Stevanović, M.
1996 *The Age of Clay: The Social Dynamics of House Destruction.* Ph.D. Dissertation. Department of Anthropology, University of California, Berkeley. University Microfilms, Ann Arbor.
1997a The Age of Clay: The Social Dynamics of House Destruction. *Journal of Anthropological Archaeology* 16:334–395.
1997b Report on Experimental Archaeology at Çatal: Manufacturing Bricks for the House Replica. http://catal.arch.cam.ac.uk/catal/archive%5Frep97/stevano97.html, pp 1–6, Electronic Document accessed December 3, 2002.

Stevanović, M., and B. Jovanović
1996 Stratigraphy of Vinča-Belo Brdo Reconsidered. *Starinar* 47:194–204.

Stevanović, M., and R. Tringham
1997 The Significance of Neolithic Houses in the Archaeological Record of Southeast Europe. In *Uzdarje Dragoslavu Srejoviću*, edited by M. Lazić, pp. 195–207. Centre for Archaeological Research, Faculty of Philosophy, University of Belgrade, Belgrade.

Stojšić, M., and N. Miljković
1992 Istorija navodnjavanja, odvodnjavanja i odbrane od poplave na slivu Dunava u Jugoslaviji (History of irrigation, drainage, and flood control in the Danube catchment area of Yugoslavia). *Vodoprivreda* 24:191–210.

Strouhal, E.
1992 *The Life of the Ancient Egyptians.* University of Oklahoma Press, Norman.

Stummer, A.
1911 Zur Urgeschichte der Rebe und des Weinbaus. *Mitteilungen der Anthropologischen Gesellschaft in Wien* 41:283–296.

Szujkó-Lacza, J.
1991 *Reconstruction of the History of the Postglacial Flora and Vegetation in the Region between Danube and Tisza and Tiszántúl.* Kiev.

Tasić, N., and J. Petrović (editors)
1988 *Gomolava: Hronologija i stratigrafija u praistoriji i antici Podunavlja i jugoistočne Evrope.* Arheološki institut, Beograd i Vojvodjanski muzej, Novi Sad.

Titow, C. E.
1972 *Winchester Yields: A Study in Medieval Agricultural Productivity.* Cambridge University Press, Cambridge.

Todorova, H., and I. Vaysov
1993 *Novo kamennata epoha v Bulgarii: krajat na sedmo-shesto hiljdoletije predi novata era* (The Neolithic in Bulgaria: the end of the seventh-sixth millennia BCE). Izdatelstvo nauka i izkustvo, Sofia.

Todorović, J., and A. Cermanović
1961 *Banjica, naselje Vinčanske kulture* (Banjica, a settlement of the Vinča culture). Muzej grada, Beograd.

Tringham, R.,
1994 Engendered Places in Prehistory. *Gender, Place and Culture* 1:169–203.
2000 Southeastern Europe and the Transition to Agriculture in Europe Bridge, Buffer, or Mosaic. In *Europe's First Farmers*, edited by T. D. Price, pp. 19–56. Cambridge University Press, Cambridge.

Tringham, R., B. Brukner, and B. Voytek
1985 The Opovo Project: A Study of Socioeconomic Change in the Balkan Neolithic. *Journal of Field Archaeology* 12:425–444.

Tringham, R. and D. Krstić (editors)
1990 *Selevac: A Neolithic Village in Yugoslavia.*

Institute of Archaeology, University of California, Los Angeles.

Tringham, R., B. Brukner, T. Kaiser, K. Borojević, Lj. Bukvić, P. Steli, N. Russell, M. Stevanović, and B. Voytek
1992 Excavations at Opovo, 1985–1987: Socioeconomic Change in the Balkan Neolithic. *Journal of Field Archaeology* 19:351–386.

Turner, M.
1994 Common Property and Property in Common: A Review of 1993 Commoners: Common Right, Enclosure and Social Change in England, 1700–1820, by J. M. Neeson. *Agricultural History Review* 42:158–162.

Valamoti, S. M.
2002 Food Remains from Bronze Age Archondiko and Mesimeriani Toumba in Northern Greece. *Vegetation History and Archaeobotany* 11:17–22.
2004 *Plants and People in Late Neolithic and Early Bronze Age Northern Greece: An Archaeobotanical Investigation.* BAR International Series 1258, Oxford.

Vankina, L. V.
1970 *Torfayanikovaya Stoyanka Sarnate* (The Sarnate peat bog site.) Zinatne, Riga.

Vasić, M.
1906 Nekoliki preistorijski nalasci iz Vinče (Some prehistoric finds from Vinča). *Starinar* 1/2:89–127.
1930 Excavations on the Neolithic Site at Vinča, *Man* XXX:197–200.
1932 *Preistorijska Vinča I: Industrija cinebarita i kosemtika u Vinči* (Prehistoric Vinča I: Industry of cinnabar and cosemetics in Vinča). Državna štamparija, Beograd.
1936 *Preistorijska Vinča II: Oblici grobova-Mistične oči-Igra na table. Datovanje Vinče* (Prehistoric Vinča II: Forms of graves-Mystic eyes-Table Games. Dating of Vinča). Državna štamparija, Beograd.
—— *Preistorijska Vinča III: Plastika* (Prehistoric Vinča III: Figurines). Državna štamparija, Beograd.
—— *Praistorisjka Vinča IV: Keramika* (Prehistoric Vinča IV: Ceramics). Državna štamparija, Beograd.

Vračarić, B., J. Bakić, D. Čolić, V. Lintner, M. Micković, R. Rajšić, D. Stevanović, and M. Uvalin
1977 *Ishrana u prirodi* (Nutrition in nature). Vojnoizdavački zavod i Narodna knjiga, Beograd.

Währen, M.
1963 *Brot und Gebäck im Leben und Glauben der alten Ägypter.* Schweizerischer Bäcker-Konditorenmeister-Verband, Bern.

Wasylikowa, K.
1981 The Role of Fossil Weeds for the Study of Former Agriculture. *Zeitschrift für Archaeologia* 15:11–23.

Waterbolk, H.T.
1988 C-14 Datierungen von Gomolava. In *Gomolava: Hronologija i stratigrafija u praistoriji i antici Podunavlja i jugoistočne Evrope*, edited by N. Tasić and J. Petrović, pp. 117–121. Arheološki institut, Beograd i Vojvodjanski muzej, Novi Sad.

Watson, J. P. N.
1979 The Estimation of the Relative Frequencies of Mammalian Species: Khirokitia 1972. *Journal of Archaeological Science* 6:127–137.

Watson, P. J.
1979 *Archaeological Ethnography in Western Iran.* The University of Arizona Press, Tucson.

White, K. D.
1976 Food Requirements and Food Supplies in Classical Times in Relation to Diet of the Various Classes. *Progress in Food and Nutrition Science* 2:143–191.

White, T. A., and Runge, C. F.
1994 Common Property and Collective Action: Lessons from Cooperative Watershed Management in Haiti. *Economic Development and Cultural Change* 43:1–41.

Whittle, A.
1996 *Europe in the Neolithic: The Creation of New Worlds.* Cambridge University Press, Cambridge.

Willis, K.
1995 The Pollen-Sedimentological Evidence for the Beginning of Agriculture in Southeast Europe and Anatolia. In The Transition to Farming in South-East Europe and Anatolia, edited by M. Budja, pp. 9–25. *Poročilo o razikovanju paleolitika, neolitka in eneolitika v Sloveniji* XXII. Filozofska fakulteta, Ljubljana.
1997 The Impact of Early Agriculture upon the Hungarian Landscape. In *Landscapes in Flux: Central and Eastern Europe in Antiquity*, edited by J. Chapman and P. Dolukhanov, pp. 193–209. Oxbow Books, Oxford.

Willis, K., and K. D. Bennett
1994 The Neolithic Transition – Fact or Fiction? Palaeoecological Evidence from the Balkans. *The Holocene* 4:32–330.

Willis, K. J., M. Braun, P. Sümegi, and A. Tóth
1998 Does Soil Change Cause Vegetation Change or Vice Versa? A Temporal Perspective from Hungary. *Ecology* 78:740–750.

Winterhalder, B., and C. Goland
1997 An Evolutionary Ecology: Perspective on Diet Choice, Risk, and Plant Domestication. In *People, Plants, and Landscapes: Studies in Paleoethnobotany*, edited by K. J. Gremillion, pp. 123–161. The University of Alabama Press, Tuscaloosa.

Wylie, A.
2002 *Thinking from Things: Essays in the Philosophy of Archaeology*. University of California Press, Berkeley.

Zamanis, A., S. Samaras, N. Stavropoulos
1988 *Report of an Expedition to Rescue Germplasm of Wild Species of Wheat and Relatives in Greece*. Greek Gene Bank Scientific Bulletin. North Greece Agricultural Research Centre.

Zeist, W. van
1970 Prehistoric and Early Historic Food Plants in the Netherlands. *Palaeohistoria* 14, 1968 (1970):41–173.
1975 Preliminary Report on the Botany of Gomolava. *Journal of Archaeological Science* 2:315–325.
1978 Ugljenisani biljni ostaci na višeslojnom nalazištu Gomolava. *Rad vojvodjanskih muzeja* 23/24:5–18. Novi Sad.
1988 Some Aspects of Early Neolithic Plant Husbandry in the Near East. *Anatolica* 15:49–67.
2001/2002 Plant Husbandry of Tell Gomolova, Vojvodina, Yugoslavia. *Palaeohistoria* 43/44:87–115.

Zeist, W. van, and S. Bottema
1971 Plant Husbandry in Early Neolithic Nea Nikomedia, Greece. *Acta Botanica Needrlandica* 20:524–538.

Zeist, W. van, G. J. Roller, R. M. Palfenier-Vegter, O. H. Harsema, and H. During
1986 Plant Remains from Medieval Sites in Drenthe, The Netherlands. *Helinium* 26:226–274.

Zeist, W. van, and M. R. van der Spoel-Walvius
1980 A Palynological Study of the Late-glacial and the Postglacial in the Paris Basin. *Paleohistoria* 22:76–109.

Zohary, D., and M. Hopf
1993 *Domestication of Plants in the Old World*. 2nd revised ed. Oxford Science Publication, Clarendon Press, Oxford.
2000 *Domestication of Plants in the Old World*. 3rd revised ed. Oxford Science Publication, Clarendon Press, Oxford.

Zolyomi, B.
1957 Der Tatarenahorn-Eichen-Losswald der zonalen Waldsteppe (Acereto tatarici-Quercetum). *Acta Botanica Academiae Sci. Hung.* 3:361–395.

Zvelebil, M.
1994 Plant Use in the Mesolithic and its Role in the Transition to Farming. *Proceedings of the Prehistoric Society* 60:35–74.
1996 The Agricultural Frontier and the Transition to Farming in the Circum-Baltic Region. In *Origins of the Spread of Agriculture and Pastoralism in Eurasia*, edited by D. R. Harris, pp. 323–346. Smithsonian Institution Press, Washington, D. C.

www.ingramcontent.com/pod-product-compliance
Lightning Source LLC
LaVergne TN
LVHW071209210926
848482LV00015B/765